Improving Regulation

Cases in Environment, Health, and Safety

Edited by
Paul S. Fischbeck and
R. Scott Farrow

Resources for the Future

New York • London

An RFF Press book
Published 2001 by Resources for the Future
711 Third Avenue, New York, NY 10017
2 Park Square, Milton Park, Abingdon, Oxfordshire OX14 4RN

First issued in hardback 2017

Library of Congress Cataloging-in-Publication Data

Improving regulation : cases in environment, health, and safety / edited by Paul S. Fischbeck and R. Scott Farrow.

p. cm.

Includes bibliographical references and index.

ISBN 1–891853–10–4 (lib.bdg) — ISBN 1–891853–11–2 (pbk.)

1. Environmental sciences—Decision making. 2. Environmental policy—Government policy. 3. Risk assessment. 4. Health risk assessment. 5. Administrative law—Economic aspects—United States. I. Fischbeck, Paul S. (Paul Stelling) II. Farrow, R. Scott.

GE105 .I56 2001
363.1′056—dc21 2001019430

This book was designed and typeset in Minion by Betsy Kulamer. It was copyedited by Pamela Angulo. The cover was designed by Debra Naylor Design.

ISBN 13: 978-1-138-16373-7 (hbk)
ISBN 13: 978-1-891853-11-1 (pbk)

About Resources for the Future and RFF Press

Founded in 1952, **Resources for the Future** (RFF) contributes to environmental and natural resource policymaking worldwide by performing independent social science research.

RFF pioneered the application of economics as a tool to develop more effective policy about the use and conservation of natural resources. Its scholars continue to employ social science methods to analyze critical issues concerning pollution control, energy policy, land and water use, hazardous waste, climate change, biodiversity, and the environmental challenges of developing countries.

RFF Press supports the mission of RFF by publishing book-length works that present a broad range of approaches to the study of natural resources and the environment. Its authors and editors include RFF staff, researchers from the larger academic and policy communities, and journalists. Audiences for RFF publications include all of the participants in the policymaking process—scholars, the media, advocacy groups, nongovernment organizations, professionals in business and government, and the general public.

Contents

Contributors

SCOTT M. BARTELL is currently pursuing a doctorate in epidemiology at the University of California, Davis, as a U.S. Environmental Protection Agency STAR Fellow. His research interests are probabilistic models for biomarker-based health risk assessment.

JAMES BOYD is a senior fellow at Resources for the Future. He received his Ph.D. from the University of Pennsylvania's Wharton Business School and works principally in the field of law and economics, with an emphasis on the economic analysis of environmental regulation, institutions, and policy.

TIMOTHY J. BRENNAN is a senior fellow at Resources for the Future and professor of policy sciences and economics at University of Maryland, Baltimore County. His research interests include regulatory design, competition policy, and intellectual property.

JARED N. DAY, a historian, is a research associate at the Center for the Study and Improvement of Regulation in the Department of Engineering and Public Policy at Carnegie Mellon University. Previously, he was a historical consultant specializing in Superfund litigation and other urban and environmental issues. He is the author of *Urban Castles: Tenement Housing and Landlord Activism in New York City, 1890–1943,* a study of the social and policy roots of rent control.

MICHAEL L. DEKAY is an assistant professor in the Department of Engineering and Public Policy and in the H. John Heinz III School of Public Policy and Management at Carnegie Mellon University. A social psychologist, DeKay studies judgement and decisionmaking in the environmental and medical domains, currently focusing on risk ranking, ecological risk perception, and precautionary reasoning.

ALEX FARRELL is the executive director of the Carnegie Mellon Electricity Industry Center and a research engineer in the Department of Engineering and Public Policy at Carnegie Mellon University. His research focuses on the scientific and policy implications of energy systems.

R. SCOTT FARROW is the director of the Center for the Study and Improvement of Regulation and a principal research economist in the Department of Engineering and Public Policy and in the H. John Heinz III School of Public Policy and Management at Carnegie Mellon University. Previously, he served as the associate director for pollution control and prevention and the lead economist for the White House Council on Environmental Quality.

ELAINE M. FAUSTMAN is a professor and director at the Institute for Risk Analyses and Risk Communication in the School of Public Health and Community Medicine at the University of Washington. She is also the director of a child health center, funded by the U.S. Environmental Protection Agency and the National Institute of Environmental Health Sciences, which is evaluating key mechanisms that define children's susceptibility to pesticides.

PAUL S. FISCHBECK is an associate professor in the Department of Engineering and Public Policy and the Department of Social and Decision Sciences at Carnegie Mellon University. His research involves normative and descriptive risk analysis and has included the development of a risk index to prioritize inspections of offshore oil production platforms; an engineering and economic policy analysis of air pollution from U.S. and international shipping; a large-scale probabilistic risk assessment of the space shuttle's tile protection system; and a geographic information system designed to evaluate the environmental risk, economic potential, and political factors of abandoned industrial sites.

BARUCH FISCHHOFF is University Professor in the Department of Engineering and Public Policy and the Department of Social and Decision Sciences at Carnegie Mellon University. His research focuses on the assessment, evaluation, and management of health, safety, and environmental risks.

H. KEITH FLORIG is a senior research engineer in the Department of Engineering and Public Policy at Carnegie Mellon University. He pursues applied research in risk analysis and environmental policy. His areas of specialization include environment and development in China, ionizing and non-ionizing radiation protection, and risk communication.

JAMES N. FOLLIN works for the Bettis Atomic Power Laboratory in the Environmental Affairs Activity, where he supports efforts toward the proposed repository for spent nuclear fuel and high level waste at Yucca Mountain. His current work includes human reliability, criticality analyses, and public interactions. Previous projects included nuclear plant operations and training, environmental impact statements involving the storage and transportation of spent nuclear fuel, and perceptions of environmental impacts.

PATRICK GURIAN is an assistant professor of civil engineering at the University of Texas, El Paso. His research interests include the regulation of arsenic in drinking water, Bayesian risk analysis, and the history of water regulation.

KAREN E. JENNI is currently a principal at Geomatrix Consultants in Denver, Colorado. She specializes in the application of decision analysis to large-scale risk management and environmental policy problems. She teaches decision analysis, multiattribute utility analysis, and priority setting for budgeting.

LESTER B. LAVE is University Professor, Higgins Professor of Economics, and professor of engineering and public policy at Carnegie Mellon University. He is also director of the Green Design Initiative, a universitywide effort to make pollution prevention practical and to make the economy more environmentally sustainable.

MOLLY K. MACAULEY is a senior fellow at Resources for the Future and a visiting professor at Johns Hopkins University. Her research interests are the economics of outer space, the value of information, and the economic and policy issues related to new technologies.

DEANNA H. MATTHEWS is a post-doctoral researcher in the Department of Civil and Environmental Engineering at Carnegie Mellon University. For the past three years, she has worked with the Green Design Initiative at Carnegie Mellon University, researching environmental management in organizations.

H. SCOTT MATTHEWS is the research director of the Green Design Initiative and a research faculty member in the Graduate School of Industrial Administration at Carnegie Mellon University. His research, teaching, and consulting activities involve systems engineering, benefit–cost analysis, environmental management, and the socioeconomic implications of information technology.

KARA M. MORGAN is a research environmental scientist at Research Triangle Institute in Washington, DC. A decision analyst, she works to improve the tools and information used for decisionmaking and to develop guidance and training on obtaining better information for decisions.

M. GRANGER MORGAN is Lord Chair in Engineering, professor and head of the Department of Engineering and Public Policy, and a professor in the

Department of Electrical and Computer Engineering and the H. John Heinz III School of Public Policy and Management at Carnegie Mellon University. He is interested in the integrated assessment of complex policy problems involving science and technology; the treatment of uncertainty in quantitative policy analysis; risk analysis, management, and communication; and the improvement of health, safety, and environmental regulation.

KATHY NOTARIANNI leads the Integrated Performance Assessment Group at the Building and Fire Research Laboratory of the National Institute of Standards and Technology. She served on a task group that developed guidelines for performance-based fire protection analysis and design of buildings. She conducts research in areas of smoke detection, automatic suppression, fire modeling, and performance-based design.

ALAIN NADAÏ is a research economist at the Center for the Study and the Improvement of Regulation at Carnegie Mellon University. Previously, he spent eight years as a research economist at CERNA, the Center of Industrial Economics at the School of Mining in Paris, France. His research has focused on the economics of environmental regulation as applied to the European Union greenhouse-gas abatement policy, pesticide registrations, the regulation of genetically modified organisms, product eco-labeling, and voluntary approaches to regulation.

RAFAEL A. PONCE is a senior scientist at Zymogenetics, with an affiliation to the Department of Health and Community Medicine at the University of Washington. His research interests are in developmental neurotoxicology, immunotoxicology, and human health risk assessment.

DONNA M. RILEY is assistant professor in the Picker Engineering Program at Smith College. Previously, she spent a year as an American Association for the Advancement of Science Fellow at the U.S. Environmental Protection Agency, as well as two years at Princeton University as the Clayton Postdoctoral Fellow in Industrial Ecology.

MARTIN T. SCHULTZ is a Ph.D. candidate in the Department of Engineering and Public Policy at Carnegie Mellon University. His research addresses problems in evaluation and decision support for environmental and natural resource policy. He has worked on problems related to pollution trading, water quality indices, valuation of water rights, and the evaluation of water conservation programs.

JOHN R. SHULTZ is an engineer in the Nuclear Materials Control and Accountability Program, Office of Safeguards and Security, Headquarters, U.S. Department of Energy. Previously, he worked as a lead project engineer at the National Energy Technology Laboratory and consulted on risk related topics for the Minerals Management Service, U.S. Department of Interior.

MITCHELL J. SMALL is the H. John Heinz III Professor of Environmental Engineering in the Departments of Civil and Environmental Engineering and of Engineering and Public Policy at Carnegie Mellon University. His research involves mathematical modeling and statistical evaluation for environmental decisionmaking. He currently serves as the chair of the Environmental Models Subcommittee of EPA's Science Advisory Board.

DAVID STIKKERS is a Ph.D. student in the Department of Engineering and Public Policy at Carnegie Mellon University. Previously, he worked as a civil engineer for Sierra Pacific Power Company.

TIMOTHY K. TAKARO is a clinical assistant professor in the Occupational and Environmental Medicine Program at the University of Washington and director of the program's Center for Chemically Related Illness. His research and teaching interests emphasize lung disease and include risk assessment and management, medical surveillance, and the validation and field application of biomarkers.

JOEL A. TARR is the Richard S. Caliguiri Professor of Urban and Environmental History and Policy at Carnegie Mellon University. He has written extensively regarding the history of environmental pollution and currently is editing a book on Pittsburgh's environmental history.

EVA WONG is a research scientist with the Institute for Risk Analysis and Risk Communication, housed in the Department of Environmental Health at the University of Washington. Her current research is in quantitative risk analysis.

RICHARD O. ZERBE is a professor in the Evans School of Public Affairs and the Law School at the University of Washington. His specialty areas are economic regulations, benefit–cost analysis, antitrust, environmental policy and economic history.

Acknowledgements

The research reported in this book took place at Carnegie Mellon University, Resources for the Future, and the University of Washington. The book's concept grew out of support from the A.W. Mellon Foundation for research on how regulations actually work. The effort of individual researchers was supported by many sources, including the U.S. Environmental Protection Agency, the National Science Foundation, the National Institute for Standards and Technology, the A.W. Mellon Foundation, the ExxonMobil Corporation, Ford Motor Company, the American Chemistry Council, and financial supporters of the Center for the Study and Improvement of Regulation at Carnegie Mellon University and the University of Washington.

We especially thank Catherine Ribarchak for secretarial support; Resources for the Future, its referees, and its publications staff (headed by Don Reisman) for patient and useful feedback; and our student and faculty colleagues in the Departments of Engineering and Public Policy, Social and Decision Sciences, and Environmental Health and in the H. John Heinz III School of Public Policy Analysis and Management.

Dedication

To our wives, Katie Greeno and Elaine King,
for their intellectual and emotional support and
ongoing efforts at husband management,
and to our family members of all ages.

Improving Regulation

Cases in Environment, Health, and Safety

1

Introduction: The Challenge of Improving Regulation

Paul S. Fischbeck, R. Scott Farrow, and M. Granger Morgan

The regulatory system is associated with much of the dramatic progress that we have made in controlling the most obvious environmental, health, and safety problems in the United States. Not only have the emissions of important pollutants declined compared with those that would have been expected based on emission levels of decades ago, but other pollutants also have declined absolutely even with economic growth. Although population increased 33% from 1970 to 1999 and (real) gross domestic product by 147%, aggregate criteria air emissions declined by almost one-third compared with 1970 levels (U.S. EPA 1998a). Lead emissions declined by as much as 98%, and particulate emissions of 10 micrometers or smaller declined by 73%. Industrial conventional water pollution did not even appear among the top eight causes of river water quality impairment in the late 1990s (U.S. EPA 1998b). Accidental death rates were down more than 35% from 1970 (National Safety Council 2000).

The Emerging Challenge

Our evolving knowledge of natural science and engineering opens new doors to measuring and understanding environmental issues. Evolving social science and preferences identify new approaches and demands for improving regulation. To improve the regulatory system, we now often deal with problems that cannot be readily detected with our senses, whose observable effects, if any, may be seen only statistically after a significant time lag and cover a geographic area ranging from a neighborhood to the entire planet. How these problems should be regulated also has undergone an evolution in thinking. Rising costs of control; the potential involvement of previously unregulated sectors; and concerns about fairness, participation, and the appropriate level for government action have led to new opportunities for and barriers to improving regulation.

Numerous stakeholders believe that one or another of these problems requires improvements to the existing system. Various "big think" documents emerged in the 1990s with suggestions for regulatory improvement under the auspices of well-respected institutions: the Carnegie Commission (1993), the Center for Strategic and International Studies/Enterprise for the Environment (1998), the National Academy of Public Administration (1997), the Office of Technology Assessment (U.S. Congress, OTA 1995), the President's Commission on Risk Assessment and Risk Management (1997a, 1997b), Resources for the Future (Davies and Mazurek 1998), and Yale University (Certow and Esty 1997).

Our review of these and other proposals, the academic literature, and our own experience leads to a view that substantial improvements in regulation require an integration of the "top-down" big picture and the "bottom-up" proof of concept that connects in detail to what really exists in the field. The studies in this book are examples of the bottom-up approach. Ultimately, they are both specific suggestions for regulatory improvement derived from policy-focused research and a proof of concept for methods to improve policy research. The chapters demonstrate that easy answers are difficult to obtain when one confronts details, but that answers to some questions exist by dint of hard work. If the answers to problems of environmental, health, and safety regulation were as easy to achieve as traveling to the Moon, we might reasonably expect to have solved them by now. Instead, we are mid-flight, and course corrections and opportunities for improvement exist.

What Direction Reform?

Suggestions for improving regulation are not consistently targeted at the same performance objectives. It is no surprise to find that different parties want different things (for example, Sexton and others 1999). Some parties primarily wish to increase the benefits, or efficacy, of a regulation, such as improved environmental quality or safety. Other parties want to include costs by comparing costs with benefits or to find the least expensive (most cost-effective) way of achieving a goal. In this view, the direction for regulatory improvement is that of finding the level of action that maximizes the sum of benefits minus costs, where both are measured in economic terms that typically ignore distributional consequences.[1] Many stakeholders find this prescription overly narrow. Although they care about efficiency, they also care about other objectives, such as equity and consent (Slovic, Fischhoff, and Lichtenstein 1980; National Research Council 1989, 1996; Lave 1996; Farrow 1998; Sexton and others 1999; Morgan 2000). As a result, other parties are more concerned with matters of distributional equity and procedural fairness. The direction of improvement is itself the subject of debate.

Added to concern about the direction of improvement is concern about various kinds of uncertainty and the very notion of trade-offs. Political or other aspects of regulatory decisions can influence decisionmakers who may not want to make choices strictly on the basis of an average or expected level of satisfaction. Under some circumstances, other decision rules—such as minimizing the worst possible outcome or cautiously waiting for new information—make good sense. Beyond such decision rules that remain fundamentally economic, rights-based formulations do not frame the question in terms of benefits and costs that are fungible but rather in terms of right and wrong, or rights and duties. Rights-based advocates argue that some issues, such as lead pollution that reduces the IQs of young children, simply may not be traded off for economic or other benefits and that such trade-offs should never be contemplated, let alone allowed.

Many economists or business-trained analysts, steeped in utility-based (that is, satisfaction-based) thinking, find rights-based formulations frustrating. They argue that anything can be traded off if the price is right and that, in principle, any rights-based formulations can be recast in terms of utility functions with large nonlinearities. While perhaps true, much of our legal, political, and religious discourse is framed in terms of rights and responsibilities, not benefits and costs. People who think in terms of rights and responsibilities typically will oppose economic-based reformulations of their beliefs.

Ultimately, a regulation is a codification of a resolution of the above concerns. The level and type of performance can be embedded in a regulatory design in various ways. Some ways, such as standard setting based on "best available technology" or the "minimum detectable level," duck the problems of estimating benefits and costs and fall back on physical constraints that often are based on intended benefits. However, in the face of rapidly evolving technology, such constraints are rarely stable. Moreover, terms such as "best available" must be given meaning, which can result in costs becoming an implicit consideration. Because the treatment is not explicit, it often varies by case or by regulation. This variation is one of the major challenges in moving toward more explicitly performance-based regulation. Several chapters in this volume (Chapters 11, 16, and 19) demonstrate broad-based analytical tools that help decisionmakers choose a level of action.

After careful reflection, our society may conclude that there are good reasons not to use the same regulatory design and performance criteria in all risk-management legislation. We may want to choose rights-based formulations in some cases and economic-based formulations in other cases. We may want to vary the level of control systematically, depending on the social context. However, much of the current variability in existing legislation does not reflect any broad consideration of national values and policy objectives. It arose from a plethora of incremental political and legislative decisions, most made without any overall philosophical consideration of how our society wants to manage risks and various levels of scientific input. Several chapters attest to this history, particularly Chapters 2, 3, and 4.

However, once a regulatory agency exists, it develops a culture of its own, along with a momentum for self-preservation and expansion. It continues to try to do what it has learned how to do, resisting fundamental change. Most regulatory agencies grew during an era in which risk problems were relatively easy to see, taste, and smell. As a consequence, the big problems lay in enforcement, not in choosing whether and how to act. The result was that legal and institutional considerations shaped the modern culture of regulatory organizations. The principal problem-solving strategy became adversarial. Insights from natural and social science took a back seat. Science and analysis came to be viewed more as weapons to be deployed in the adversarial process than as vehicles for gaining insight on whether and how to proceed. As a result, the substantive issues often became secondary to the adversarial process—to the objective of "getting" a polluter, to "scoring points," to winning the argument, and to pushing the standard up or down. Once this happens, goals of social efficiency and equity can get lost in the shuffle. The organizational dynamic is true of other parties, such as environmental advocacy groups and trade associations, which are regular players in the regulatory process. Institutional issues appear throughout this book but appear most strongly in Part 1.

Even in the best of worlds, good science is rarely sufficient for informed regulatory decisionmaking. Not understanding or choosing to ignore this basic fact, many people adopt a simple two-stage model in which science produces facts that are fed to decisionmakers who set policies. Such a model underlaid proposals several decades ago for the establishment of "science courts." In reality, what usually is needed is a three-step process involving science, other information, and values. Raw scientific findings are rarely directly applicable to regulatory decisionmaking. For the results of research to be useful, they must be combined with other knowledge and value judgments and then analyzed. This process of analysis often can be quite demanding, requiring every bit as much effort, care, and formal training as more discipline-focused scientific research.

Because of specific legislation, a short-term adversarial mind-set, institutional momentum, and inadequate research, regulators often adopt a narrow outlook on the risks that they manage. Similarly, industry, environmental groups, or others being regulated have their own organizational procedures and perceptions to go along with their self-interest. Consequently, regulatory actions frequently have imposed unintended risks and costs. In some cases—such as the Consumer Product Safety Commission's requirement that children's sleepwear be treated with a flame retardant that was later found to be a carcinogen—a more-understood risk was replaced by one that was less understood. A similar evolution has taken place with fuel additives in automobiles, as described in Chapter 4. However, the requirements of enabling legislation and the Administrative Procedure Act often transform even minor efforts at change into major activities that lead to substantial inertia to the system.

Proof of Concept: Improving Regulation and Improving Regulatory Analysis

The challenge of improving regulation is problem-driven, both by new problems and by new opportunities to revisit old problems. The chapters in this book provide proof of concept for improving regulation and regulatory analysis. Innovative approaches are presented for the design of performance-based regulation for the workplace exposure of genetically susceptible individuals, the design of fire building codes, eco-labels and consumer warning labels, highway construction, safety inspections, air emissions, and more. As important, methods and techniques of analysis for improving regulation are demonstrated. Innovative proof-of-concept approaches include historical and institutional analysis to understand how unintended consequences can creep into regulation; life-cycle analysis as a complement to understanding the potential for unintended consequences; simulation methods to address uncertainty in natural science, engineering, behavior, and valuation; technology assessment to understand current constraints and opportunities and to forecast into the future; optimization models to develop regulatory designs; and the use of psychometric "mental models" and surveys to hear what the public and other stakeholders may be saying and to communicate regulatory desiderata.

The chapters of this book are organized into four parts, each with a thematic introduction. The themes we chose highlight both substantive topics and issues in regulatory analysis that we feel lag behind the more standard techniques of impact, benefit–cost, and survey-based analysis. The thematic areas are institutions and performance (Part 1), behavior and perception (Part 2), uncertainty and technology (Part 3), and design and performance (Part 4). Chapters in these sections raise general questions for improving regulation such as the following:

- What is the role of scientific expertise compared with place-based or public expertise in the development of environmental standards?
- What process or processes can help avoid unintentional consequences of regulation?
- What are the roles of experts, advocates, and public opinion?
- Instructions and warning labels can hold only so much information. How does one decide that one piece of information is important and another is not?
- Ranking risks is not the same as ranking risk-mitigation alternatives. How could risk rankings be used by policymakers trying to establish regulations?
- Suppose that people rank risks differently on the basis of the display and type of information provided. Which is the "correct" ranking?
- When are performance-based regulations easy to verify? What types of problems are difficult to verify? When are computer simulations good enough?
- How should performance standards be written when uncertainty about human behavior is significant? Should they assume typical behavior of an average person?
- Is it appropriate to use genetic testing to screen job applicants on the basis of potential occupational health risks?
- Should emissions control technologies or emission standards be mandated? Which would cost less, which would be easier to monitor, which would be more effective?
- Should analysts consider only direct impacts or impacts throughout a chain of economic activity?
- Can analysts distinguish the impact of alternative regulatory designs and programs on measures of performance such as net present value? What are the difficulties?
- What role can tools such as extended input–output analysis and computerized benefit–cost analysis play in regulatory development and stakeholder discussion? What role should these tools play?

The chapters address these and other questions in the context of specific regulatory issues. Water pollution regulation from 1914 to the present forms the core of Part 1 on institutions; historical and case study methods drive the analytic approaches. The regulatory histories by Day (Chapter 2) on the interaction between the Safe Drinking Water Act and Superfund, Gurian and Tarr (Chapter 3) on the development and evolution of the first federal drinking water laws, and Stikkers (Chapter 4) on the interaction of air and water regulation in the debate about methyl-*tert*-butyl ether (MTBE) as a gasoline additive are all studies of institutional and stakeholder behavior. Boyd (Chapter 5) then investigates whether there is a consistent bias against decisions regarding pollution prevention in corporations, whereas Farrell (Chapter 6) reviews the rapidly changing face of the electric power sector and the way in which environmental issues are, or are not, incorporated into economic regulation in a multistate setting.

Part 2, on behavior and perception, begins with two chapters on labeling. Riley and others (Chapter 7) show how careful research into how paint strippers are used can improve warning labels that will reduce the risk to many users. Nadaï (Chapter 8) demonstrates the complexity of the process for establishing an international eco-label and how the interaction of varied stakeholders brings about some unintended results. The remaining two chapters in Part 2 provide examples of how public concerns can be assessed using mental model and survey approaches. Follin and Fischbeck (Chapter 9) show that different subgroups of the population differ on what they find important and underestimate the variety of opinions held by others. DeKay and others (Chapter 10) demonstrate a robust process by which risk rankings can be assessed from focus groups of laypeople.

The chapters on uncertainty and technology in Part 3 cover issues as diverse as fire performance standards for buildings (Notariani and Fischbeck in Chapter 11) and health regulation as linked to genetic testing (Ponce and others in Chapter 12). Going further afield, Corbett and Fischbeck (Chapter 13) survey their work identifying ocean shipping as an overlooked source of pollution and discuss various technologies and policies to reduce ship emissions; Macauley and Brennan (Chapter 14) investigate the emerging technologies of satellite imaging in support of environmental compliance objectives. Analytical methods include simulation, value of information from statistical analysis, and technological assessment.

Part 4, on design and performance, addresses the substance of safety in the workplace as well as water and air quality. Analytical methods include the statistical analysis of surveys, augmented input–output analysis, quantitative optimization, and simulation. Designing inspections to reduce accidents in a heavy industrial environment is the subject of work by Shultz and Fischbeck (Chapter 15), which is complemented by the multi-industry analysis of accidents and their costs by Matthews and Lave (Chapter 16). The challenges of designing a water-quality trading program are investigated by Schultz and

Small (Chapter 17). The section concludes with two chapters on air pollution control. Matthews (Chapter 18) extends the work of EPA to isolate the net benefits of mobile and stationary source control. In the concluding chapter, Farrow and others (Chapter 19) test the sensitivity of EPA results to alternative health response functions. Chapter 19 can be read on its own; however, the full version of the flexibile risk and benefit–cost template used in the chapter—the Fast Environmental Regulatory Evaluation Tool (FERET)—is supplied on the compact disk accompanying this volume. FERET is a Windows-based program and is provided for academic and research purposes. It operates using Microsoft Excel, augmented with a commercial add-in, Decisioneering Crystal Ball, for full functionality.

Alternative Reading Paths

The thematic sections of the book are focused on issues and techniques to improve policy-focused research through context-rich illustrations. However, the chapters can be combined in various ways to follow a more traditional taxonomy of regulatory issues. Below, we illustrate several ways in which the chapters can be organized for different courses or interests.

The following reading sets, which focus on air, water, toxics and land, and health and safety, generally proceed from older, larger issues to newer, more detailed issues.

AIR

Chapter and Author(s)		*Title*
18	Matthews	Analysis of the Benefits and Costs of Clean Air
4	Stikkers	The Unintended Consequence of Reformulated Gasoline
19	Farrow and others	Facilitating Regulatory Design and Stakeholder Participation
6	Farrell	The Political Economy of Interstate Public Policy
13	Corbett and Fischbeck	International Technology Policy
14	Macauley and Brennan	Private Eyes in the Sky
5	Boyd	The Barriers to Corporate Pollution Prevention
8	Nadaï	The Impact of Industrial Strategy and Expert Information on Eco-Labels
9	Follin and Fischbeck	Trade-Offs among Environmental, Human Health, and Quality-of-Life Impacts
10	DeKay and others	The Use of Public Risk Ranking in Regulatory Development
7	Riley and others	Behaviorally Realistic Regulation

WATER

Chapter and Author(s)		Title
2	Day	Safe Drinking Water—Safe Sites
3	Gurian and Tarr	The First Federal Drinking Water Quality Standards and Their Evolution
4	Stikkers	The Unintended Consequence of Reformulated Gasoline
5	Boyd	The Barriers to Corporate Pollution Prevention
9	Follin and Fischbeck	Trade-Offs among Environmental, Human Health, and Quality-of-Life Impacts
10	DeKay and others	The Use of Public Risk Ranking in Regulatory Development
17	Schultz and Small	Integrating Performance in the Design of a Water Pollution Trading Program

TOXICS AND LAND

Chapter and Author(s)		Title
2	Day	Safe Drinking Water—Safe Sites
4	Stikkers	The Unintended Consequence of Reformulated Gasoline
5	Boyd	The Barriers to Corporate Pollution Prevention
8	Nadaï	The Impact of Industrial Strategy and Expert Information on Eco-Labels
9	Follin and Fischbeck	Trade-Offs among Environmental, Human Health, and Quality-of-Life Impacts
10	DeKay and others	The Use of Public Risk Ranking in Regulatory Development
7	Riley and others	Behaviorally Realistic Regulation
17	Schultz and Small	Integrating Performance in the Design of a Water Pollution Trading Program
12	Ponce and others	Genetic Testing and the Workplace

HEALTH AND SAFETY

Chapter and Author(s)		Title
16	Matthews and Lave	Evaluating Occupational Safety Costs and Policy in an Input–Output Framework
3	Gurian and Tarr	The First Federal Drinking Water Quality Standards and Their Evolution
15	Shultz and Fischbeck	Workplace Accident and Compliance Monitoring
9	Follin and Fischbeck	Trade-Offs among Environmental, Human Health, and Quality-of-Life Impacts
10	DeKay and others	The Use of Public Risk Ranking in Regulatory Development

11	Notarianni and Fischbeck	Performance with Uncertainty
12	Ponce and others	Genetic Testing and the Workplace

The chapters also may be viewed in more traditional categories of policy and social science research, although part of our message is the usefulness of blending insights from various approaches. Following are several different "disciplinary" packages ranging from history to engineering.

HISTORY AND PUBLIC ADMINISTRATION

Chapter and Author(s)		*Title*
2	Day	Safe Drinking Water—Safe Sites
3	Gurian and Tarr	The First Federal Drinking Water Quality Standards and Their Evolution
4	Stikkers	The Unintended Consequence of Reformulated Gasoline
6	Farrell	The Political Economy of Interstate Public Policy
8	Nadaï	The Impact of Industrial Strategy and Expert Information on Eco-Labels
11	Notarianni and Fischbeck	Performance with Uncertainty
15	Shultz and Fischbeck	Workplace Accident and Compliance Monitoring

ECONOMICS

Chapter and Author(s)		*Title*
18	Matthews	Analysis of the Benefits and Costs of Clean Air
19	Farrow and others	Facilitating Regulatory Design and Stakeholder Participation
16	Matthews and Lave	Evaluating Occupational Safety Costs and Policy in an Input–Output Framework
17	Schultz and Small	Integrating Performance in the Design of a Water Pollution Trading Program
5	Boyd	The Barriers to Corporate Pollution Prevention
8	Nadaï	The Impact of Industrial Strategy and Expert Information on Eco-Labels

POLICY ANALYSIS

All chapters.

PSYCHOLOGY AND BEHAVIOR

Chapter and Author(s)		*Title*
5	Boyd	The Barriers to Corporate Pollution Prevention

8	Nadaï	The Impact of Industrial Strategy and Expert Information on Eco-Labels
9	Follin and Fischbeck	Trade-Offs among Environmental, Human Health, and Quality-of-Life Impacts
10	DeKay and others	The Use of Public Risk Ranking in Regulatory Development
7	Riley and others	Behaviorally Realistic Regulation
15	Shultz and Fischbeck	Workplace Accident and Compliance Monitoring
11	Notarianni and Fischbeck	Performance with Uncertainty

ENGINEERING, TECHNOLOGY, AND SYSTEM DESIGN

Chapter and Author(s)		*Title*
13	Corbett and Fischbeck	International Technology Policy
14	Macauley and Brennan	Private Eyes in the Sky
5	Boyd	The Barriers to Corporate Pollution Prevention
11	Notarianni and Fischbeck	Performance with Uncertainty
15	Shultz and Fischbeck	Workplace Accident and Compliance Monitoring
16	Matthews and Lave	Evaluating Occupational Safety Costs and Policy in an Input–Output Framework
18	Matthews	Analysis of the Benefits and Costs of Clean Air
19	Farrow and others	Facilitating Regulatory Design and Stakeholder Participation
17	Schultz and Small	Integrating Performance in the Design of a Water Pollution Trading Program

STATISTICS AND QUANTITATIVE MODELING

Chapter and Author(s)		*Title*
12	Ponce and others	Genetic Testing and the Workplace
16	Matthews and Lave	Evaluating Occupational Safety Costs and Policy in an Input–Output Framework
15	Shultz and Fischbeck	Workplace Accident and Compliance Monitoring
9	Follin and Fischbeck	Trade-Offs among Environmental, Human Health, and Quality-of-Life Impacts
11	Notarianni and Fischbeck	Performance with Uncertainty
18	Matthews	Analysis of the Benefits and Costs of Clean Air
19	Farrow and others	Facilitating Regulatory Design and Stakeholder Participation
17	Schultz and Small	Integrating Performance in the Design of a Water Pollution Trading Program

Note

1. Advanced practice, such as that suggested in the Office of Management and Budget 1996 and in Farrow 1998, include distributional issues in efficiency calculations.

References

Carnegie Commission, The. 1993. *Risk and the Environment: Improving Regulatory Decision Making.* New York: The Carnegie Commission.

Center for Strategic and International Studies/Enterprise for the Environment. 1998. *The Environmental Protection System in Transition: Toward a More Desirable Future.* Washington, DC: Center for Strategic and International Studies.

Certow, M., and D. Esty (ed.). 1997. *Thinking Ecologically: The Next Generation of Environmental Policy.* New Haven, CT: Yale University.

Davies, J.C., and J. Mazurek. 1998. *Pollution Control in the United States.* Washington, DC: Resources for the Future.

Farrow, S. 1998. Environmental Equity and Sustainability: Rejecting the Kaldor–Hicks Criteria. *Ecological Economics* 27(2): 183–188.

Lave, Lester. 1996. Benefit–Cost Analysis. Do the Benefits Exceed the Costs? In *Risks, Costs, and Lives Saved: Getting Better Results from Regulation,* edited by Robert Hahn. Oxford, U.K.: Oxford University Press.

Morgan, G.M. 2000. Risk Management Should Be about Efficiency and Equity. *Environmental Science and Technology* 34(1): 32A–34A.

National Academy of Public Administration. 1997. *Resolving the Paradox of Environmental Protection.* Washington, DC: National Academy of Public Administration.

National Research Council. 1989. *Improving Risk Communication.* Committee on Risk Perception and Communication. National Research Council: Washington, DC.

———. 1996. *Understanding Risk: Informing Decisions in a Democratic Society.* National Research Council: Washington, DC.

National Safety Council. 2000. *Injury Facts.* Itasca, IL: National Safety Council.

Office of Management and Budget. 1996. *Economic Analysis of Federal Regulations under Executive Order 12866.* http://www.whitehouse.gov/OMB/inforeg/riaguide.html/#iii (accessed March 5, 2001).

President's Commission on Risk Assessment and Risk Management. 1997a. *Framework for Environmental Health Risk Assessment.* Washington, DC: President's Commission on Risk Assessment and Risk Management.

———. 1997b. *Risk Assessment and Risk Management in Regulatory Decision-Making.* Washington, DC: President's Commission on Risk Assessment and Risk Management.

Sexton, Ken, A. Marcus, W. Easter, and T. Burkhardt. 1999. *Better Environmental Decisions.* Washington, DC: Island Press.

Slovic, P., B. Fischhoff, and S. Lichtenstein. 1980. Facts and Fears: Understanding Perceived Risk. In *Societal Risk Assessment,* edited by Richard Schwing and Walter Albers. New York: Plenum, 181–216.

U.S. Congress, OTA (Office of Technology Assessment). 1995. *Environmental Policy Tools: A User's Guide.* Washington, DC: OTA.

U.S. EPA (Environmental Protection Agency). 1998a. *National Air Quality and Emissions Trends Report.* January. Washington, DC: U.S. EPA, Office of Air Quality Planning and Standards.

———. 1998b. *National Water Quality Inventory.* March. Washington, DC: U.S. EPA, Office of Water.

PART I

Institutions and Performance

Quantitative analysts and advocates of rational government often forget that real people in very different institutions play roles in the regulatory drama. The people have their own interests and training; each institution has its own constituency, culture, process, and objectives. The very human cast of the regulatory system is composed of federal, state, and local regulators; environmental, citizen, and other public groups; industry trade associations, individual firms, and not-for-profit entities such as municipal water utilities; and others. What is desirable performance for one group or individual often has a different weight in or runs counter to the objectives of another group. Many stakeholders in the regulatory arena, including economists, claim to represent the interests of the general public.

What are the actual strategies and actions of stakeholders? Where do organizational politics, science, technology, and performance enter the activity of improving regulation? The chapters in this section address such questions in the context of specific regulatory issues.

Water regulation is one strand of investigation in the first three papers. In Chapter 2, Day investigates the intended and unintended links between drinking water regulation and cleanup levels at federal hazardous waste sites. In Chapter 3, Gurian and Tarr take us farther back with a look at the interstate commerce basis of federal drinking water regulation. This historical perspective echoes many contemporary issues such as federalism, the use of science, and the structure of the regulatory development process. In Chapter 4, Stikkers builds on the theme of water regulation with a still-unfolding story on the use of methyl-*tert*-butyl ether (MTBE) as a gasoline additive. He describes the various institutions focused on improving air quality through MTBE use and the resulting backlash as its impact on water quality became apparent.

Boyd opens a different line of institutional research in Chapter 5. In light of concerns that corporations systematically ignore opportunities to prevent pollution at the source instead of through end-of-pipe treatments, Boyd investigates three cases of failed corporate pollution prevention efforts. In general, he finds no "smoking gun" of ignored opportunity but instead finds the complexity of large organizations seeking the best investment projects in the face of market, regulatory, and technological uncertainty. In Chapter 6, Farrell concludes this section with a view of conflicting governmental institutions and an emerging institutional hybrid, the multistate and multistakeholder group. His context is the deregulation of the electric power market, and his competing institutions are primarily the Federal Energy Regulatory Commission and the U.S. Environmental Protection Agency.

These chapters raise numerous questions for research and discussion. While reading, you might consider questions such as these:

- Do regulations exist independent of one another, or are they part of an intentionally or unintentionally integrated system?
- What is the role of scientific expertise compared with place-based or public expertise in the development of environmental standards?
- What process or processes can help avoid unintentional consequences of regulation? Are the consequences unintentional? Are they beneficial?
- How do for-profit firms, environmental groups, and regulators behave in the development of and the response to regulation?
- What is the role of science and technology?
- What are the roles of experts, advocates, and public opinion? What other groups participate in regulatory development and implementation?

2

Safe Drinking Water—Safe Sites

Interaction between the Safe Drinking Water Act and Superfund, 1968–1995

JARED N. DAY

On December 16, 1974, President Gerald Ford signed into law the Safe Drinking Water Act (SDWA). This legislation granted the U.S. Environmental Protection Agency (EPA) authority to regulate levels of contamination in the nation's drinking water supplies. Built on the loosely enforced and incomplete standards of the U.S. Public Health Service (PHS) from 1962 (see Chapter 3), EPA gradually developed a more sophisticated "multiple barrier" approach to water protection that required different actions: assessing drinking water sources, monitoring wells and collection systems, and improving the qualifications of operators and technical staff at the state and local levels.

As their primary measure of contamination, the U.S. Congress and EPA established a series of National Drinking Water Regulations—legally enforceable standards that applied to all drinking water supplies. These regulations usually took the form of maximum contaminant levels (MCLs), a set of strict numeric values established by EPA that included radionuclides as well as organic, inorganic, and microbial contaminants. With these regulations, Congress significantly expanded EPA's powers to regulate toxic contaminants compared with the narrow 1962 PHS standards, which applied only to bacteriological contaminants. Along with the Clean Air Act (1970) and the Clean Water Act (1972), SDWA became part of the central legislative bedrock on which EPA attempted to regulate environmental threats to public health.

Later environmental legislation built on the regulatory procedures and rules established under SDWA. Laws such as the Resource Conservation and Recovery Act (1976); the Toxic Substances Control Act (1976); the Comprehensive Environmental Response, Compensation, and Liability Act (CERCLA) (1980); and important reauthorizations of CERCLA and SDWA in 1986 and later further expanded EPA's regulatory agenda. The laws routinely taxed the agency's scientific, technical, and administrative capacities. Additionally, the inherent tension between the SWDA's legislative goals set by Congress and EPA's administrative and scientific resources and capabilities created a series of unintended consequences within the drinking water program itself and in other programs and agencies. The origins and implementation of SDWA showcase these tensions as MCLs—the complex regulatory standards designed strictly as water regulations in the 1970s—became central regulatory standards for EPA's largest toxic substance program, Superfund.

The information in this chapter is gleaned from congressional hearings, EPA records, federal archive materials, interviews, and related primary documents as well as secondary sources. The first section is a review of the historical context in which the safe drinking water program was created. In the next section, I explore the primary measurements developed by EPA to determine drinking water safety—the MCLs—and the agency's initial efforts at developing them. In subsequent sections, I examine the creation of the Superfund program in 1980, the conflicting ideological views of Congress and President Ronald Reagan's EPA, and the eventual clash over cleanup standards and timetables. Also explored are the landmark 1986 reauthorizations of SDWA and Superfund, in which Congress imposed strict rules, schedules, and mandates on both programs. One of the central elements of the Superfund reauthorization was the linkage of Superfund cleanup standards to the MCLs proscribed by SDWA. This linkage had important long-term consequences for both programs from the late 1980s through the 1990s.

Origins of the SDWA Standards Regime

Congress grappled with safe drinking water legislation throughout the late 1960s. The PHS water quality standards proved inadequate in the face of rising public concern about drinking water quality and the expanding complexity of America's water contamination problems. Although the PHS standards set limits on many known bacteriological, organic, and inorganic contaminants, the regulations were binding on only about 700 interstate drinking water carriers, fewer than 2% of the nation's water supply systems (Train 1974, 562; see Chapter 3). Even though the standards had a broad effect because all 50 states adopted them as regulations or guidelines, enforcement on the federal and state levels was abysmal. In 1969, PHS began a comprehensive review of 969 water supplies to inform a revision of its standards. In their final report, PHS officials found that 41% of the systems studied did not meet state or federal standards (U.S. PHS 1970; U.S. EPA 1971, 20–21). In addition, states seldom monitored their drinking water supplies, especially in small communities (Cotruvo and Vogt 1990, 7).

Despite these shortcomings, Congress's goals in 1968 and 1969 were quite modest: to revise and update the PHS standards. For example, the PHS administrator described the proposed Safe Drinking Water Act of 1969 as "proposals limited to minor, technical, or clarifying amendments" (Johnson 1968).

A series of federal studies from 1972 through 1974 quickly brought national attention to the weaknesses of the PHS standards. EPA Region VI's landmark 1972 publication, *Industrial Pollution of the Lower Mississippi River in Louisiana,* and an equally important 1974 EPA follow-up were key to setting in motion a series of similar public and private investigations into drinking water quality. The Mississippi River study, which included studies of contamination in the drinking water supplies for the city of New Orleans, Louisiana, found high concentrations of 86 contaminants, including 66 organic compounds from one supplier (U.S. EPA 1972; *Environmental News* 1974, 1–3; Pontius 1999a). In 1973, a General Accounting Office study of water supplies in Maryland, Massachusetts, Oregon, Vermont, Washington, and West Virginia found that of the 446 systems studied, only 60 fully complied with the PHS bacteriological standards (Symons 1984, 34).

These federal studies were followed in quick succession by independent investigations that exposed similar (or worse) problems. *Consumer Reports* published a damning three-part report on water quality standards based on its investigation. The Environmental Defense Fund (EDF) conducted another study of the New Orleans water supply and found that residents drinking treated Mississippi River water had a higher probability of developing cancer than people in surrounding areas who drank water from groundwater sources. In addition, on December 5, 1974, CBS television aired a program with Dan Rather entitled "Caution, Drinking Water May Be Dangerous to Your Health" (Harris and Brecher 1974; Pontius and Clark 1999).

After these and other investigations (and the resulting wave of negative publicity) surrounding drinking water contamination, the pressure for meaningful change became inexorable. After more than four years of off-and-on efforts, Congress finally enacted legislation to develop a national program to protect the quality of the nation's public drinking water systems. The architects of the 1974 Safe Drinking Water legislation tried to establish a cooperative program among local, state, and federal agencies and require the creation of primary drinking water regulations that would ensure safe drinking water for consumers. These regulations were the first to cover both chemical and microbial contaminants and the first to apply to all public water systems in the United States. The House of Representatives passed a Senate-amended bill on December 3, 1974, and President Ford signed the SDWA 13 days later (Pontius 1999a).

The 1974 act specified the process by which EPA was to adopt national drinking water regulations. Interim regulations, later renamed National Interim Primary Drinking Water Regulations, were to be adopted within six months of enactment of SDWA. In this chapter, most references to SDWA "standards" refer specifically to these and later interim standards. By March 1977, two and a half years after SDWA's passage, EPA proposed Revised National Drinking Water Regulations based on a National Academy of Science (NAS) study of health effects of contaminants in drinking water. EPA went through a two-step process. First, it needed to publish recommended MCLs (RMCLs), later renamed MCL goals (MCLGs), for contaminants thought to have adverse health effects on the basis of the NAS study. The RMCLs needed to be set at a level, with an adequate margin of safety, such that no known or anticipated health effect would occur. These levels were to act only as health goals and were not federally enforceable (Pontius and Clark 1999).

In a second step, EPA officials established MCLs as close to the RMCLs as the agency thought technologically or economically feasible. If it was not feasible to determine the level of a contaminant, SDWA authorized EPA to establish a required treatment technique instead of an MCL. Together, the MCLs and treatment techniques would eventually comprise the final, federally enforceable National Primary Drinking Water Regulations (Pontius 1999a).

The Challenge of Establishing MCLs

EPA took on a complex task that would severely test the agency's administrative, scientific, and financial resources over the next 26 years. Both government studies and private researchers in the early 1970s routinely noted that water systems were contaminated with substances whose toxicity or risk levels to human health were unknown (Harris and Brecher 1974, 440). In their efforts to impress legislators, SDWA proponents routinely noted that public water supplies contained up to 12,000 contaminants whose effects on human health were unknown (McDermott, Kabler, and Wolf 1970, 11). For example, the concentration levels at which mercury and molybdenum caused adverse health effects had yet to be determined (McDermott 1971, 13–14). Highlighting the inadequacy of existing scientific methods concerning organic contaminants, James McDermott, director of EPA's Water Hygiene Division, noted in June 1972 that "there is a need to develop simple rapid methods for detection and quantification of bacteria, chemicals, and toxic agents in water. Many of the current assay methods are complicated and time-consuming and, therefore, are not applicable to routine monitoring of drinking water supplies" (McDermott and Cofrancesco 1972, 21–22).

Federal environmental officials noted that the very idea of identifying contaminant levels for thousands or even hundreds of contaminants was fantasy and that many other aspects of the drinking water legislation were equally unrealistic (Johnson 1971, 2–7). Officials at EPA viewed the problem in much smaller terms. Throughout the legislative debates on safe drinking water in the early 1970s, EPA officials planned to simply use the existing PHS standards, update them if necessary, and conduct new research on a small set of contaminants, such as trihalomethanes (THMs).

Instead of addressing hundreds or thousands of contaminants, EPA officials planned immediate implementation for 20–30 contaminants (Reizen 1972, 2). Even this task was more complicated than it appeared on the surface because some of the PHS standards—notably, the standards for arsenic and selenium—were suspect, and years of research were needed to establish their standards with greater certainty (McDermott, Kabler, and Wolf 1970, 12–13). The true health significance of the whole category of synthetic organic chemicals in drinking water was not known. Many other questions regarding the health significance of low concentrations of organic chemicals and disinfection by-products were unanswered. EPA officials had long recognized that establishing MCLs as absolute numbers would require more extensive scientific and administrative work than that required by the PHS (Lee 1972). In August 1973, EPA officials noted that the agency's authority "will be broadened, not only in depth but in the scope of the systems included. We will move from being primarily concerned with standards governing the prevention of communicable diseases and the annual classification of 650 water supply systems to the establishment of much broader standards applicable to at least 40,000 [water systems] and potentially as many as 250,000" (U.S. EPA n.d.).

Some officials looked beyond the initial group of 20–30 contaminants and saw clearly that, in the future, EPA needed significant resources to identify and assess suspected toxic substances as well as new chemical compounds. In 1972, Ralph Nader noted, "What is required is a systematic policy which produces adequate [drinking water standards] with sufficient authority and funds to enforce them, and prompt revision of these standards with adequate planning of water resources for the future. The emphasis must be on prevention, on foreseeing and forestalling the long-term as well as short-term risks and ravages on human health from the chemicals, bacteria, pesticides, toxic metals, radioactive substances, hormones, viruses, and many other wastes which are making our water heavy with silent violence" (U.S. Senate 1972, 5). Officials from the Conference of State Sanitary Engineers stated succinctly in May 1973 that "the success of a national water program is no stronger than its weakest link, so attention must be directed to ample provisions and funding for research; training of operating, teaching, and regulatory personnel; Federal, State, and local staffing; and demonstration of new methodology" (Policy Statement 1973).

The need for a systematic process and adequate funding was further underscored by the expanding nature of the contamination problem. In the early 1970s, with an average of 500 new chemical compounds coming to the market each year from chemical manufacturers, the need to have institutional procedures in place to assess them and set standards seemed obvious (McDermott, Kabler, and Wolf 1970, 11). Indeed, some recognized that without a strong, sustained effort at keeping the standards current, the program might be crippled. In 1973, one EPA official noted that "given no added resources, it will not be possible [under the proposed SDWA legislation] to even promulgate the new drinking water standards by the date required in the pending legislation" (U.S. EPA n.d., 2).

For most of the years following the passage of SDWA, the resources and commitment to fulfill EPA's mission were notably lacking. By 1977, after officials from EPA and NAS had updated the PHS standards, the standards-setting process became bogged down in the pursuit of data and standards for high-profile contaminants showcased in the New Orleans studies such as THMs and synthetic organic chemicals (Cotruvo and Vogt 1990, 9). Setting MCLs, and establishing a rational process for setting priorities and standards for other common contaminants in general, moved at an extremely slow pace. After the establishment of MCLs for the initial PHS group of 22 contaminants in 1976, EPA set an MCL for only one contaminant between 1976 and 1986 (Table 2-1). For some observers, it was no surprise. Shortly after the passage of SDWA, EPA Administrator Russell Train noted, in specific reference to SDWA, "There will be little support [from the next Congress] for regulatory programs involving additional costs or threatening employment" (*Clean Water Report* 1974, 2). Through the late 1970s and into the 1980s, environmental groups such as EDF attacked EPA's unwillingness to vigorously pursue the setting of standards.

EPA's slow pace of setting standards was often caused by issues beyond the agency's control. For example, Congress had authorized NAS to advise EPA about specific numeric levels for standards during the 1970s and 1980s. As Rosenthal points out, even though the NAS received material and funding for this purpose, it declined to comment on proposed levels, choosing instead to assess the quality of the data on which standards were based. As Rosenthal notes, the NAS "perceived standard setting as a fundamentally regulatory function, and as a scientific body it was uncomfortable reducing complex toxicological information to a single number upon which EPA would base its standards." EPA attorneys, in turn, felt they could not avoid

TABLE 2-1. Total Number of Interim Contaminant Standards Set by Type, 1976–1998

Contaminant	*1976*	*1979*	*1987*	*1989*	*1991*	*1992*	*1998*
Organics	6	7	14	14	38	55	60
Microbials	2	2	2	6	6	6	7
Inorganics	10	10	10	10	14	19	19
Radionuclides	4	4	4	4	4	4	4
Total	22	23	30	34	62	84	90

Source: Data courtesy of U.S. EPA's Office of Ground Water and Drinking Water (OGWDW), 2000.

legal challenges without specific guidance from the NAS. The resulting conflict, according to one industry source, left the drinking water program "in a state of paralysis" during the 1970s and 1980s (Rosenthal 1991, 162). Additionally, EPA officials channeled resources toward high-profile contaminants identified in the public investigations of the early 1970s (such as THMs), and for many years, neither Congress nor the Executive Branch made standard setting a priority.

Despite the shortcomings of contaminant identification and the setting of new standards, by most accounts, SDWA effectively regulated the small group of contaminants for which MCLs had been set. State and federal enforcement procedures generated little controversy, and local authorities tended to accept the scientific and technical underpinnings of individual MCLs, at least initially. At times, modest opposition to elements of SDWA and the MCL regime manifested itself from state and municipal water suppliers through their organization, the American Water Works Association (AWWA). Even so, AWWA usually worked cooperatively with EPA, and its influence was clearly circumscribed. With the exceptions of vociferous criticisms from EDF, Ralph Nader's organizations, and a handful of others, the program was widely viewed as a successful model of state and federal cooperation. All of these strengths tended to de-politicize the program and mask for a time the slow pace at which new standards were being set. These strengths came into even sharper focus when Congress and the Carter and Reagan administrations took on the infinitely more complex and politically charged task of cleaning up large-scale toxic waste sites when Superfund, a program that was to develop regulatory ties to the SDWA MCLs, was created in 1980.

Origins of Superfund

Beginning in August 1978, widespread public concern over Love Canal and similar so-called toxic waste time bombs prompted the Carter administration to begin crafting a comprehensive program to clean up toxic waste. Officials from the U.S. Department of Transportation, the agency that historically had had the most dealings with hazardous waste sites, led an interagency task force to develop a program to deal with such situations. EPA Administrator Douglas Costle, however, wanted to use toxic waste site management to establish EPA's credentials in the field of public health (Landy, Roberts, and Thomas 1994, 140–142).

Before other agencies or Congress had concrete proposals, an EPA preimplementation planning group completed a draft of the National Contingency Plan (NCP). It was the major blueprint guiding Superfund response actions and a waste site management plan that outlined the decision points and general phases required for long-term cleanups (Stafford 1981, 10; Cohen and Tipermas 1983, 48). Under the leadership of James Florio (D-NJ), the House proposed a waste cleanup program that accepted the broad outlines provided by Costle's staff and left the program's technical and regulatory details to EPA. When the House and Senate took up the matter during a lame-duck session after the 1980 presidential elections, the drafting and the hurried deliberations over the bill's final language largely focused on funding and liability issues and largely overlooked the technical, scientific, and regulatory elements raised by a large federal cleanup program (Greenberg n.d., 7–37). On December 3, 1980, the House passed the Senate's Superfund bill, and on December 31, President Jimmy Carter signed CERCLA into law.

Because of congressional haste in passing the bill and EPA's still largely untarnished public reputation for technical expertise, Congress and the President left cleanup standards and many other aspects of program design to the agency. In its final draft, CERCLA did not contain any detailed statutory standards for cleanup. Section 104(c)(4) authorized EPA to select appropriate remedial actions, "which are to the extent practicable in accordance with the national contingency plan and which provide for that cost-effective response which provides a balance between the need for the protection of public health and welfare and the environment" (Environmental Law Institute 1986, 10). As Florio later stated, "CERCLA gave EPA virtually unlimited discretion in carrying out its provisions. For example, the statute's central cleanup provision gave the agency open-ended authority to respond to a release or threatened release of any 'hazardous substance, pollutant, or contaminant' that 'may present an imminent and substantial danger to the public health.' Each of these key terms was defined in the broadest possible manner" (Florio 1986, 356–357).

By the close of 1980, Costle had largely achieved his goal of locating an important new public health program within EPA. Congress also had given the agency substantial discretion to administer the program, set its priorities, and establish a strong federal presence in the area of toxic waste cleanup (Cohen and Tipermas 1983, 57).

Reagan's EPA, Superfund, and the Slow Pace of Toxic Waste Cleanup

Ronald Reagan's victory over Jimmy Carter in the 1980 presidential election signaled a fundamental shift in the focus of CERCLA. As conservative Reagan appointees took the program in hand, CERCLA was recast using very different ideological and administrative priorities that were strongly influenced by the highly polarized political environment of the times. In addition to the ideological differences, this environment was created largely by the mutual distrust between the group made up of environmental lobbyists, many EPA officials, and congressional staffers on the one hand, and conservative legislators and industry lobbyists on the other. Within the narrow group of congressional staffers and Carter EPA officials who wrote up the original legislation, there was a clear ideological disposition for a strong federal program. As Landy, Roberts, and Thomas (1994) argue, these staffers routinely displayed contempt for the capabilities of states and municipalities. In the view of one former state official, these staffers suffered from the so-called Mississippi Syndrome—focusing on the worst examples of state government behavior and assuming that such behavior is the norm. Compounding this contempt for state administration, staffers and environmental lobbyists also were extremely suspicious of chemical industry representatives, often assuming that the dangers linked with hazardous waste were as life threatening as many extreme environmentalists argued. They viewed the Superfund debate as simply the latest skirmish in an ongoing battle to make polluters responsible for the toxic messes they created. For policymakers and legislators who shared these and similar views, the main purpose of Superfund was to establish new environmental rights and to establish clear lines of corporate responsibility. Establishing efficient, rational, and feasible cleanup levels was of secondary importance (Landy, Roberts, and Thomas 1994, 165).

The ideological stridence of many CERCLA supporters was equally matched by the partisan views of Superfund's conservative critics. Among conservative state activists, industry lobbyists, and particularly within Reagan's policymaking circle, environmentalists and EPA were viewed with deep suspicion. Reagan, in assessing the goals of environmentalists and EPA stated, "What they believe in is a return to a society in which there wouldn't be the need for industrial concerns" (Cannon 1982, 287). Many conservative policymakers within the new administration viewed the environmental movement as a sham dominated by anti-American radicals greedy for power. For example, Anne Gorsuch, Reagan's choice to lead EPA, stated in February 1981 that, "In my experience with Washington-based environmental lobbyists, their main concern is seeing how much money they can raise for their organizations by scaring the American public half to death. The truth about the vast majority of them is that they are not interested in the environment at all. They are interested in power, political power, and the environment is just a platform for them." EPA staffers were suspected as well, as she argued, "In my opinion [EPA staff] are simply anti-industry and anti-business. They want to make controls as costly as possible, purposely, almost as if they think it is somehow possible to go back to nature.... I believe that if they could, they would happily make regulations so costly that they would have the effect of cutting off all economic growth. They would be happy to deindustrialize the United States" (Burford with Greenya 1986, 98).

The professional backgrounds and orientations of many Reagan appointees contributed to the suspicion of environmentalists and EPA. Many top-level officials and advisors came from corporate backgrounds, and they routinely viewed EPA as a bloated agency rife with mismanagement and waste. They viewed the agency as a bureaucratic and regulatory lodestone around the country's economy. Norman Livermore, the top environmental official in Reagan's administration while he was governor of California, noted that Reagan "really hasn't had any environmental advisors.... So things are left up to [Joseph] Coors and people like that, who feel they've been strictured [sic] by environmental laws and regulations." For many Reagan policymakers, the central environmental problem was EPA's overregulation of U.S. corporations, and their solution was, in part, to rein in the agency (Cannon 1982, 369). In January 1981, Fred Khedouri, the budget director of the Office of Management and Budget, interviewed John Hernandez for the post of Gorsuch's deputy administrator. At one point, Khedouri leaned over in his chair, and in all seriousness, asked Hernandez, "Would you be willing to bring EPA to its knees?" (Burford with Greenya 1986, 84).

These ideological undercurrents significantly influenced the administration of the Superfund program. Top-level decisionmakers such as Gorsuch and David Stockman, the director of the Office of Management and Budget under the Reagan administration, decided that Superfund should end when it came up for reauthorization. They also decided that the $1.6 billion allotted by Congress in 1980 was the only money the program was going to get for cleaning up toxic waste sites—a prospect most of the original framers of the program had not intended. Stockman believed that EPA and the Carter administration's justifications for Superfund were vastly overstated from the start and that EPA and environmental groups had used excessive scare tactics to push the bill through in 1980 (Burford with Greenya 1986, 28).

According to William Hedeman, director of EPA's Office of Emergency and Remedial Response under Gorsuch, and Gene Lucero, director of Hazardous Waste Enforcement, this decision to end Superfund in 1985 led to most of the delays experienced at various Superfund sites during the early 1980s. At Senate committee hearings in 1983, Maine Senator George Mitchell asked Hedeman and Lucero, "Was there any policy explicitly or implicitly to slow down expenditures from Superfund so that there would be money left in the fund when the law expires in order to support the preconceived notion that no extension of the law was necessary and to minimize the problem?" Both men, under oath, agreed that there was (Lash, Gillman, and Sheridan 1984, 83). Hedeman said he believed that there was "an implicit policy to at least curtail the progress of the program." It was explicit, said Hedeman, that there was "a view that the agency should not support Son of Superfund" (U.S. House of Representatives 1983, 134; Lucero 1988).

Most EPA staffers charged with administering Superfund knew that completing the job with this goal in mind was impossible. Even the nation's 419 priority sites could not have been cleaned up within those limits, let alone the 15,000 less notorious sites, some of which might have turned out to be as bad as or worse than the priority sites. In addition, with each passing year, the National Priority List and the lists of lesser sites were expanding. Although Congress had originally intended Superfund to continue beyond 1985, many lawmakers and environmentalists had only just started to appreciate that the original authors of Superfund had considerably underestimated the scope of the problems. For top-level EPA officials, however, the decision still stood: no reauthorization (Lash, Gillman, and Sheridan 1984, 83).

Many critics also started to suspect that the leadership at EPA purposefully prolonged negotiations with polluters and that these officials were indifferent about the real or perceived toxic impact on local populations at Superfund sites. As Hedeman later noted, "I always felt that [Rita] Lavelle [the assistant administrator for Solid Waste and Emergency Response] was a friend of business. She was advocating the approach that from the standpoint of risk, and in particular risk management, one had to live with risk" (Burford with Greenya 1986, 105).

According to Lucero, director of Hazardous Waste Enforcement, Lavelle did not see a health threat as an "emergency" until it involved immediate sickness or death. In her view, a long-term health threat, such as a toxic chemical in a water supply that might later produce nerve damage or cancer, was not an emergency. Indeed, to Lucero and other critics, it seemed that Lavelle (and many other high-ranking officials in the administration) insisted on a "body count" before conceding that an emergency existed. Evidence of a health threat had to be persuasive and overwhelming. In the Superfund program, Lavelle required more studies, more risk assessments, or in Lucero's view, "an additional quantum of proof to show a health problem." These high standards of proof caused major delays because EPA lacked the personnel to do the research. Officials had to shuffle scientists and staffers from one project to another, which caused delays in other programs (Lash, Gillman, and Sheridan 1984, 94). A Hazardous Waste Treatment Council Report later said of Superfund, "For the first four or five years it accomplished absolutely nothing except creating a huge new industry of environmental consultants who put that $1.5 billion into their pockets by studying sites, taking literally years to figure out what the site is like and what kind of remedies might be used" (Greenberg n.d.).

Long before Congress acted to change Gorsuch's case-by-case approach to cleanup standards, both environmental organizations and the courts began to force important changes. In 1982, shortly after Gorsuch had the NCP and case-by-case cleanup strategy published in the *Federal Register*, EDF—one of the most vigorous lobbies overseeing environmental law enforcement—and the state of New Jersey sued EPA to compel the agency to develop uniform national cleanup standards. The case, Environmental Defense Fund v. United States Environmental Protection Agency (No. 82-2234, DC Circ. 1984), presented a two-pronged legal argument. First, EDF argued that EPA's failure to promulgate uniform cleanup regulations that used the standards contained in other environmental laws was a violation of the Administrative Procedure Act, which barred capricious and arbitrary administrative actions. EDF also argued that, because CERCLA did not explicitly repeal the articulated standards laid out in other federal environmental laws, Superfund was legally bound to comply with the provisions contained in other major environmental laws. Thus, if regulations mandated under SDWA contained binding maximum contamination levels for certain substances at the tap, EPA had to ensure that the substance levels in drinking water supplies affected by Superfund sites were brought down at least to the maximum required standard.

The main solution sought by the lawsuit was the promulgation of uniform regulations used in other major environmental laws (*BNA Environment Reporter* 1985, 1595; Florio 1986, 365). In effect, EDF officials wanted the MCLs designed for a drinking water program to apply to a toxic waste cleanup program. In addition, as William Hedeman noted, EDF "argued that [Gorsuch's NCP] was so flexible that it wasn't a plan of any sort; it completely eliminated any public involvement in agency decision-making; it had no standards on how clean is clean or the direction that the agency should go in cleaning up sites; and it prohibited the listing of federal facilities on the national priority list, among other things" (Burford with Greenya 1986, 103).

In January 1984, EDF settled its lawsuit with EPA. The settlement agreement required that EPA would issue regulations that applied "relevant quantitative health and environmental standards and criteria" developed by the agency in other programs and to apply them at Superfund sites. A year later, the agency proposed a series of changes to the NCP in response to the settlement agreement (*Federal Register* 50: 5862, 1985). Even then, however, the agency retained some elements of their discretionary authority. EPA stated that "the requirements of other Federal environmental and public health laws, *while not legally applicable to CERCLA response actors*, [author's italics] will generally guide EPA in determining the appropriate extent of cleanup" (*Federal Register* 50: 47912, 1985).

When these revisions were promulgated in their final form in November 1985, EPA stated that it was adopting the standards as a matter of policy and not because it was legally obligated to do so. Moreover, EPA officials retained flexibility in their application of standards. Agency officials refused to concede some of the central legal issues of the EDF lawsuit (*BNA Environment Reporter* 1985, 1595; Florio 1986, 366).

Locking in Strict Standards with the CERCLA and SDWA Reauthorizations of 1986

While Superfund was becoming increasingly controversial by 1984 and 1985, SDWA remained a popular program widely supported by both political parties. However, by 1985, EPA's reputation for inaction in the toxic waste arena began to spill over into the drinking water program, where the agency's slow pace of setting MCLs appeared to confirm for many critics that the agency was guilty of deliberate inaction in their key programs. Congress had been looking to revise SDWA for some time and, after lengthy hearings, chose to limit EPA's discretionary authority by imposing mandatory schedules, standards, and rules for water supplies (Florio 1986, 367–368).

By the mid-1980s, many congressional leaders realized they had greatly underestimated the time and resources EPA needed to develop reputable SDWA regulations in a timely fashion. EPA's sluggishness in regulating contaminants and its failure to require certain types of treatment for organic contaminants served as focal points of debate over the law's revision. Reports in the early 1980s of drinking water contamination by organic contaminants, chemicals, and microbes such as *Giardia lamblia* prompted congressional concern over the adequacy of SDWA. From 1982 to 1985, both houses of Congress conducted hearings and reviewed the various proposals that helped to shape the SDWA amendments enacted in 1986 and that emerged in Senate Bill 124. The House passed this legislation on May 13, 1986, and the Senate followed suit on May 21, 1986. President Reagan signed the SDWA amendments into law on June 19, 1986 (Pontius 1999a).

To strengthen SDWA's regulation-setting process, most of the original 1974 SDWA was amended to include tougher schedules, standards, and rules and to limit EPA's discretion. For example, the 1986 amendments included mandatory standards for 83 contaminants by June 1989, mandatory regulation of 25 new contaminants every three years, and the monitoring of unregulated contaminants. The 1986 amendments significantly altered the speed with which EPA had to set drinking water standards and monitoring requirements (Dyksen, Hiltebrand, and Raczko 1988; Gray and Koorse 1988). For example, the amendments required EPA to set standards for 9 contaminants by June 1987, 40 by June 1988, and the total 83 by June 1989. EPA also had to develop a priority list of contaminants that posed a health risk in drinking water. This list formed the basis for standard-setting priorities in the 1990s (Cotruvo and Vogt 1990, 14).

With Superfund, Congress and EPA remained mired in the program's costs, its slow pace of cleanup, and the lack of clear standards. Lavelle was fired and Gorsuch resigned in early 1983; they were replaced by William Ruckelshaus and Lee Thomas. Ruckelshaus and Thomas tried to placate congressional anger over CERCLA by relying more on court litigation as an enforcement tool (Light 1985). Together with Assistant Attorney General Henry Habicht of the Land and Natural Resources Division at the Department of Justice (like Gorsuch before them), Ruckelshaus and Thomas concluded that EPA should promote policies that encouraged voluntary settlements and cleanups. The agency promulgated this approach in EPA's interim CERCLA Settlement Policy on December 5, 1984. During congressional hearings on the administration's proposals for CERCLA reauthorization in 1985, EPA officials expressed great concern about the program's legal and administrative costs. An agency study estimated that 30% of Superfund spending would be consumed by these costs. During the Senate Environment and Public Works Committee hearings on the Superfund Improvement Act of 1985, a case in which litigation expenses would actually exceed cleanup expenditure was discussed (Environmental Law Institute 1986, 6–7; *Environmental Science and Technology* 1995).

In response to the 1984 EDF lawsuit settlement and to forestall congressional action, EPA officials began to develop more stringent Superfund standards and enforcement policies. One of the key elements of these new standards was to link Superfund site cleanup levels to standards already established in other existing environmental programs, such as the SDWA MCLs. Beginning with the NCP of 1985, EPA recognized "applicable" or "relevant and appropriate" requirements as guides to cleanup (*BNA Environment Reporter* 1985, 1595); these requirements eventually became abbreviated as "ARARs."

In October 1985, EPA Assistant Administrator J. Winston Porter set forth the agency's policy on the applicability of the various state and federal standards, criteria, and advisories as they related to Superfund. Porter informed the 10 regional EPA administrators that "as a general rule, the Agency's policy is to attain or exceed *applicable* or *relevant and appropriate* [author's italics] federal environmental and public health requirements in CERCLA response actions" (Porter 1985, 2). By framing EPA's policy around applicable laws and regulations as well as relevant and appropriate requirements, Porter and federal authorities laid out very high standards. The applicability determination was fundamentally a legal one; determinations were based on the existing set of state and federal requirements, criteria, or limitations that dealt with a specific hazardous substance, contaminant, type of remedial action, location, or other circumstance at a CERCLA site. In contrast, relevant and appropriate requirements relied on professional judgement and consideration of environmental and technical factors at each Superfund site. As a result, relevant and appropriate determinations were more flexible. At any given site, regulators might judge that only portions of a regulation or standard were relevant and appropriate. In contrast, if a regulation was applicable, all substantive parts had to be followed (U.S. EPA 1989, 1). Under this restriction, SDWA's standards and MCLs in particular became cleanup standards for Superfund sites. In promulgating standards that were both applicable or relevant and appropriate in 1985, agency officials tried to make the case that EPA took its enforcement responsibilities seriously.

In the eyes of many members of Congress, these actions represented too little too late. Frustrated by scandals surrounding Gorsuch and Lavelle and by Superfund's slow pace of cleanup, congressional critics vented their anger by attacking and eventually limiting EPA's authority in ways that had long-term consequences for CERCLA and for SDWA. Long before the 1985 NCP, lawmakers had moved to include stricter standards, as evidenced by the House consideration of a reauthorization bill in 1984. "The legislation," reads the House report, "would accomplish the crucial goal [of establishing uniform national cleanup guidelines] by applying appropriate standards and criteria developed under the other major environmental laws to Superfund remedial actions. These provisions will ensure that once cleanup at a Superfund site is completed, the environment will not be contaminated at levels in excess of these permitted under … other federal laws" (U.S. House of Representatives 1984, 28).

As Congress took up the reauthorization in earnest, EPA's general lack of credibility on environmental issues and its continued unwillingness to rigorously enforce strict standards in other programs crippled the agency as an advocate for EPA discretion. In July 1986, Senate and House conferees agreed on programmatic changes; President Reagan, despite his strong opposition to many elements of the bill, signed the Superfund Amendments and Reauthorization Act (SARA) into law on October 17, 1986 (U.S. EPA 1989, 199–204).

Advocates of an expanded Superfund saw SARA as "the most significant piece of environmental legislation this decade" (*BNA Environment Reporter* 1986, 955).

New Standards and New Conflicts in the 1990s

SDWA and CERCLA reauthorizations of 1986 placed new schedules and rules on both programs. These strict mandates affected each program somewhat differently. For SDWA, EPA was able to set MCLs for most of the contaminants demanded by Congress. They particularly excelled in establishing standards for organic chemicals (Tables 2-1 and 2-2).

To meet these mandates, the agency paid a clear price. Competing priorities within EPA and lack of resources prevented the agency from fully complying with the mandates of the 1986 amendments (Pontius 1999a). Because of pressure from Congress and the courts to meet date-specific sum totals (such as the 83 contaminants by 1989 mandated in SDWA amendments), EPA officials tended to devote precious resources to contaminants that were easiest to determine rather than to ones that scientists considered the greatest threats to human health. For example, whereas EPA did well in establishing standards for organic chemicals, the agency made less progress for inorganic contaminants, such as chemicals that are difficult to remove from groundwater (where they usually occur) and that are relatively expensive on a per capita basis, particularly for small public water suppliers.

EPA continued to confront problems with arsenic, barium, lead, natural fluoride, and, most notably, nitrate and sulfate (Cotruvo and Vogt 1990, 13). According to Robert Perciasepe, former assistant administrator of EPA's Office of Water, EPA recognized that the many contaminant-specific standard-setting deadlines created a "regulatory treadmill [that] dilutes limited resources on lower-priority contaminants and as a consequence may hinder more rapid progress on high-priority contaminants" (Safe Drinking Water Act Report No. 104-632, quoted in U.S. EPA 1999, 10; Pontius 1999a). In addition, several reports were released challenging the standards program and raising specific issues about how realistic SDWA implementation schedules were and whether adequate funding existed, among other things.

The 1986 reauthorizations also incited heated controversy to the ARARs and the linkage of MCLs with Superfund cleanup standards. Before the passage of SARA, the chief critics and outside commentators on MCLs and many other elements of SDWA were water suppliers, AWWA, and state officials. Although these critics were influential, their influence was constrained because EPA routinely sought out their input, and they tended to work cooperatively with EPA's Office of Water. Most of these critics shared a common point of reference with EPA officials: that SDWA was, at its core, a drinking water program.

Once SARA linked MCLs with Superfund cleanup standards, however, it ceased to be just a water program. A new set of less cooperative and more influential critics emerged that included lobbies from chemical manufacturers, the mining industry, and petroleum companies—groups whose members were, in their view, getting financially "hammered" by Superfund's strict joint-and-several liability requirements. One way industrialists sought to limit their cleanup costs was to attack the cleanup standards themselves. When the sulfate standard came up for outside comment in 1990, the normal contingent of SDWA critics and researchers was joined by the National Coal Association, the American Mining Congress (which two organizations later merged into the National Mining Association), Asarco, and other groups. These mining interests challenged the standards because "EPA fails to consider [the effect] that adopting a sulfate MCL of 500 mg/l will have under other federal and state environmental programs. MCLGs and/or MCLs *automatically* [original italics] become enforceable cleanup and/or performance standards under numerous other federal or state programs," specifically including CERCLA. "Increased remediation costs triggered by adoption of a sulfate MCL at 500 mg/l could range from $600 million to $12 billion annually, depending on the method of estimation used. EPA has made absolutely no effort to quantify these costs" (National Mining Association 1995, 4–5, 36–37; Reid and others 1995). In 1990, 1992, 1994, 1995, and later, EPA tried to set the sulfate standard at 500 mg/l. In each instance, the standards were shot down in the review process by industry lobbies and other opponents. To date, the standard still has not been set.

EPA's efforts to update existing SDWA standards also may delay and drive up costs on Superfund cleanups. For example, in May 2000, EPA announced that it was lowering arsenic's MCL from 50 parts per billion (ppb) to between 3 and 20 ppb. This shift, many industry critics charged, would significantly increase cleanup costs at mining sites. At the Bunker Hill site in northern Idaho, for example, where the contaminant (lead) drives local cleanup efforts, arsenic's new standard would make it the predominant contaminant, forcing a thorough review of ongoing cleanup strategies there. More broadly, EPA and industry sources agree that the new arsenic standard will cause the suspension of ongoing cleanups where arsenic is involved as EPA and the cleanup managers review remedy agreements and make changes. Even under the old standard of 50 ppb, many industry critics viewed the goal as economically unfeasible. One source noted, "If the agency says 'pump and treat [the groundwater for arsenic]' … the current drinking water standard is so tough you can't get there." Another industry critic described meeting the current standard "impossibly expensive to treat" (*Environmental Policy Alert* 2000, 13).

TABLE 2-2. Specific Contaminants for Which Interim Standards Were Set by EPA, by Year (1976–1998)

1976	*1979*	*1987*	*1989*	*1991*	*1992*	*1998*
Organics						
Endrin	Trihalomethanes	Benzene[a]		Acylamide[ay]	Adipate di(2-ethylhexyl)	Bromate
Lindane		Carbon tetrachloride[a]		Alalchlor[a]	Benzo[*a*]pyrene	Chlorine dioxide
Methoxychlor		1,2-Dichloroethane[a]		Aldicarb	Dalapon[a]	Chlorite
Toxaphene		*p*-Dichloroenzene[a]		Aldicarb sulfone	Dichloromethane	Chloromine
2,4-D		1,1-Dichloroethylene[a]		Aldicarb sulfoxide	Dinoseb[a]	Haloacetic acids
2,4,5-TP (Silvex)		1,1,1-Trichloroethane[a]		Atrazine[a]	Diquat[a]	
		Vinyl chloride		Carbofuran[a]	Endothall[a]	
				Chlordane[a]	Glyphosate[a]	
				(mono)-Chlorobenzene	Hexachlorobenzene[a]	
				Dibromochloropropane [a]	Hexachloro-cyclopentadiene[a]	
				o-Dichlorobenzene	Picloram[a]	
				cis-1,2-Dichloroethylene	Pthalate, di(2-ethylhexyl)	
				trans-1,2-Dichloroethylene	Oxamyl[a]	
				1,2-Dichloropropane	Simazine	
				Epichlorohydrin	2,3,7,8-Tetrachlorodibenzo-	
				Ethylbenzene	*p*-dioxin[a]	
				Ethylene dibromide[a]	1,2,4-Trichlorobenzene	
				Heptachlor[a]	1,1,2-Trichloroethane	
				Heptachlor epoxide[a]		
				Pentachlorphenol[a]		
				Polychlorinated biphenyls		
				Tetrachlorethylene		
				Toluene		
				Xylenes		

Microbials			
Coliform bacteria	*Giardia*		
Turbidity	Heterotrophic plate count bacteria		
	Legionella		
	Viruses		
Inorganics			
Arsenic		Asbestos	Antimony
Barium		Copper	Beryllium
Cadmium		Nitrite	Cyanide
Chromium		Total nitrate/nitrite	Nickel
Lead			Thallium
Mercury			
Nitrate			
Selenium			
Silver			
Fluoride			
Radionuclides			
Radium-226			
Radium-228			
Gross alpha			
Gross beta			

[a] Pesticide.

Sources: Data courtesy of U.S. EPA's Office of Ground Water and Drinking Water (OGWDW), 2000; parts were published in U.S. EPA 1999 and in Shelton 2000.

In reference to the link between MCLs and Superfund, Joel Robinson, director of Environmental Sciences at Unocal Corporation, representing the American Petroleum Institute (API), stated, "We have found that the attempts to employ standards from other laws and regulations are often not achievable and more importantly are not appropriate to Superfund" (U.S. House of Representatives 1991, 330). The more conservative Congress of the mid-1990s tried to address some of these concerns in the SDWA reauthorization of 1996 by requiring that EPA conduct cost–benefit analyses on proposed MCLs to determine whether they were realistic and appropriate. Many critics from industry were still left out of the process because EPA officials refused to include studies of toxic waste cleanup costs in their assessments. According to one industry source, "EPA took the position that the costs of complying with MCLs that are enforceable under other programs should not be attributed to SDWA, but instead should be considered in promulgating the regulations under the other programs." With Superfund, however, legislative statutes passed by Congress required the presumptive use of MCLs as toxic waste cleanup standards, thus these critics from industry and elsewhere had no chance to challenge the standards during CERCLA's normal procedures (National Mining Association 1995, 1–4).

Many critics also attacked the scientific and technical underpinnings of ARARs. Some scientists and public health specialists argued that Superfund, written in reaction to popular concerns about toxic waste dumps and asbestos in schools, was based on little sound research about the actual nature of the threats. Critics argued in a 1993 *New York Times* article, for example, that "thousands of regulations were written to restrict compounds that had caused cancer in rats and mice, even though these animal studies often fail to predict how the compounds might affect humans" (Passell 1991; Schneider 1993). Both critics and supporters of the program questioned whether the remedies EPA selected were justified by the risks. Although a desire to return sites to a preindustrial condition was commendable, critics argued that a narrower focus on public health and saving lives was a far more economical use of limited environmental cleanup resources. The critics pointed to EPA estimates that some 1,000 cases of cancer annually could be linked to hazardous waste exposure compared with 5,000–20,000 lung cancer deaths from indoor exposure to radon. Critics suggested that more realistic Superfund cleanup goals would free up funds to substantially address problems such as radon (Passell 1991). EPA officials responded that they wanted to make sites safe enough to be used for any purpose, including houses or playgrounds. They did not want to revisit a site that had, in theory, already been "cleaned up" (personal communication from Robin Anderson, Crystal City, VA, October 15, 1999).

The ARARs continued to have many defenders, both within EPA and among environmental groups. EPA officials have argued that because 85% of

Superfund sites have contaminated groundwater and because many regions, particularly in the West, rely on groundwater as a primary source of drinking water, then high standards—and the drinking water requirements specifically—are appropriate. As Robin Anderson, a policy analyst in EPA's Office of Emergency Response stated in defense of the ARARs–MCLS linkage, "It comes down to 'do the people believe that groundwater is a national resource that needs to be preserved?' Is this something that we need to protect...? This [Clinton] administration says we do" (personal communication, October 15, 1999).

"We don't need a new paradigm," said David D. Doniger, a senior lawyer with the Natural Resources Defense Council. "For 35 years, the policy of the government has been that when there is uncertainty about a threat, it is better to be safe than sorry. When you are operating at the limits of what science knows, the big mistake would be to underestimate the real danger and leave people unprotected" (Schneider 1993).

Conclusions

In 1986, James Florio criticized the Reagan administration's enforcement of Superfund. He argued that EPA's cleanup standards reflected "the agency's disregard for the goals of the Superfund program" and that the "approach allowed EPA to ignore the environmental safety standards carefully developed under other federal environmental statutes to protect human health. Under EPA regulations, for example, maximum contamination levels for drinking water were established under SDWA. Thus residents near 'cleaned up' Superfund sites might be exposed to concentrations of chemicals found dangerous under the drinking water law" (Florio 1986, 365). Many other legislators and influential representatives of lobbies, including EDF, shared Florio's criticism and made tougher standards the centerpiece of their efforts to reauthorize Superfund that same year. The standards they advocated required that hazardous waste site cleanups meet the ARARs (*Washington Post* 1986).

Seven years later, however, even staunch advocates of the tougher standards and deadlines questioned whether these mandates drove up cleanup costs unnecessarily. Even Florio, one of the bill's strongest original supporters, stated, "It doesn't pay to clean up a rail yard in downtown Newark so it can be a drinking water reservoir" (Landy, Roberts, and Thomas 1994). In 1998, Senate Bill 8 proposed revising Superfund's reliance on Superfund by replacing the rigid statutory presumption in favor of treatment and the automatic adoption of all ARARs with a new, more flexible balancing test. In recent years, environmental and industry groups as well as states have consistently supported revising or eliminating ARARs in favor of more flexible or simplified standards (U.S. Senate 1998, 1–3). Although many of these efforts have stalled because of a lack of congressional consensus on reauthorizing CERCLA, federal and state authorities have found significant relief in programs such as the Brownfields Initiative put forward by EPA Administrator Carol Browner in 1995, the Superfund Redevelopment Initiative of 1999, and similar efforts that have evolved out of state-run voluntary cleanup programs (Hanley 1995; Moyer and Trimarche 1997; Cutler, Friedland, and Heavener 2000). A detailed discussion of these programs is beyond the scope of this chapter.

This overview of SDWA's regulations and their relationship with Superfund suggests some broader conclusions about the way SDWA regulations were created, the way they were applied to other programs, and the general behavior of various stakeholders. The 1986 reauthorizations of SDWA and CERCLA represented critical watersheds in EPA's history as Congress chose to no longer leave critical regulatory standards, rules, and deadlines to the agency. Instead, as Florio later noted, Congress became increasingly comfortable with making highly detailed scientific, technical, and regulatory determinations, some of which had profound, unforeseen consequences for EPA's central environmental programs. From Congress's point of view, the reauthorizations were used to send EPA (and the Executive Branch more generally) a message: ignore legislative intentions at your own peril.

Despite the overall critical tone of this analysis, it is difficult not to conclude that, since SDWA's passage in 1974, EPA has been poorly served by both Congress and the agency's overseers in the White House. Lack of funding and unrealistic deadlines have undermined the agency's ability to produce credible, timely standards. From mid-1980s through the 1990s, in an era of growing public skepticism concerning EPA's scientific and technical expertise, Congress imposed schedules and mandates that left EPA in some cases scrambling to set standards that used whatever data were available when the due date came. Many of the flaws in the process can be attributed to a necessary "learning curve" that agency officials endured in the face of large-scale change. Indeed, recent activities such as the setting of a standard for microbial and disinfectant by-products have effectively combined sound science and timeliness with meaningful stakeholder involvement (Pontius 1999b).

Despite these welcomed developments, one is still left with the overall impression of a standard-setting process in which researchers and administrators are constantly reacting to outside influences. The degree to which high-level EPA officials shaped this process in the late 1980s and the 1990s awaits further research. Additionally, more detailed analysis of the EPA Office of Water's increasingly complex, congressionally mandated review procedures would fill a pronounced need in the current literature. During the 1990s, Congress and many stakeholders remained deeply dissatisfied with the pace of standard setting, the science behind the standards, and ARARs. Also during the 1990s, particularly with the 1996 SDWA reauthorization, Congress imposed new deadlines, rules, and administrative structures to increase the pace of standard setting, create more transparency and public accountability, and foster "reasonable" and "appropriate" responses to specific environmental problems.

If the past is any indicator, EPA will be facing more of these mandates in the near future. One can hope that compared with the 1986 reauthorizations and more recent legislation, a broader cross-program and interregulatory perspective will inform its decisions.

References

BNA Environment Reporter. 1985. EPA Proposes Flexible Scheme for Choosing Cleanup Remedies under Superfund Statute. February 2, 1595–1596.

———. 1986. Superfund Update. October 24: 955.

Burford, Anne M., with John Greenya. 1986. *Are You Tough Enough? An Insider's View of Washington Power Politics.* New York: McGraw-Hill.

Cannon, Lou. 1982. *Reagan.* New York: Putnam's.

Clean Water Report. 1974. Slants and Trends. 12(December 20): 2.

Cohen, Steven, and Marc Tipermas. 1983. Superfund: Preimplementation Planning and Bureaucratic Politics. In *The Politics of Hazardous Waste Management,* edited by James P. Lester and Ann O. Bowman. Durham, NC: Duke University Press, 42–59.

Cotruvo, Joseph A., and Craig D. Vogt. 1990. Rationale for Water Quality Standards and Goals. In *Water Quality and Treatment: A Handbook of Community Water Supplies,* edited by Frederick W. Pontius. New York: McGraw-Hill.

Cutler, William, Melissa Friedland, and Stephen Heavener. 2000. Everybody Wins: The Superfund Redevelopment Initiative. Paper presented at Brownfields 2000 Research and Regionalism: Revitalizing the American Community, October 12, Atlantic City, NJ.

Dyksen, J.E., D.J. Hiltebrand, and R.F. Raczko. 1988. SDWA Amendments: Effects on the Water Industry. *Journal of the American Water Works Association* 80(1): 30–35.

Environmental Law Institute. 1986. *Superfund Deskbook.* Washington DC: Environmental Law Institute, 10.

Environmental News. 1974. Organic Chemicals Found in New Orleans Water Supply. November 8: 1–3.

Environmental Policy Alert. 2000. EPA Water Arsenic Rule Could Boost Cost of Superfund Cleanups. May 31: 13.

Environmental Science and Technology. 1995. EPA at 25. 11: 509A.

Florio, James J. 1986. Congress as Reluctant Regulator: Hazardous Waste Policy in the 1980s. *Yale Journal on Regulation* 3(2): 351–382.

Gray, K.F., and S.J. Koorse. 1988. Enforcement: U.S. EPA Turns Up the Heat. *Journal of the American Water Works Association* 80(January): 47–49.

Greenberg, Eric. n.d. *Toxic Temptations: The Environmental Protection Agency's Superfund Program.* Washington DC: Center for Public Integrity.

Hanley, Cassandra M. 1995. Developing Brownfields: An Overview. *The Journal of Urban Technology* 2(Spring): 1–20.

Harris, Robert H., and Edward M. Brecher. 1974. Is the Water Safe to Drink? *Consumer Reports* June: 440.

Johnson, Charles C. 1968. Memorandum from Charles C. Johnson, Administrator, U.S. Public Health Service, to Regional Administrators. July 30. Miscellaneous File Folder. Box 1. Office of Drinking Water—Records Relating to Program Establishment, 1968–1978. EPA Record Group 412. National Archives II, College Park, MD.

———. 1971. Statement of Charles C. Johnson, Associate Executive Director, Public Health Service, before the House Public Works Committee.

Landy, Marc K., Marc J. Roberts, and Stephen R. Thomas. 1994. *Environmental Protection Agency: Asking the Wrong Questions.* New York: Oxford University Press.

Lash, Jonathan, Katherine Gillman, and David Sheridan. 1984. *A Season of Spoils: The Reagan Administration's Attack on the Environment.* New York: Pantheon.

Lee, Roger D. 1972. Memorandum from Roger D. Lee, Chief, Community Water Supply Branch, Water Supply Division, U.S. Environmental Protection Agency, to James H. McDermott, Director, Water Supply Division. June 5.

Light, Albert R. 1985. A Defense Counsel's Perspective on Superfund. *Environmental Law Reporter* 15(July):10203–10207.

Lucero, Gene A. 1988. Son of Superfund: Can the Program Meet Expectations? *Environmental Forum* March/April: 5–12.

McDermott, James H. 1971. Drinking Water Standards, Past, Present, and Future. Paper presented at the Seventh American Water Resources Conference, October 25.

McDermott, James H., and John A. Cofrancesco. 1972. EPA's Drinking Water Supply Program. Paper presented at the American Water Works 92nd Annual Conference, June 6.

McDermott, James H., P.W. Kabler, and H.W. Wolf. 1970. Health Aspects of Toxic Materials in Drinking Water. Paper presented at the American Public Health Association Meeting, October 26–30, Houston, TX.

Moyer, C.A., and G.D. Trimarche. 1997. *Brownfields: A Practical Guide to the Cleanup, Transfer, and Redevelopment of Contamination Property.* Foresthill, CA: Argent Communications.

National Mining Association. 1995. Comments of the National Mining Association on the Proposed Maximum Contaminant Level. *Federal Register* 59(March 21): 65578.

Passell, Peter. 1991. Experts Question Staggering Cost of Toxic Cleanups. *New York Times,* September 1.

Policy Statement. 1973. Conference of State Sanitary Engineers on Federal Legislation Regarding Drinking Water Supplies. Strategy Paper Folder. Box 2. Office of Drinking Water—Records Relating to Program Establishment, 1968–1978. EPA Record Group 412. National Archives II, College Park, MD.

Pontius, Frederick W. 1999a. *History of the Safe Drinking Water Act (SDWA)*. Denver, CO: American Water Works Association.

———. 1999b. Negotiating Regulations for Microorganisms and Disinfection By-products. *Journal of the American Water Works Association* October: 14–28.

Pontius, Frederick W., and Stephen W. Clark. 1999. Drinking Water Quality, Standards, Regulations, and Goals. In *Water Quality and Treatment* (Fifth Edition), edited by Raymond D. Letterman. New York: McGraw-Hill, 1.4–1.6.

Porter, J. Winston. 1985. Memorandum from J. Winston Porter, Assistant Administrator, U.S. Environmental Protection Agency (EPA), to EPA Regional Administrators. October 2.

Reid, Jon, and others. 1995. *Final Report: A Review of the U.S. EPA's Proposal to Set an MCL for Sulfates in Drinking Water.* Washington, DC: American Mining Congress.

Reizen, Maurice S. 1972. Memorandum from Maurice S. Reizen, Director, Department of Public Health, State of Michigan, to Philip A. Hart, Chairman, Subcommittee of the Environment, U.S. Senate. August 29.

Rosenthal, Alon. 1991. Nitrates in Drinking Water. In *Harnessing Science for Environmental Regulation,* edited by John D. Graham. New York: Praeger, 159–196.

Schneider, Keith. 1993. New View Calls Environmental Policy Misguided. *New York Times* March 21.

Shelton, Theodore B. 2000. *Interpreting Drinking Water Quality Analysis: What Do the Numbers Mean?* New Brunswick, NJ: Rutgers Cooperative Extension. http://www.wairus.com/~gatherer/interpret.html (accessed November 20, 2000).

Stafford, Robert T. 1981. Why Superfund Was Needed. *EPA Journal* June: 10.

Symons, G.E. 1984. That GAO Report. *Journal of the American Water Works Association* 66(August): 34.

Train, Russell S. 1974. Facing the Real Cost of Clean Water. *Journal of the American Water Works Association* 66(October): 562.

U.S. EPA (Environmental Protection Agency). 1971. Drinking Water Supply Issue Paper. Issue 30 and Strategy Papers Folder. Box 4. Office of Drinking Water—Records Relating to Program Establishment, 1968–1978. EPA Record Group 412. National Archives II, College Park, MD.

———. 1972. *Industrial Pollution in the Lower Mississippi River in Louisiana.* Dallas, TX: EPA Region VI Office.

———. 1989. *ARARs Q's and A's: General Policy RCRA, CWA, SDWA.* OSWER 9234.2-01FS. May. Washington, DC: U.S. EPA, 1.

———. 1999. *EPA, 25 Years of the Safe Drinking Water Act: History and Trends.* Washington DC: U.S. EPA. http://www.epa.gov/safewater/sdwa/trends.html (accessed November 17, 2000).

———. n.d. Strategy Document: Implementation of the Safe Drinking Water Act. Attachment 2. Misc. File Folder. Box 3. Office of Drinking Water—Records Relating to Program Establishment, 1968–1978. EPA Record Group 412. National Archives II, College Park, MD.

U.S. House of Representatives. 1983. House Energy and Commerce Committee. The Administration of Superfund, the Withholding of Files, and the Role of OMB in EPA Rulemaking. Hearings before the Subcommittee on Investigations and Oversight. September 28.

———. 1984. Committee on Energy and Commerce. *Superfund Expansion and Protection Act of 1984 (Part I and 2)*. 98th Congress, 2nd Session. July 24.

———. 1991. Committee on Public Works and Transportation. *The Administration of the Federal Superfund Program: Hearings before the Subcommittee on Investigations and Oversight of the Committee on Public Works and Transportation*. 102nd Congress, 2nd Session. October 3 and November 12 and 13.

U.S. PHS (Public Health Service). 1970. *Community Water Supply Study: Significance of National Findings*. Springfield, VA: U.S. PHS.

U.S. Senate. 1972. Senate Commerce Committee. Statement of Ralph Nader on Drinking Water Policy Before the Senate Subcommittee on the Environment. Untitled Folder. Box 1. Office of Drinking Water—Records Relating to Program Establishment, 1968–1978. EPA Record Group 412. National Archives II, College Park, MD.

———. 1998. *Superfund Cleanup Acceleration Act of 1998*. Senate Report 105-192. 105th Congress, 2nd Session. May 9.

3

The First Federal Drinking Water Quality Standards and Their Evolution

A History from 1914 to 1974

PATRICK GURIAN AND JOEL A. TARR

Americans today live in a world bounded by environmental standards largely set at the federal level. Although environmental regulation at the federal level is a relatively new development, dating primarily from the 1970s, several regulatory milestones established earlier provided important precedents. In this chapter, we focus on the development and impact of one of these milestones: the first federal drinking water quality standards, established by the U.S. Public Health Service (PHS) in 1914.[1] We explore the subject from two perspectives. The first perspective considers the scientific and engineering advances that laid the technical foundation for the standards: improved understanding of the causality of waterborne disease, the formulation of analytical and measurement techniques, and the development of treatment technologies. The second perspective examines the process of setting standards, including their legal authority and institutional context as well as society's reaction to the standards themselves.

During the standard-setting process, many issues arose relevant to current environmental regulation. They included concern about the difficulty of providing absolute protection from environmental hazards and the need to determine an acceptable level of risk; debate over the proper role of the federal government in local decisionmaking; and concern about unnecessarily broad or strict regulation producing unintended consequences, such as increasing public anxiety and inducing unnecessary expenditures. In several features, however, the regulatory process differed from current practice. The most critical differences included the amount of public participation and stakeholder involvement in the regulatory process and the extent of reliance on formal analytical approaches such as benefit–cost analysis.

Scientific and Engineering Foundations for Standards

The Etiology of Waterborne Disease

John Snow is generally credited with having made the first scientific association of water quality with disease when he determined that water from the Broad Street Pump in London, England, was the source of a cholera infection. Although Snow made an important contribution by showing an association between water quality and disease, he did not determine the causal mechanisms for the transmission of waterborne diseases (Hamlin 1990). Discovery of the germ theory of disease and the method of transmission of waterborne disease was still decades away, and the development of a science of water analysis was beset by conflicting theories and controversy among experts (Brock and Madigan 1988; Tomes 1998). Even after some diseases had been shown to result from microorganisms, skeptics pointed out that the causes of waterborne diseases had not yet been experimentally demonstrated (Hamlin 1990).

The development of the compound microscope made it possible to identify microorganisms in water and provided the technology for additional advances in understanding the etiology of waterborne disease. In 1880, Robert Koch and Karl Eberth identified the microorganism responsible for typhoid, and in 1883 Koch identified the causative agent of cholera as the bacteria *Vibrio cholera* (Gorham 1921). William Sedgwick and others at the Massachusetts Board of Health's Lawrence Experiment Station confirmed the role of sewage pollution in the spread of typhoid in the 1890s (Tarr 1996). Most researchers accepted the waterborne nature of typhoid and cholera by the 1890s, but controversy persisted among the advocates of chemical analysis versus bacterial analysis (Hill 1894; Hamlin 1990; Gossel 1992). However, it is difficult to establish a precise date for the recognition of the role of waterborne bacteria in disease. Skepticism persisted into the early twentieth century, well beyond when such doubts appear to be reasonable from a contemporary viewpoint (Duffy 1990; Hamlin 1990; Tomes 1998).

Analytical Methods

The determination of water standards required the development of analytical methods to measure water quality. Water analysis originated in attempts to understand the chemical composition of mineral waters. Early analytical protocols were aimed at determining the concentrations of various salts in waters and were not well suited to assessing the potability of polluted water (Hamlin 1990). In the 1850s, as physicians and sanitarians embraced the anticontagionist theory of disease etiology, the focus of analyses shifted toward organic material, because water with decaying organic matter was believed to contribute to poor health. Several surrogate parameters were evaluated as being indicative of organic compounds, such as the mass lost by an evaporative residue on ignition, the presence of nitrogen, and the amount of chemically oxidizable material. Whereas these analyses were capable of detecting gross pollution, they could not detect the small quantities of contamination capable of transmitting disease (Hamlin 1990). Some sanitary authorities recognized the limitations of these tests, but the tests may have created a false sense of confidence in the safety of water supplies and led to the widespread belief in the rapid self-purification of running water (Hyde n.d.; Hamlin 1987; Tarr 1996).

One reason for the continued reliance on chemical tests, even though they were inadequate, was that early bacteriological tests were not very informative. In the 1880s, water analysts began quantifying the number of bacteria in water by using new culturing methods developed by Koch and other microbiologists. Although these techniques gave better quantitative results than microscopic analysis, they did not address the fundamental problem that had plagued bacterial analysis: how to distinguish harmful microorganisms from the harmless varieties found in all water sources (Hamlin 1990).

Reliably analyzing water for pathogenic organisms had serious practical problems. A wide variety of pathogenic organisms potentially could be present, and to be useful, a test would have to be capable of detecting all of those pathogens. In addition, pathogenic organisms account for only a tiny fraction of the total microorganisms present in a water sample; however, pathogenic bacteria are capable of causing disease even in small numbers. Because the ingestion of a single typhoid bacteria has a 1–2% chance of causing the disease, tests for individual pathogens would have to be extremely sensitive (Gainey and Lord 1952; Fair and Geyer 1956).

The determination of pathogenic organisms was not only impractical but also an insufficient test of the safety of a water supply over time. A water polluted with sewage might not have any pathogens at a particular time because the population from which the sewage pollution derived was then healthy. However, the water might still be capable of transmitting disease should any of the upstream population contract a waterborne disease. This line of reasoning led to the development of the indicator–organism concept, which involves testing for microorganisms that indicate the presence of sewage without regard to whether the particular organisms are pathogenic (Gossel 1992).

Coliform bacteria proved to be such an indicator organism. These bacteria grow at body temperature and are capable of forming gas by lactose fermentation. Coliforms occur naturally in the intestines of humans and other animals, but like pathogenic bacteria, they do not live for an extended time in water supplies. Thus, the presence of coliforms indicates that a water supply has been polluted with sewage recently enough for the bacteriological contamination to still be harmful (Brock and Madigan 1988). Theobald Smith developed the experimental method for the coliform test and proposed its use as a measure of water quality in 1892. The New York State Board of Health adopted Smith's proposal and used the method to study pollution in the Mohawk and Hudson Rivers. Researchers at the Lawrence Experiment Station in Massachusetts applied the test as a measure of filtered water quality in 1896–1897. C.E.A. Winslow in 1901 and the distinguished Harvard sanitary engineer George Whipple in 1903 proposed rough quantitative guidelines for acceptable levels of coliforms, although debates over the acceptable levels of coliforms continued even after the establishment of the 1914 standard (Draffin 1939; AWWA 1940, 1950).[2]

The analytical methods used by basic researchers such as Koch, Smith, and Sedgwick had to be standardized and publicized so that different analysts could obtain consistent results. By following the history of these standardization efforts, one can observe the shift from chemical to more sensitive bacteriological analysis that occurred gradually around the turn of the century. In the 1880s, the American Association for the Advancement of Science organized a committee on methods for the "sanitary examination of water." The 1889 report of this committee included chemical tests for nitrogen species and oxygen demand but no bacteriological tests. However, the growing importance of bacteriological testing and the various methods used by laboratories led to the first formal gathering of bacteriologists in the United States, held at the New York Academy of Medicine in June 1895. The discussions at the conference made apparent the "myriad sources of technical variation in bacteriology" and led to the appointment of a committee to compile a set of standard methods (Gossell 1992).

In 1898, the American Public Health Association (APHA) created the Laboratory Committee, renamed in 1900 as the Section of Bacteriology and Chemistry, the first permanent section of the APHA. The section's main task was to devise a set of standard methods including bacteriological analyses as well as other types of water analyses (Gossell 1992). In 1905, the section issued its first publication, *Report of the Committee on Standard Methods of Water Analysis.* This report recommended that several tests—bacteriological, chemical, physical, and microscopic—be used to assess the quality of water supplies. The subsequent various editions of *Standard Methods of Water Analysis,* formulated by the APHA, the American Water Works Association

(AWWA), and other groups, came to furnish the basic techniques of determining water quality for the nation (Tarr 1996).[3]

The 1905 *Standard Methods* report noted that its recommendations constituted a shift toward more use of the bacteriological analyses, particularly the coliform test and the total bacteria test (APHA 1905; Franson 1985). As Whipple (1904) noted, "improved methods of bacterial examinations, however, have shifted this [analytic] emphasis somewhat, and, at the present time, greater weight is perhaps given to the biological and physical examinations, than to the chemical analysis so far as sanitary questions are concerned . . . ." The degree of confidence that had developed in the total coliform test is evident in the report of the commission that established the 1914 standards: "The test is an extremely delicate one, showing traces of pollution not detectable by any other means" (Monfort 1915; Rosenau 1927).

Treatment Techniques

One final technical requirement for water quality standards was the development of effective treatment techniques. One could not mandate the provision of high-quality water before practical and economical methods for providing such water were available. The 1914 PHS standard limited coliform bacteria to no more than 2 coliforms/100 milliliters. In an era when coliform counts in the water supplies of many major cities ran into the hundreds of coliforms per 100 milliliters (values for several major cities are shown in Table 3-1), such a restrictive standard clearly necessitated treatment by all suppliers except those with the most protected source waters.

The two technologies that made large-scale water treatment feasible were filtration and chlorine disinfection. Although filtration has been used since antiquity to purify water, many of the early filters probably served only to strain out large particles and would not have removed pathogenic bacteria to any appreciable degree (AWWA 1926; Baker 1948; Melosi 2000). Not until the development of a method known as slow-sand filtration did filters became capable of reliably removing small particles such as bacteria.

The first slow-sand filters were developed in Great Britain (the "English" filter) beginning in the late eighteenth century and were intended to clarify water, not improve its bacteriological quality (Baker 1948). Thus, the improved bacteriological quality of water was an unforeseen consequence of technological change. In fact, the germ theory of disease had not even been established when the earliest filters were designed. Slow-sand filters consist of a bed of fine sand supported by layers of coarser material through which water is passed at a low flow rate. By the 1890s, it was understood that the process produced the growth of a mat of harmless bacteria in the upper portion of the filter bed that strained out pathogenic

TABLE 3-1. Performance of Water Treatment Methods

		Raw water		*Treated water*			
City	*Year*	*1-ml samples positive (%)*	*Presumptive coliforms/100 ml*[a]	*1-ml samples positive (%)*	*Presumptive coliforms/100 ml*[a]	*Coliform removal (%)*	*Compliant with 1914 standard?*
Filtration							
Birmingham, AL	1910	84	180	8.3	8.7	95	No
	1911	78	150	9.7	10	93	No
Lawrence, MA	1910	100	—	7.8	8.1	—	No
	1911	100	—	8.4	8.8	—	No
	1912	100	—	16.9	19	—	No
Washington, DC	1910	26.9	31	0	—	—	Yes
	1911	16.9	18.5	0.8	0.8	96	Yes
	1912	43.5	57	0	—	—	Yes
New Orleans, LA	1913	39	49	3.1	3.1	94	Possible
Albany, NY	1912	100	—	5	5.1	—	Possible
Wilmington, DE	1912	99.2	480	11.2	12	98	No
Filtration and disinfection							
Harrisburg, PA	1910	55	80	0.17	0.17	99.8	Yes
	1911	77	150	0.61	0.61	99.6	Yes
	1912	47	63	0.81	0.81	99	Yes

Columbus, OH	1913	69	117	0.5	0.5	99.6	Yes
Cincinnati, OH	1910	91	240	0.8	0.8	99.7	Yes
	1911	85	190	5.2	5.3	97	Possible
	1912	93	270	3	3	99	Possible
Bangor, ME	1913	86	200	0.89	0.89	99.6	Yes
Clarksburg, WV	1911–1913	67	110	0.3	0.3	99.7	Yes
Filtration and partial disinfection							
Birmingham, AL	1912	89	220	5.7	5.9	97	Possible
	1913	78	150	0	—	—	Yes
Little Falls	1911–1912	89	220	6.1	6.3	97	Possible
Philadelphia, PA	1912	—	—	4.1	4.2	—	Possible
Disinfection							
New York, NY	1911	65–21	—	29.8	35	0–65	No
	1912	42–35	—	10	11	69–74	No
Cleveland, OH	1911	11.1	12	0	—	—	Yes

Note: ml = milliliters.

[a] The percentage of positive tests is converted to the expected number of coliforms per 100 milliliters (ml) using a maximum-likelihood estimation procedure.

Source: Fuller 1915.

bacteria and other small particles that would otherwise pass through the sand. Even after the role of water in disease was better understood, the confirmation of the ability of filters to remove bacteria had to wait until the development of analytical methods to quantify the bacteriological quality of water.

In 1872, the first slow-sand filtration plant in the United States was built in Poughkeepsie, New York, under the direction of engineer J.P. Kirkwood. However, the technology did not diffuse rapidly, and by 1890, only a small fraction of the urban population was supplied with water treated by slow-sand filtration (Hazen 1904; Baker 1948). However, in 1893 engineer Hiram Mills, chairman of the committee on water control and sewage of the Massachusetts Board of Health, constructed a slow-sand filtration plant for Lawrence that reduced the local incidence of typhoid fever; the typhoid death rate fell 79% and the overall death rate by 10% over a five-year period (Hazen 1907; Melosi 2000). Several cities proceeded to install slow-sand filtration plants, and by 1914, such plants supplied water to more than 5,400,000 urbanites (Melosi 2000).

A competing technology was the rapid-sand filter, also known as the mechanical or "American" filter. It resulted from attempts to settle and clarify very turbid water by use of a coagulant. Originally, it was feared that the high rate of filtration of the technology would not result in bacterial purification, but experiments in Louisville, Kentucky; Cincinnati, Ohio; and Washington, District of Columbia; in the late 1890s demonstrated the effectiveness of rapid-sand filters. Rapid-sand filtration soon overtook slow-sand filtration as the preferred method of protecting water supplies in the United States, and by 1914 these filters served 11,900,000 urbanites (Johnson 1914; Turneaure and Russell 1924).

Once filtration technology showed major demonstrable public health effects, American public and private water suppliers adopted it rapidly in the early twentieth century. The population served rose from 1.9 million in 1890 to more than 17 million in 1914, and the percentage of the urban population supplied by filtered water rose from 6% to 41% (Johnson 1914; Melosi 2000). A study of 13 American cities found that typhoid mortality dropped by an average of 65% after the adoption of water filtration, and reductions in individual cities ranged from 34% to 92% (Johnson 1914).

Table 3-1 shows the coliform levels of raw and treated water for six cities using filtration but not disinfection between 1911 and 1914. The rates of removal of coliform bacteria by filtration range from 93% to 98%. These data imply that filtration of a raw water with hundreds of coliforms per 100 milliliters would produce finished water in the range of 2–70 coliforms/100 milliliters. Many suppliers, however, would have been unable to comply with the 1914 coliform standard of 2 coliforms/100 milliliters through the use of filtration alone (confirmed by an examination of the data in Table 3-1 for treated water).

The second important water treatment technique was chlorination. Research in the 1890s demonstrated chlorine's efficacy as a disinfectant, and it occasionally was used as a temporary measure in response to epidemics. In 1908, Jersey City, New Jersey, became the first American municipality to adopt chlorination, using hypochlorite of lime (Turneaure and Russell 1924; AWWA 1926; Laubusch 1971). Shortly afterward, liquid chlorine was produced by the electrolysis of common salt, and this method rapidly became preferred for treating water supplies. The technique of chlorine disinfection diffused much more rapidly than filtration. In 1911, only three years after its introduction, more than 800 million gallons of drinking water were being chlorinated in the United States each day, supplying roughly 4 million people or 10% of the urban population. By 1915, about 11 million people—one-quarter of the urban population—were supplied with chlorinated drinking water (AWWA 1926). The rapid adoption of chlorination can be attributed to its cheapness (chlorination cost about $0.01 per capita per year compared to $0.40–0.50 per capita per year for filtration) and the fact that it presented few technical problems (Johnson 1914; AWWA 1926).

Chlorine used by itself as a water supply disinfectant was relatively effective in disease protection, but for many cities, chlorination alone would not have achieved the 1914 standard. A 1939 study by PHS sanitary engineer H.W. Streeter estimated that chlorination, when used alone, could reliably reduce coliform levels from 50 coliforms/100 milliliters to 1 coliform/100 milliliters, a reduction of 98% (Phelps 1944). Chlorination of a source water with hundreds of coliforms per 100 milliliters would produce a treated water with 2–20 coliforms/100 milliliters. Much more effective was a combination of chlorination and filtration, with coliform removals ranging from 97% to 99.8% (as shown in Table 3-1). Of the five cities in Table 3-1 using filtration and full disinfection, four clearly achieved coliform levels below the 1914 standard.[4] Removals of 97–99.8% from a source water with hundreds of coliforms would produce treated water with concentrations in the range of 0.2–30 coliforms/100 milliliters. Thus, filtration and disinfection allowed many suppliers to achieve compliance with the 1914 standards, but suppliers with a combination of highly polluted source waters and poor treatment performance might not.

The Legal and Institutional Context

In addition to the technological preconditions for water standards, legal authority is required to establish drinking water standards. In an era when the U.S. Constitution was interpreted narrowly, it was necessary for any PHS action to be clearly within the constitutional authority of the federal government. The legal basis of the 1914 standards began with the authority given to Congress in the U.S. Constitution to regulate interstate commerce. The interstate commerce clause of the Constitution frequently has been used as justification for the expansion of federal powers. Because so many aspects of life involve commercial transactions, this clause may be used to justify federal involvement in a wide range of activities.

In 1893 Congress passed the Interstate Quarantine Act to reduce the spread of communicable diseases through interstate commerce. The act gave the Department of the Treasury broad powers to establish regulations preventing the spread of disease from one state to another in the following clause (Cumming 1932; Kraut 1994):

> The Secretary of the Treasury shall, if in his judgement it is necessary and proper, make such additional rules and regulations as are necessary to prevent the introduction of such diseases (communicable) into the United States from foreign countries, or into one State or Territory or the District of Columbia from another State or Territory or the District of Columbia....

This clause was not immediately perceived as requiring any regulations relating to drinking water. In fact, methods of bacteriological analysis and water treatment were not sufficiently developed at this time for the establishment of quantitative standards.

In the years after the passage of the Interstate Quarantine Act, the prevailing political sentiment of the Progressive Era began to favor an expanded federal role in the protection of public health. This trend is seen in the passage of laws such as the Pure Food and Drug Act and the Meat Inspection Act in 1906 (Williams 1993–1996a; 1993–1996b). In 1912 and 1913, PHS promulgated regulations containing very detailed procedural requirements for drinking water on interstate carriers. Regulations were issued forbidding the use of common drinking cups (Cotruvo and Vogt 1990) and requiring weekly steam cleanings of all water containers used on common carriers (Treasury Department 1913). There was also a requirement for certifying water quality issued as part of the Amendment to Interstate Quarantine Regulations Number 6 on January 25, 1913 (Treasury Department 1913):

> Water shall be certified by the state or municipal health authority within whose jurisdictions it is obtained as incapable of conveying disease: Provided, that water in regard to the safety of which a reasonable doubt exists may be used if the same has been treated in

> such manner as to render it incapable of conveying disease, and the fact of such treatment is certified by the aforesaid health officer.

This clause met practical problems in implementation. The basis on which the local health authorities would judge the quality of water was not specified. Taken literally, the requirement that water be "incapable of conveying disease" was an impractical standard. Even treated water supplies that were generally considered to be safe occasionally had measurable amounts of coliform bacteria, and a source with coliform bacteria was in principle capable of conveying disease. Given that local health authorities could not be expected to insist on the complete absence of coliforms, the problem was standardizing an acceptable level so that the various local authorities could judge water quality consistently.

These problems were recognized, and work on the issue was begun concurrently with the establishment of the certification requirement. On January 22, 1913, the secretary of the treasury appointed a commission to begin the development of quantitative standards for drinking water quality on interstate carriers (Anderson 1914). The following year, this commission's recommendations were adopted as the first federal drinking water standards. Thus, the first federal standards derived their legal authority from and were the natural product of efforts to reduce the spread of disease through interstate commerce. The standards to be developed would be legally binding only for common carriers involved in interstate commerce.

It was immediately recognized that the standards would have a widespread impact. When Surgeon General Rupert Blue (1913a) requested authorization for the formation of the commission from the secretary of the treasury, he pointed out that "the adoption of such a standard by the Department would be practically the creation of an official standard of purity for all drinking waters." The letter of invitation sent to prospective committee members did not indicate that the standard to be developed was for interstate commerce but instead stressed the need for a general standard of water purity: "In connection with the enforcement of regulations relative to pure drinking water, the necessity for a federal standard of purity for drinking water has arisen."

The name chosen for the committee, Commission for the Determination of a Standard of Purity for Drinking Water, also emphasized the goal of developing a generally applicable standard. The surgeon general described it as a history-making event in a letter to Professor Charles Hyde of Berkeley (Blue 1913b):

> I sincerely hope that you will find it possible to add this duty to your already crowded existence, not only because I feel that the commission will receive great benefit from your services thereon, but also because I

> feel that the body is going to make history. The standard which they will set will probably be the official standard of the United States for many years to come....

What began as an effort to develop a standard for a specific class of water supplies broadened to include developing appropriate standards for all public water supplies, encouraging people to improve their water supplies, and improving the funding of state and local health departments (Olsen 1979). PHS appears to have planned a de facto "unfunded mandate" in the hope that this would spur greater local action and awareness. In a memorandum, Surgeon General Blue (1913c) wrote

> If this regulation is enforced to its fullest extent, it will mean that eventually the water supply of every portion of the country traversed by common carriers must be examined. It will readily be seen, that if the water supply of a community cannot receive the certificate of the health authorities, that the people living in the community will be stimulated to bring about an improvement of their ice and water supplies. It is true that many of the state and municipal health authorities are not provided with the machinery and funds to carry out the said work, but the constant requests for these certificates will create, it is believed, a condition which will aid in securing larger appropriations from state and local governments for this important public health work.

Although the strict interpretation of the Constitution and lack of specific authorizing legislation constrained the actions that the federal government could take on health and environmental issues, they did not close the door completely. When the opportunity arose to develop drinking water standards for common carriers, PHS seized on it to further the more general agenda of improving the nation's drinking water quality.[5]

The Standard-Setting Process

The disciplines represented on the commission that set the first federal drinking water standards ranged from the basic sciences of biology and chemistry to applied fields such as sanitary engineering and medicine. Of the 15 members, 7 had their primary affiliations with academic institutions, 3 with PHS, 4 with state or local government, and 1 with a nongovernmental organization. In contrast to current regulatory standard setting, there was no opportunity for public participation in the process, and in fact, no representatives of the regulated parties—the interstate carriers and water suppliers—

were included in the process. Also in contrast to current regulatory efforts, the commission had no budget. Its members served without compensation, and all work was conducted by correspondence (Blue 1913a). Considering the importance of this commission and the complexity of its mission, it is striking that they produced a standard without a single face-to-face meeting.

The commission did not set the standard on the basis of formal analytical approaches such as benefit–cost analysis; instead, the members relied on their informed judgement. The subcommittee charged with proposing limits for inorganic components reported, "We have found it very difficult to get any authoritative information on the subject or even to find anyone who acknowledged that he was competent to advise us in regard to it. We have, therefore, been compelled to exercise our best judgement" (Anderson 1913a).

The situation with the bacteriological standards was better in terms of scientific knowledge. The potential for transmission of disease by contaminated water was well understood. Judgement was still required in the setting of bacteriological standards but for a different reason. Although theoretically, any amount of bacteriological contamination could result in illness, it was not feasible to set a standard of zero for coliform bacteria. Members of the commission were called on to make a risk-based decision, long before the science of risk analysis had been formally developed.

Given the state of technical knowledge at the time and the variety of professional orientations of its members, it is not surprising that the commission's initial opinions varied widely. After 10 months of work, Chair John Anderson reported in a letter to Assistant Surgeon General W.C. Rucker that "while in the beginning of the work of the commission there was a very great variance in the opinions of the different members ... they have gotten to be in pretty fair agreement and I believe that, with the exception of Prof. Bartow, we will probably be able to get out a unanimous report" (Anderson 1913b). The letter contains an interim report summarizing each member's opinions on the bacteriological standards in late 1913. The final report was submitted September 21, 1914. Thus, Anderson's interim report gives a snapshot of various commission members' views at roughly the halfway point in the standard-setting process.

Two bacteriological standards were being considered. The first was a standard for total bacteria. In the interim report, six commission members recommended a standard of 100 bacteria/milliliter. Three recommended stricter standards (10–50 bacteria/milliliter), two made no quantitative recommendation, and four recommended that the standard be omitted. The divergence of opinion as to whether there even should be a standard is because of the fact that the total bacteria test is an imprecise indicator of sewage pollution. The bacteria measured by this test may occur naturally in the water supply. The use of culture conditions that favored sewage-derived bacteria over aquatic bacteria minimized but did not eliminate this problem. Of the six members recommending a standard of 100 bacteria/milliliter, two indicated that the standard should be used only as an indication of whether further investigation into the quality of the water was necessary, not as sufficient basis to judge the water unsafe (Anderson 1913a).

The second bacteriological standard was for coliform bacteria. The test used for this did not directly determine the concentration of coliforms in a sample. It indicated only the presence or absence of coliforms in the sample. To estimate the coliform concentration, multiple tests were made. The estimated number of coliforms was a function of both the percentage of tests that were positive and the amount of water sampled in each test. Larger sample volumes tended to detect coliform bacteria more frequently.

Of the 15 commission members, 9 recommended that five 10-milliliter samples be tested and that no more than one of the test results be positive for coliforms. This standard corresponds to a concentration of approximately 2 coliforms/100 milliliters. Charles Bass was the only commission member to advocate a stricter standard at this point. He recommended that none of the five samples be allowed to test positive. Bass's standard implies a most probable number of zero for coliform bacteria, perhaps reflecting a desire for absolute protection against waterborne disease. The rest of the commission considered such absolute protection infeasible and wished instead to set what in today's terms would be considered an allowable level of risk. Three commission members recommended slightly more lenient standards for coliforms: that no more than one of three 10-milliliter samples test positive (Rosenau and Winslow) and that no more than two of five 10-milliliter samples test positive (Whipple) (these recommendations correspond to 4 and 5 coliforms/100 milliliters, respectively). Edward Bartow recommended a substantially less stringent standard, proposing that no more than one of five 1-milliliter samples test positive, which corresponds to a concentration of approximately 20 coliforms/100 milliliters.[6]

On September 21, 1914, the committee submitted its final report to the surgeon general. Despite Anderson's earlier concerns, the members had reached a unanimous decision on appropriate standards. The most common views expressed at the end of 1913 had been adopted by the commission as a whole. Total bacteria were limited to 100/milliliter, which the commission described as "very liberal limits" in the final report; this limit was justified by noting that bacteria detected by the test were not necessarily present due to sewage pollution, especially when the water had been stored for a time after treatment, as would be the case on common carriers (U.S. PHS 1914).

In contrast, the coliform standard of 2 coliform/100 milliliters was relatively stringent (U.S. PHS 1914). Several major cities would not have been able to comply with the standard despite having filtration plants that were generally regarded as providing safe water (Jordan 1915). However, the combination of filtration and chlorination, if applied properly, would have permitted all but the cities with the most polluted source waters to meet this standard. The availability of economical treatment methods capable of attaining the standard is explicitly cited by the commission as justification of the stringent standard for coliforms (U.S. PHS 1914):

> Compliance with the requirements herein recommended will insure a quality of water equal to that of municipal supplies which have been demonstrated by experience to be entirely safe and satisfactory and will at the same time impose no great burden upon common carriers since it is entirely practicable, with moderate expense and pains, to purify water to the degree required.

This criterion is similar to the current requirement in the Safe Drinking Water Act that drinking water standards be "feasible." As was the case with the 1914 standards, the contemporary requirement [*U.S. Code,* Section 300, Title 42, g.1 (b)(4)(D)] explicitly includes both technical and economic considerations:

> The term "feasible" means feasible with the use of the best technology, treatment techniques and other means which the Administrator finds, after examination for efficacy under field conditions and not solely under laboratory conditions, are available (taking cost into consideration).

The commission's report indicates that no standards were set for inorganic constituents, noting that "it is impossible in the present state of our knowledge to definitely specify the ill effects, if any, which given amounts of these substances produce." These constituents would have been difficult to regulate on the basis of a "feasible" treatment because "it is generally impracticable to remove the mineral salts present in waters by measures of practical application...." The "chemical impurities" of water supplies were left for additional discussion, and the report dealt with only bacteriological impurities (U.S. PHS 1914).

In contrast to the original interpretation of the committee's mission, the final report explicitly indicated that the standards were intended for interstate carriers and disavowed any intentions to set a general "standard of purity" for drinking water (U.S. PHS 1914). The latter can be interpreted as a political position, intended to diffuse initial criticism. Jack Hinman (1920), a water bacteriologist and chemist from the Iowa State Board of Health, noted that "the effect of the posting of notices in stations and the condemnation of local waters has been to force the standard upon the local plants through the action of governmental prestige and public opinion."

Reaction to the Standards

Several major professional journals of the time noted the establishment of the standards. *American Journal of Public Health* (1915) and *Engineering and Contracting* (1915) limited their coverage to factual reporting on the standards. *Journal of the American Water Works Association* (Monfort 1915) and *Engineering Record* (1914a, 1914b) published generally favorable commentaries on the standards, as well as factual descriptions of the standards. One notable exception is *Journal of the New England Water Works Association,* which did not report on the standards in the 1914 and 1915 issues.

In contrast to the reaction of the national publications, some local suppliers and state boards of health, which feared they would be expected to meet the new standards, reacted negatively. Water analyst Charles Hoover (1915), for instance, criticized the standards during a discussion period at a 1914 AWWA meeting as leading to an inaccurate public perception of risk:

> The bacteriological standard for water adopted by the Treasury Department for the drinking water supplied to the public by common carriers in interstate commerce may work an injustice upon well regulated municipal water purification plants, for undue suspicion is apt to be thrown upon the plant, and the minds of the citizens may become unduly prejudiced against their public water supply if it becomes known that the water has at some time failed to meet the government requirements.

H.E. Jordan (1915), superintendent of filtration at the Indianapolis Water Company, presented a paper at the February 1915 meeting of the Indiana Sanitary and Water Supply Association that criticized the standards as unnecessarily strict: "The chief objection to the standard as it now stands [is] that public water supplies have been produced from year to year, not conforming to the percentage of *B. coli* [coliform bacteria], yet having a good effect upon the health of the community." Jordan cited examples of cities that did not achieve the standard of 2 coliforms/100 milliliters but were generally considered to have safe water. Lawrence, Massachusetts, averaged around 11 coliforms/100 milliliters, and his own plant in Indianapolis averaged 3 coliforms/100 milliliters. He stated that for these cities, "It is the consensus of opinion that the public water supply as a factor in typhoid causation has been eliminated" and referred to the "mistaken idea on the part of so many, that *B. coli* found at all is a bad indication." Jordan was particularly critical of the standards because he believed that they would be applied to all water suppliers, not only interstate carriers: "The fact that there is this government standard—and the word 'government' or 'United States' is a fetish to many—no matter if the committee does not intend its application in local cases, will in fact result in that very thing being done."

Earle Phelps of PHS's Hygienic Laboratory, a chemist who trained at the Massachusetts Institute of Technology and had devised the method commonly used for calculating the *B. coli* index, attended the meeting where Jordan presented his paper. Phelps (1915a) reported that "in conversation with other water works men I found [Jordan's views] to be the prevailing sentiment in Indiana." He replied to Jordan's criticism during the discussion after his paper by presenting two contradictory tracts. On the one hand he argued, "It is stated as emphatically as the commission can state, and in three distinct places that these standards apply only to the water supplies of common carriers. They have nothing to do with city water supplies" (Phelps 1915b). He indicated that a stringent standard had been set because railroads had ready access to many supplies and should, therefore, be encouraged to choose the best source of water. The commission had been obliged to set the standard based on laboratory results alone, not detailed knowledge of the water supply in question. City water supplies where the source was well-known and found to be safe might still be appropriate for public use even if the water did not meet the new standards.

However, Phelps did not consistently portray the standards as irrelevant to city water supplies. He suggested that water suppliers not currently meeting the standards could explain to their customers the additional cost of meeting the standards and allow them to decide whether the standards should be met. Phelps (1915c) also disputed Jordan's assertion that the transmission of typhoid by water had been completely eliminated in communities that did not meet the standard. He pointed out that in Lawrence, Massachusetts, in 1899, coliform levels of 26/100 milliliters were associated with an elevated typhoid death rate and concluded that there probably did not exist a coliform concentration at which protection from disease was absolute.

Phelps's view of the problem as a risk-based decision explains the apparent contradictions in his reply. On the one hand, he wished to encourage communities to consider improving their water supplies, because he believed that a reduction in coliforms would reduce the risk of disease, even if coliform levels were already fairly low. On the other hand, he appreciated that it was not feasible to reduce the risk of waterborne disease to zero. Even compliance with the standards would not completely eliminate the risk of waterborne disease, and so some latitude should be allowed for communities to determine what level of risk they could feasibly attain, given the characteristics of their water supply and their financial resources.

Municipal acceptance of the standard was gradual. At the 1920 AWWA annual convention, Hinman reported the results of a standards survey that he had conducted of municipal water works. Hinman noted that of 168 plants that had replied to his survey, 46 said that they operated "continuously" under the PHS standard, 4 followed a "more rigid standard," 16 followed a "less rigid" standard, 45 used no standard, and 44 used standards "whose position with respect to the Treasury Department [PHS] standard is difficult to determine."[7] Of the set, 47 considered the PHS standard "reasonable," 19 were in favor of it but "with reservations"; 21 opposed it, and 81 had no opinion. Several survey respondents called for flexibility rather than absolutes, and Hinman (1920) concluded that standards "must have a certain amount of flexibility and depend in large part upon the sanitary survey."

At the state level also, acceptance was mixed. Although the standard was enforced for water used in interstate commerce, in 1917 only California, Georgia, Michigan, and Wisconsin enforced the PHS standard for other water supplies, and 22 states used no general standards at all. Again, the response reflected the belief that the standards were too rigid (Morse and Wolman 1918).

In a paper critical of PHS standards, Morse and Wolman (1918) reported that many "analysts and sanitary engineers expressed the opinion rather vehemently that no general standard could be devised to fit the peculiar conditions existing in that particular section of the country." In addition, they observed, most official sanitarians had mixed opinions about "standards of quality." Morse and Wolman specifically criticized PHS standards as not providing "an accurate index" to the quality of a water supply: "If inconsistency reigns in the determination of the fundamental units, such as the total count, the *B. coli* content, the chemical constituents, and the sanitary survey, then the general standard of quality, a derived unit composed of basic measures, becomes of extremely little value."

For Morse and Wolman (1918), the standards were simply a "convenient mode of analysis to be used with considerable caution for comparative purposes" and of "little value" until "basic or fundamental units for measurement of quality" were established.

In contrast to the reaction of municipal water suppliers and of state public health authorities, which were not subject to the regulation, the railroads—which were subject to it—accepted it without protest. The regulation was a minor constraint on their operations without significant cost. State departments of health used the federal water quality standard in certification of railroad water supplies almost without exception, even if not in broader situations. Thus, by narrowing the stated applicability of the standards to interstate carriers, PHS diffused its most vocal critics (Morse and Wolman 1918).

In retrospect, the criticism of the standards seems overly harsh. From a scientific perspective, the standards reflected an attempt to form a consensus among a disparate group of professionals in a field characterized by an evolving state of knowledge. From a policy perspective, many of the negative responses seem to have been driven by concerns over control of turf and the fear of dictation from a higher authority. Such control from above eventually came, but not for more than half a century.

Subsequent History of the Standards

After 1914, PHS periodically reviewed the standards and revised them in response to criticisms and new findings concerning water quality. Revised standards were issued in 1925, 1942, 1946, and 1962 (Borchardt and Walton 1971). In 1974, Congress passed the Safe Drinking Water Act, and responsibility for drinking water standards moved to the U.S. Environmental Protection Agency (EPA) (Cotruvo and Vogt 1990). Revisions to the drinking water standards have followed three trends over time: the standards have become stricter, have been applied to more contaminants, and have been applied to more types of water supplies.

In accordance with the first trend toward stricter standards, the coliform concentration standard was revised downward to 1 coliform/100 milliliters in 1925. This standard was adopted as a National Interim Primary Drinking Water Standard in 1975 under the Safe Drinking Water Act. In 1989, the Total Coliform Rule lowered the standard dramatically to 0.05 coliform/100 milliliters (Cotruvo and Vogt 1990). This dramatic reduction in the standards was made possible by gradual improvements in filtration and disinfection that allowed increased and more consistent removal of pathogens. It also was made possible by the widespread adoption of sewage treatment, which improved the quality of many drinking water sources.

The second trend toward regulating larger numbers of contaminants can be seen in the addition of standards for three inorganic constituents (lead, copper, and zinc) in 1925. PHS added 10 more standards in 1942. Nine of these standards were for additional inorganic constituents: fluoride, arsenic, selenium, magnesium, chloride, sulfate, total solids, alkalinity, and the sum of iron and manganese. In addition, the first standard for an organic constituent (phenolic compounds) was added. In 1946, a standard for hexavalent chromium was added, and in 1962, seven more standards were added: five for inorganics (barium, cadmium, cyanide, nitrate, and silver) and two for organics (alkyl benzene sulfonate and carbon chloroform extract). The regulation of more components reflected the accumulating knowledge of the harmful effects of many substances and the analytical methods that allowed the quantification of trace amounts of such substances. In addition, the acceptance and success of the higher priority bacteriological standard enabled PHS to focus its efforts on other constituents (Tarr 1996).

The total bacteria standard was an exception to the trend of regulating additional parameters. The value of this parameter had been questioned even in 1914. By 1925, confidence in the coliform test as a sufficient measure of bacteriological quality was such that the total bacteria standard was omitted from the 1925 standards and never reinstated.

The third trend is that standards have become applicable to a much wider group of water supplies. The 1914 standards stated that they were applicable only to drinking water supplied on interstate carriers. The 1925 standards also stated that they were intended only for water used in interstate commerce, but they included requirements clearly intended for the system supplying the water, such as the requirement that the supplier complete a sanitary survey of the water source and that the distribution system be appropriately maintained (U.S. PHS 1925). Beginning with the 1946 revision, the standards were explicitly "applicable to all water supplies in the United States" but still legally binding only for water used on interstate carriers. The AWWA endorsed this broadening of the intended scope of the standards (U.S. PHS 1946). In practice, an entire city's supply had to conform to the standards if any portion of it was used in interstate commerce.

By the early 1970s, about 800 public water supplies were regulated under the federal standards. In addition, many other water supplies were regulated under state standards that were often based on the federal standards (Borchardt and Walton 1971). With the promulgation of National Interim Primary Drinking Water Regulations under the Safe Drinking Water Act in 1975, water quality standards became legally binding for all U.S. public water supplies (Cotruvo and Vogt 1990).

Conclusions

In this chapter, we have considered the development and impact of the first federal water quality standards in the context of changes in water quality analysis and water treatment technologies. The history suggests several conclusions about the development of science-based standards in relationship to public health. In the standards literature, standards are viewed as "specified codified knowledge elements with high normative significance [that] ... emerge when technical knowledge has entered the stage of consolidation." They possess "stabilizing effects," emerging from processes that involve various actors with different values, knowledge bases, and interests (Schmidt and

Werle 1998). This description fits the process involved in formation of the 1914 PHS water quality standards well.

The 1914 standards came at a time when the scientific uncertainty over the etiology of waterborne disease had been largely resolved. In addition, the technical capacity to quantify the extent of bacteriological contamination and to remove the contamination through treatment had been convincingly demonstrated. The standard-setting effort could be undertaken only when consensus existed on these fundamental scientific issues. Nevertheless, PHS established the standard with continuing scientific uncertainty as to acceptable levels of coliform bacteria in drinking water. Thus, it appears to have been an attempt to create a consensus in favor of higher water quality through the adoption of new treatment techniques by using the prestige of the federal government to endorse stricter standards.

The development of a standard for drinking water quality required PHS to balance the health risks of polluted water against the costs of improving water quality. In such a situation, it is not possible to develop a single unambiguous threshold for acceptable water quality. The acceptability of a given water source instead would depend on an evaluation of the risks it presented and the alternatives available. The committee did not explicitly include such considerations in their standard. However, by establishing a strict standard and emphasizing its narrow applicability, they were able to encourage local decisionmakers to strive for purer water while avoiding mandating compliance with a somewhat arbitrary criterion.

How much of the improved water quality in the United States during the first quarter of the twentieth century can be considered attributable to the PHS standards? Only a limited amount. Typhoid death rates had already declined by about half before the standards were established, driven by new water treatment technologies, principally rapid-sand filtration and chlorine disinfection. The technologies had become available during this time and were adopted in response to local demands for pure water—not federal requirements. However, one can argue that by establishing a strict standard for water quality, PHS helped strengthen the existing trend toward higher water quality. That is, the standards had both a moral and symbolic importance aside from their practical significance.

Compliance with the standards (although not necessarily because of the standards) gradually became the norm among major urban water suppliers. Of 19 large U.S. cities, only two (Milwaukee, Wisconsin, and Providence, Rhode Island) exceeded the standard in 1922 (Wolman 1923). One factor motivating change may have been the desire of some municipal officials to remain competitive with other cities. In accordance with Blue's hopes and Jordan's fears, the standard was influential far beyond its legal scope of water on interstate carriers. By setting a stricter standard than was normal water works practice, the PHS standard may have driven water suppliers to improve the quality of the water they provided. This goal might be called the "secret agenda" of PHS and of most members of the standard-setting committee. The eventual acceptance of the standards as generally applicable is an example of how the "unintended" effects of regulation may be clearly foreseeable, may be explicitly desired by some parties, and may prove to be beneficial (Whipple and Pincus 1922).

An additional conclusion involves the importance of the larger political context on the timing of the regulatory process. The political climate in the Progressive Era favored government actions to ensure public health, reflecting the growing belief of Americans that science and bureaucracy together could improve the nation's health. On the federal level, historians have recognized the Meat Inspection Act and the Pure Food and Drugs Act of 1906, approved by Congress after long and contentious battles, as most significant in regard to regulatory precedent. Following constitutional interpretations of the time, these acts applied only to food, meat, and drugs in interstate commerce but recognized the necessity of dealing with some issues on a national level (Young 1989; Andrews 1999). These precedents undoubtedly stimulated PHS to undertake the establishment of drinking water standards based on general authority given by the Interstate Quarantine Act, even though lacking a specific congressional mandate. On the other hand, the narrow interpretation of the Constitution during this time period limited the scope of the standards. Although the interpretation of the Constitution has broadened greatly since 1914, the conflicting desires for universal standards of public health and for local, decentralized decisionmaking remain very much alive (Nieminski 2000).[8]

Whereas scientific understanding and institutional roles have evolved since PHS established the 1914 standards, this standard-setting process may provide some important perspective on contemporary policy issues. In particular, the PHS standards for drinking water show the strengths and weaknesses of voluntary standards. The strengths of the voluntary approach include the flexibility to proceed without the delay and contentious debate often involved in obtaining enforcement authority and the lack of administrative monitoring and enforcement costs. The ability to inform local debates while preserving local control may encourage a more economically efficient outcome, because local decisionmakers may have specific information about the particular situation that is not available to regulators on the national level. However, the primary weakness of the voluntary approach is the inability to achieve complete compliance. Local control may not be an advantage if the local decisionmakers are not well informed and capable of making complex technical decisions. Given the public health risks associated with drinking water contamination, it is not surprising that voluntary standards were replaced by enforceable standards.

Acknowledgements

This research was supported by the Heinz Family Foundation through the grant of a Teresa Heinz Scholarship for Environmental Research (to Patrick Gurian).

Lauren MacWilliams assisted in the location of numerous important sources and shared detailed notes on the history of water supply. Mitchell Small provided valuable comments on a draft of the manuscript.

Notes

1. For a brief survey of the history of U.S. drinking water standards, see Taylor 1977.

2. For a more detailed discussion of the bacteriological standards development, see MacWilliams 2000.

3. The 1905 report was the first edition of *Standard Methods*, which remains the standard handbook of chemical and bacteriological analysis with updated editions published every 5–10 years.

4. The fifth city (Cincinnati, Ohio) met the 1914 standard in one year, but its status for the remaining two years is unclear.

5. This method of furthering a policy agenda is in keeping with Kingdon's (1995) model of policy "windows."

6. It is interesting to contrast the disciplinary backgrounds of the two commission members with the most extreme positions on the coliform standard. Charles Bass was affiliated with a clinical medical institution, whereas Edward Bartow's position in a state water survey was more engineering-oriented. These affiliations parallel the disciplinary divisions in a debate taking place during this same time period in which sanitary engineers defended the practice of discharging of untreated sewage to waterways as being economical and a minimal threat to public health if downstream cities treated their drinking water properly. In contrast, medical authorities advocated the treatment of sewage to provide more complete protection of the environment and human health. (See Tarr 1996 for a full discussion of this debate.)

7. The numbers are given here as originally reported by Hinman, even though in the first sentence, they add up to only 155 (not 168). A total of 168 plants are accounted for in the second sentence.

8. Steve Gordon (1999), during his tenure as president of the American Water Works Association, asked, "Isn't it odd how everyone else thinks they know how to run our systems better than we do?"

References

American Journal of Public Health. 1915. Bacteriological Standard for Water on Transportation Lines. 5(1): 79.

Anderson, John. 1913a. Attachment to a Letter from John Anderson, Chair of the Commission, to Commission Members. November 24. Records of the U.S. Public Health Service, Record Group 90. Washington, DC: National Archives.

———. 1913b. Letter from John Anderson to Assistant Surgeon General W.C. Rucker. November 25. Records of the U.S. Public Health Service, Record Group 90. Washington, DC: National Archives.

———. 1914. Letter of Transmittal from John Anderson to the Surgeon General for the Commission's Final Report. September 21. Records of the U.S. Public Health Service, Record Group 90. Washington, DC: National Archives.

Andrews, Richard N.L. *Managing the Environment, Managing Ourselves: A History of American Environmental Policy.* New Haven, CT: Yale University Press.

APHA (American Public Health Association). 1905. Report of Committee on Standard Methods of Water Analysis for the Laboratory Section of the American Public Health Association. *Journal of Infectious Disease* May(Suppl. 1): 6–8.

AWWA (American Water Works Association). 1926. *Water Works Practice.* Baltimore, MD: Williams & Wilkins.

———. 1940. *Manual of Water Quality and Treatment* (First Edition). New York: AWWA.

———. 1950. *Water Quality and Treatment, A Manual* (Second Edition). New York: AWWA, 32–33.

Baker, M.N. 1948. *The Quest for Pure Water.* New York: American Water Works Association.

Blue, Rupert. 1913a. Letter from Surgeon General Rupert Blue to the Secretary of the Treasury. January 10. Records of the U.S. Public Health Service, Record Group 90. Washington, DC: National Archives.

———. 1913b. Letter from Surgeon General Rupert Blue to Professor Charles Gilman Hyde. February 12. Records of the U.S. Public Health Service, Record Group 90. Washington, DC: National Archives.

———. 1913c. Memorandum from Surgeon General Rupert Blue to Assistant Secretary Allen. April 29, Records of the U.S. Public Health Service, Record Group 90. Washington, DC: National Archives.

Borchardt, J.A., and G. Walton. 1971. Water Quality. In *Water Quality and Treatment* (Third Edition). New York: McGraw-Hill.

Brock, T.D., and M.T. Madigan. 1988. *The Biology of Microorganisms* (Fifth Edition). New York: Prentice-Hall.

Cotruvo, J.A., and C.D. Vogt. 1990. Rationale for Water Quality Standards and Goals. In *Water Quality and Treatment* (Fourth Edition). New York: McGraw-Hill.

Cumming, H.S. 1932. Letter from Surgeon General H.S. Cumming to M.H. Thatcher, House of Representatives. March 3. Records of the U.S. Public Health Service, Record Group 90. Washington, DC: National Archives.

Draffin, J.O. 1939. *The Story of Man's Quest for Water.* Champaign, IL: Garrard Press.

Duffy, John. 1990. *The Sanitarians: A History of American Public Health.* Urbana, IL: University of Illinois Press, 192–203.

Engineering and Contracting. 1915. Bacteriological Standard for Drinking Water Supplied by Common Carriers. January 27: 77–78.

Engineering Record. 1914a. Bacteriological Standard for Drinking Water on Common Carriers. 70(21): 551.

———. 1914b. Bacteriological Standard for Water on Common Carriers. 70(23): 617–618.

Fair, G.M., and J.C. Geyer. 1956. *Water Supply and Waste-Water Disposal.* New York: John Wiley and Sons, 513.

Franson, M.A.H. (ed.). 1985. *Standard Methods for the Analysis of Water and Wastewater* (Sixteenth Edition). Washington, DC: American Public Health Association, American Water Works Association, and Water Pollution Control Federation, i.

Fuller, George W. 1915. Biochemical and Engineering Aspects of Sanitary Water Supply. *Journal of the Franklin Institute* July: 17–61.

Gainey, P.L., and T.H. Lord. 1952. *Microbiology of Water and Sewage.* New York: Prentice-Hall, 159–160.

Gordon, Steve. 1999. Monday Night Football: EPA vs. Water. *Mainstream* 43(10): 2.

Gorham, Frederic P. 1921. The History of Bacteriology and Its Contribution to Public Health Work. In *A Half-Century of Public Health,* edited by M.P. Ravenel. New York: American Public Health Association, 71–72.

Gossel, Patricia P. 1992. A Need for Standard Methods: The Case of American Bacteriology. In *The Right Tools for the Job: At Work in Twentieth Century Life Sciences,* edited by A.E. Clark and J.H. Fujimura. Princeton NJ: Princeton University Press.

Hamlin, Christopher. 1987. *What Becomes of Pollution? Adversary Science and the Controversy on the Self-Purification of Rivers in Britain, 1850–1900.* New York: Garland.

———. 1990. *A Science of Impurity.* Berkeley, CA: University of California Press.

Hazen, Allen. 1904. Purification of Water for Domestic Use: American Practice. *Transactions of the American Society of Civil Engineers* 54(Part D): 131–154.

———. 1907. *Clean Water and How to Get It.* New York: John Wiley & Sons, 70–71.

Hill, John. 1894. The Quality of Water Supplies. *Transactions of the American Society of Civil Engineers* 32(August): 130–170.

Hinman, Jack J. 1920. Standards of Quality of Water. *Journal of the American Water Works Association* 7: 835.

Hoover, Charles P. 1915. In Discussion following Some Considerations in Estimating the Sanitary Quality of Water Supplies (by W. H. Frost). *Journal of the American Water Works Association* 2: 725.

Hyde, Charles Gilman, n.d. (1922?). *Quality Standards for American Water Supplies.* Berkeley, CA: University of California, Water Resources Center Archives.

Johnson, George A. 1914. Present Day Water Filtration Practice. *Journal of the American Water Works Association* 1: 31–80.

Jordan, H.E. 1915. The Standards for Drinking Water Used on Interstate Carriers. *Proceedings of the Eighth Annual Convention of the Indiana Sanitary and Water Supply Association* 38–42.

Kingdon, John. 1995. *Agendas, Alternatives, and Public Policies.* New York: HarperCollins College Publishers.

Kraut, Alan M. 1994. *Silent Travelers: Germs, Genes, and the "Immigrant Menace."* New York: BasicBooks.

Laubusch, E.J. 1971. Chlorination and Other Disinfection Processes. In *Water Quality and Treatment* (Third Edition). New York: McGraw-Hill.

MacWilliams, Lauren. 2000. Evolution and Evaluation of the Coliform Drinking Water Standard. Independent Study Paper. Durham, NC: Duke University, Department of Civil Engineering.

Melosi, Martin V. 2000. *The Sanitary City: Urban Infrastructure in America from Colonial Times to the Present.* Baltimore, MD: Johns Hopkins University Press, 135–145.

Monfort, W.F. 1915. A Special Water Standard. *Journal of the American Water Works Association* 2: 65–73.

Morse, R.B., and A. Wolman. 1918. The Practicability of Adopting Standards of Quality for Water Supplies. *Journal of the American Water Works Association* 15: 216–217.

Nieminski, Eva. 2000. Regulatory Compliance Versus Self-Regulation. *Journal of the American Water Works Association* 92(2): 64–65.

Olsen, Johen. 1979. Reorganization as a Garbage Can. In *Ambiguity and Choice in Organizations* (Second Edition). Bergen, Norway: Universitetsforlaget, 314–337.

Phelps, Earle B. 1915a. Letter from Earle B. Phelps to the Surgeon General. March 3. Records of the U.S. Public Health Service, Record Group 90. Washington, DC: National Archives.

———. 1915b. In Discussion following The Standards for Drinking Water Used on Interstate Carriers (by H.E. Jordan). *Proceedings of the Eighth Annual Convention of the Indiana Sanitary and Water Supply Association* 42.

———. 1915c. Enclosure in letter from Earle B. Phelps to the Surgeon General, detailing his remarks at the Feb. 1915 meeting of the Indiana Sanitary and Water Supply Association. March 3. Records of the U.S. Public Health Service, Record Group 90. Washington, DC: National Archives.

———. 1944. *Stream Sanitation.* New York: John Wiley & Sons, 196–198.

Rosenau, Milton. 1927. *Preventive Medicine and Hygiene* (Fifth Edition). New York: Appleton, 959.

Schmidt, Susanne K., and Raymund Werle. 1998. *Coordinating Technology: Studies in the International Standardization of Telecommunications.* Cambridge, MA: MIT Press.

Tarr, Joel A. 1996. *The Search for the Ultimate Sink.* Akron, OH: University of Akron Press.

Taylor, Floyd B. 1977. Drinking Water Standards Principles and History 1914 to 1976. *Journal of the New England Water Works Association* 91: 237–259.

Tomes, Nancy. 1998. *The Gospel of Germs.* Cambridge, MA: Harvard University Press.

Treasury Department. 1913. *Pure Drinking Water for Passengers in Interstate Traffic Amendment to the Interstate Quarantine Regulations No. 6.* Treasury Department Memo. January 25. Records of the U.S. Public Health Service, Record Group 90. Washington, DC: National Archives.

Turneaure, F.E., and H.L. Russell. 1924. *Public Water Supplies; Requirements, Resources, and the Construction of Works* (Third Edition). New York: John Wiley & Sons, 463–475.

U.S. PHS (Public Health Service). 1914. Bacteriological Standard for Drinking Water. *Public Health Reports* 29: 45 (Reprinted in Monfort 1915).

———. 1925. Report of the Advisory Committee on Official Water Standards. *Public Health Reports* 40(15): 693–723.

———. 1946. Public Health Service Drinking Water Standards, 1946. *Public Health Reports* 61 (11): 371–384.

Whipple, George C. 1904. Discussion: Purification of Water for Domestic Use. *Transactions of the American Society of Civil Engineers* 54(Part D): 192–206.

Whipple, George C., Chair, and Sol Pincus, Secretary. 1922. *First Tentative Report of the Sub-committee on Field Survey of the U.S. Public Health Service Advisory Committee on Official Water Standards.* December. Records of the U.S. Public Health Service, Record Group 90. Washington, DC: National Archives.

Williams, Willard. 1993–1996a. Pure Food and Drug Acts. In *Encarta 97 Encyclopedia.* Redmond, WA: Microsoft Corporation.

———. 1993–1996b. Meat Packing Industry. In *Encarta 97 Encyclopedia.* Redmond, WA: Microsoft Corporation.

Wolman, Abel. 1923. *Preliminary Report of the Sub-Committee on Bacteriological Standards as Amended by Sub-Committee on Appraisal.* Records of the U.S. Public Health Service, Record Group 90. Washington, DC: National Archives.

Young, James Harvey. 1989. *Pure Food: Securing the Federal Food and Drugs Act of 1906.* Princeton, NJ: Princeton University Press.

4

The Unintended Consequence of Reformulated Gasoline

David Stikkers

On March 25, 1999, California Governor Gray Davis issued an executive order banning the use of a controversial chemical because of its environmental risks. Whereas similar actions have taken place during the history of environmental policy, this directive was unique because it represented a remarkable policy reversal.

The target of the governor's executive order was methyl-*tert*-butyl ether (MTBE), a gasoline additive that by 1999 had become one of the highest-volume end-use chemicals manufactured in the United States. What made this ban unique was that nine years earlier, Congress had mandated the use of fuel additives such as MTBE as part of the 1990 Clean Air Act Amendments. Subsequent regulations created a marketplace that decisively favored MTBE over ethanol, its chief rival. California's action set in motion a nationwide rethinking of this mandate as a means to clean the air and ultimately led the U.S. Environmental Protection Agency (EPA) to reverse policy and call for a reduction in the use of MTBE. These actions are among the important events in the regulatory history of MTBE (summarized in Table 4-1).

In this chapter, I describe the evolution of the oxygenate mandate in Congress, how regulators drafted rules that favored MTBE, and the environmental consequences that resulted. This story illustrates the important role played by institutions and the perceptions of the people within those institutions in the development of technically complex policies. From this analysis, it is evident that several opportunities existed to prevent environmental problems associated with MTBE use or to intervene in such a way as to minimize them. The MTBE story also illustrates several challenges facing the current system of environmental regulation in this country. Among them are a fragmented environmental policy system, lack of flexibility in the implementation of regulations, and a need for timely availability and use of scientific information across a broad spectrum of agencies.

The 1990 Clean Air Act and Oxygenated Fuels

One of the primary issues facing Congress before the Clean Air Act Amendments in 1990 was the continuing problem of urban air pollution caused by vehicle emissions. Two of the biggest problems were ozone pollution and the buildup of carbon monoxide (CO) levels in several cities during the winter months. Framing the debate about how to deal with these problems were proposals to mandate the use of vehicles that could use clean alternative fuels such as methanol and compressed natural gas. This strategy was strongly supported by important aides to President George Bush, high-level EPA officials, and some members of Congress (Bryner 1995; Cohen 1995).[1] These proposals were controversial, however, and faced powerful political opposition from the auto and oil industries as well as their supporters in Congress.

This situation changed on August 15, 1989, when Atlantic Richfield Company (ARCO) announced that it would introduce EC-1, an environmentally friendly reformulated fuel containing MTBE, to replace leaded gasoline in the pollution-plagued southern California market. ARCO touted the new fuel as a cost-effective way to clean up emissions of older vehicles by as much as 15% (Wald 1989). This announcement set off a chain of events that undermined the industry's arguments against strong federal fuel standards and opened the door to reformulated gasoline (RFG) as a means to clean the air. In the words of one lobbyist, "ARCO let the genie out of the bottle by saying they could make cleaner gasoline" (Kriz 1990).

After ARCO's announcement, Congressional debate shifted away from alternative fuels and focused on the technical details of gasoline formulation. Of particular importance was the debate over the oxygen levels that would be required for the new fuels. Adding oxygen to fuels achieves two objectives. First, increased oxygen in the air–fuel mixture promotes more efficient combustion and thus reduces CO emissions. Second, high-octane chemicals such as MTBE and ethanol can be used to replace other volatile organic compounds (VOCs) that had been used in place of lead as octane enhancers during the 1980s (Seymour 1992).[2,3] Some of these compounds have a high vapor pressure, and others (particularly benzene) are carcinogenic. Replacing them with substances such as MTBE and ethanol reduces the tendency for the fuel to evaporate and lowers the risk of exposure to tailpipe and evapora-

TABLE 4-1. Major Milestones in the Regulatory History of MTBE

Date	*Air pollution regulation*	*Health and environmental regulation*
1979	EPA approves MTBE for concentrations up to 7% by volume	
1981	EPA approves MTBE for concentrations up to 11% by volume	
1986		Maine State Toxicologist recommends maximum MTBE level for domestic water of 50 ppb Interagency Testing Committee recommends MTBE to be considered for priority regulation under Toxic Substances Control Act
1988	EPA approves MTBE for concentrations up to 15% by volume California Clean Air Act passes Denver (Colorado) institutes nation's first wintertime oxygenated fuel program	New York establishes action level for MTBE in groundwater
1989	Cities in Arizona, New Mexico, and Nevada establish wintertime oxygenated fuel programs ARCO introduces MTBE-rich EC-1, the first reformulated gasoline, in southern California	Residents in Phoenix (Arizona) complain about health effects associated with breathing MTBE fumes
1990	Clean Air Act Amendments pass, requiring oxygenated fuels	First case of MTBE groundwater contamination reported in California Clean Air Act lists MTBE as one of 189 hazardous air pollutants to be regulated by EPA
1991	California regulators approve oxygenated fuel requirements statewide, including a 2.2% cap on oxygen content due to concern about NO_x emissions. EPA announces its intent to formulate rules for reformulated gasoline using regulatory negotiation	California establishes action level for MTBE in groundwater of 35 ppb
1992	Clean Air Act's oxygenated fuels program begins, requiring fuel with 2.7% oxygen in CO nonattainment areas	Widespread health complaints accompany introduction of oxygenated fuel in Alaska and other states
1993	EPA completes regulatory impact analysis for reformulated gasoline	Anti-MTBE group OxyBusters is formed in New Jersey
1994	EPA proposes rule requiring that 30% of oxygenates for use in reformulated gasoline come from renewable sources such as ethanol EPA announces registration testing requirement rules for all fuel additives as required by the Clean Air Act	EPA, CDC, and state agencies undertake studies of possible health effects associated with MTBE fumes Alaska bans fuel containing MTBE

continued on next page

TABLE 4-1. Major Milestones in the Regulatory History of MTBE—*Continued*

Date	*Air pollution regulation*	*Health and environmental regulation*
1995	Clean Air Act's reformulated gasoline program begins, requiring fuel with 2% oxygen in ozone nonattainment areas First bill to outlaw MTBE introduced in California Federal court strikes down renewable oxygenate rule	City of Santa Monica (California) closes several MTBE-contaminated drinking water wells EPA asks Health Effects Institute to conduct a comprehensive review of all information available on health effects of fuel oxygenates USGS releases report describing the extent of urban groundwater contamination by MTBE
1996	California cleaner-burning gasoline program begins statewide	First bills that would exempt California from the oxygenated fuels requirements of the Clean Air Act are introduced in Congress
1997	California legislature passes bill requiring a comprehensive study of MTBE	EPA establishes advisory level for MTBE in drinking water U.S. Senate Environment and Public Works Committee holds field hearing on MTBE risk to drinking water
1998	EPA forms Blue Ribbon Panel on Oxygenates	University of California releases report recommending phaseout of MTBE because of its threats to water supplies and public health EPA places MTBE on Safe Drinking Water Act Contaminant Candidate List
1999	Governor Gray Davis of California issues executive order to phase out MTBE EPA Blue Ribbon Panel recommends "substantial reduction" in MTBE use	Maine and New Jersey end the use of oxygenated fuels because of concerns over groundwater contamination
2000	EPA reverses policy, announcing support for the increased use of ethanol and urging Congress to amend the Clean Air Act such that MTBE can be phased out Senate Environment and Public Works Committee approves a bill to phase out MTBE	New York governor issues executive order to phase out MTBE EPA announces plans to regulate MTBE under the Toxic Substances Control Act

Notes: EPA = U.S. Environmental Protection Agency; MTBE = methyl-*tert*-butyl ether; CDC = Centers for Disease Control; USGS = U.S. Geological Survey.

tive emissions. Reducing evaporative emissions was seen as a particularly important means of controlling ozone pollution, and any policy options addressing this issue received much attention (U.S. Congress, OTA 1989).

Both ethanol and MTBE were already approved for use in gasoline and therefore were the leading contenders in the significantly expanded oxygenate market that would emerge from the congressional debate over RFG. One of the most contentious issues of this debate was oxygen content and the underlying battle for market share between ethanol and MTBE. MTBE has several advantages over ethanol as an oxygenate because it can be manufactured from refinery by-products (and thus is relatively cheap to produce), mixes readily with gasoline, has a lower tendency to evaporate, and has a high octane rating. Ethanol, in contrast, is more expensive to produce than MTBE, is difficult to mix and transport, is more volatile, and contains more oxygen. Because of ethanol's higher volatility and higher oxygen content, less ethanol can be blended per gallon of gasoline to achieve the same oxygen content; thus, it does not displace the same volume of toxic components or decrease gasoline's tendency to evaporate as effectively as MTBE (California EPA 1997).

During the Congressional negotiations, every increase in oxygen content won by ethanol interests was seen as "money taken from the pockets" of the auto and oil industries (Cohen 1995, 183). Ethanol interests pushed for a law requiring gasoline oxygen content as high as possible because MTBE, with its lower oxygen content, faced a significant disadvantage if oxygen requirements were set above certain levels (Adler 1992). For example, ethanol blended with gasoline at 10% by volume produces an oxygen content of 3.5%, whereas MTBE blended at 15% by volume results in only 2.7% oxygen content (Piel and Thomas 1990). Because EPA had approved MTBE only as a gasoline additive at concentrations up to 15%, a bill that mandated oxygen content higher than 2.7% was very favorable to ethanol.[4]

The Clean Air Act Amendments signed by President Bush on November 15, 1990, included two new programs that would mandate gasoline oxygen content.[5] The first, a wintertime oxygenated fuel program, was designed to reduce CO pollution during the winter months and required the addition of 2.7% oxygen to gasoline in 39 cities classified as nonattainment areas for this criteria pollutant. This program would be implemented by the states and was scheduled to begin during the winter of 1992–1993. The second, known as the RFG program, required the addition of 2% oxygen to gasoline sold in cities that failed to achieve ambient air quality standards for ozone. This program was estimated to cost between $3 billion and $4 billion to implement and applied to the nation's nine most heavily polluted urban areas: Los Angeles and San Diego, California; New York, New York; Chicago, Illinois; Houston, Texas; Milwaukee, Wisconsin; Baltimore, Maryland; Hartford, Connecticut; and Philadelphia, Pennsylvania (*Environment Reporter 1991*).

The amended Clean Air Act also granted California, the nation's largest consumer of gasoline, permission to develop its own emissions regulations, because the state's progressive air pollution laws were in place before 1990. Under the 1988 California Clean Air Act, the California Air Resources Board (CARB) developed a reformulated fuels program of its own that required the sale of gasoline containing 2% oxygen content statewide beginning March 1, 1996. One of the unique features of this program is that refiners can comply by either producing fuel that meets content requirements for eight predefined components, varying the content of these components using a predictive model to determine emission performance, or conducting vehicle emissions testing to certify alternative blends (CARB 1998).

MTBE Becomes the Environmentally Preferred Oxygenate

Although the new clean air law included detailed specifications for new fuel formulations, Congress left many unresolved questions for EPA. Whereas the wintertime oxygenates program scheduled to begin in 1992 was administered by the states, rules for the RFG program were left to EPA. Instead of using traditional notice-and-comment rulemaking procedures, EPA opted for a collaborative regulation-and-negotiation (reg-neg) approach for the development of these rules. The goals of this process were to avoid the court challenges and lengthy litigation that typically accompanies the notice-and-comment procedure and to establish rules in time to meet the aggressive deadline of selling RFG by January 1, 1995 (Weber 1998).

To begin the reg-neg process, EPA brought together a set of 35 stakeholders directly affected by the RFG program and its implementation. These interests were grouped into seven categories: refiners, the auto industry, manufacturers of oxygenates (methanol and ethanol), marketers, state and local regulators, environmental and public interest groups, and federal regulators (*Federal Register* 1991). An important task facing this group was to draft regulations for RFG that would not increase nitrogen oxide (NO_x) or evaporative emissions. These constraints hinged on oxygen content and would prove to be key elements in determining how the market would be divided between MTBE and ethanol. As indicated earlier, ethanol is significantly more volatile than MTBE, placing it at a disadvantage when attempting to limit evaporative emissions.

Another problem facing ethanol supporters was that studies conducted at the time indicated that oxygen content levels above approximately 2.1% would increase NO_x emissions when using ethanol but that MTBE would not increase NO_x emissions even at its maximum allowable concentration of 2.7% (Adler 1992; Piel and Thomas 1990). On the basis of these findings,

EPA made an initial proposal allowing a maximum oxygen level of 2.7% for MTBE and 2.1% for all other oxygenates (*Environment Reporter* 1991). Although the final agreement did not specify lower oxygen levels for ethanol, it did establish the same volatility standard regardless of the oxygenate used, thus blunting attempts by ethanol supporters to establish a rule granting a higher volatility standard for ethanol than for MTBE. This provision provided a distinct advantage to MTBE, and ethanol interests threatened to walk out of negotiations over the issue. EPA, environmentalists, and state regulators had serious concerns over ethanol's potential to increase evaporative and NO_x emissions, however, and joined forces to keep a single volatility standard in the final agreement (Weber 1998).

Although they ultimately signed the final agreement in August 1991, ethanol interests refused to accept the single volatility standard and appealed to the Bush administration and Congress to modify the final rule by granting ethanol a higher volatility standard than MTBE (Bryner 1995). The other participants from the recently concluded reg-neg process teamed up to lobby against this effort, arguing that other industries would have to suffer under more stringent standards to compensate for the increased pollution that would result if ethanol were granted a waiver (*Environment Reporter* 1992). President Bush was not convinced by these arguments and, in the midst of his presidential campaign and looking for votes from the farm states, granted ethanol a vapor pressure waiver of one pound per square inch in October 1992 so that the additive could be used in RFG (Schneider 1992).

EPA refused to implement this rule in the waning days of the Bush administration, and ethanol backers were forced to appeal to the incoming EPA Administrator Carol Browner (Weber 1998). As the debate over the proposed volatility waiver for ethanol intensified, the first arguments were made against the use of MTBE on environmental grounds. In a fierce advertising war waged in the pages of *The New York Times*, *The Washington Post*, and *The Wall Street Journal*, ethanol backers argued that the methanol-derived MTBE posed an environmental threat (Bryner 1995). In a speech that would prove prophetic, the specter of MTBE groundwater contamination was raised by Senator Charles Grassley of Iowa when he cited a published study detailing MTBE's slow biodegradability in the subsurface environment (U.S. Senate 1993a, S16641).[6]

EPA Administrator Browner eventually issued the RFG rule without Bush's volatility waiver for ethanol in late 1993, but she made good on President Bill Clinton's campaign promises to voters in the Corn Belt and proposed a new rule that would require that 30% of the oxygenate used to produce RFG be made from renewable resources (Kriz 1994; Stone 1994). This rule promised a guaranteed market share for ethanol, the only oxygenate that could qualify as renewable.

Environmentalists joined with the oil and petrochemical industries in an intense lobbying effort to convince Congress to pass legislation rescinding the renewable oxygenate mandate (Brown 1997; Bryner 1995). One of the key arguments of the rule's opponents was that granting exemptions for ethanol would result in negative environmental consequences because of its higher volatility and propensity to increase NO_x emissions. The Sierra Club described even more dire consequences in a Senate hearing on the new rule, stating that reserving a market segment for ethanol would "potentially increase global warming, increase smog, increase air toxics" and "potentially increase water pollution and damage to erodible and sensitive habitat areas as a result of intensive farming to produce additional corn" (U.S. Senate 1994, 71).

These efforts to defeat the renewable oxygenate mandate legislatively failed, and environmentalists and the oil industry were forced to attack the proposed rule in federal court. The American Petroleum Institute (API) took the lead in this effort and sued EPA over the rule in the D.C. Circuit Court of Appeals. On April 28, 1995, the court ruled in favor of API, finding that the renewable oxygenate mandate was "not directed toward the reduction of VOCs" (Brown 1997, 1312). This action was a significant blow to ethanol and cleared the way for MTBE to dominate the emerging oxygenate market.

Ethanol was also losing ground in California. When CARB approved rules for the implementation of California's Wintertime Oxygenates Program in 1991, one of the most contentious issues was a staff proposal to seek a waiver from EPA to reduce oxygen content. CARB staff felt that they had sufficient evidence to justify reducing the oxygenate content to a range of 1.8–2.2% instead of the 2.7% minimum specified in the Clean Air Act Amendments. This position was based on oil industry and CARB studies that showed increased NO_x emissions associated with oxygen contents above approximately 2.2% (CARB 1991a). California ethanol interests argued against this rule, claiming that it would amount to a virtual mandate for MTBE and put them out of business in the state (CARB 1991c, 1992). The CARB rule had the effect of essentially prohibiting ethanol use in California after 1991, guaranteeing that MTBE would dominate the California oxygenate market (Wiley 1998).

Demand for MTBE soared as the federal court's ruling against the renewable oxygenate rule and California's action to lower the oxygen content cap sent a signal that regulators viewed MTBE as the environmentally preferred oxygenate (Figure 4-1). During the early 1990s, MTBE was considered the fastest growing petrochemical in the world. Anticipating this increased demand as the Clean Air Act Amendments moved through Congress, industry geared up to produce MTBE in large quantities and announced a 75% increase in production capacity that would soon push daily output to 175,000 barrels (Uhlman 1990).

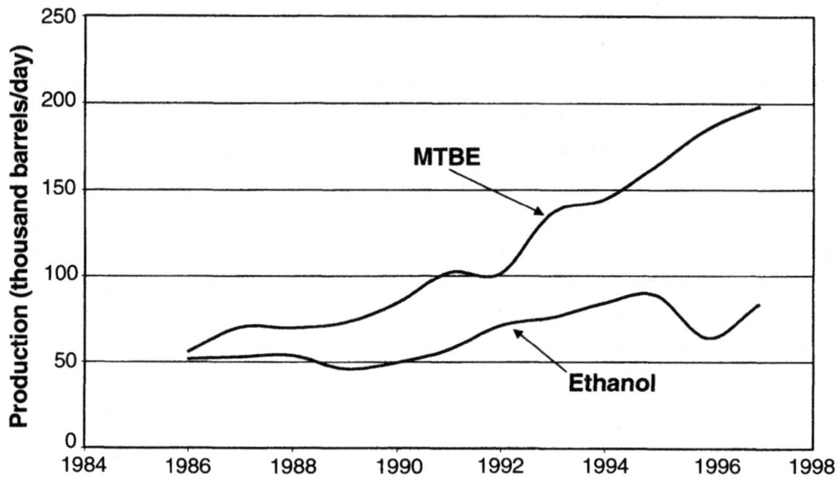

FIGURE 4-1. U.S. Oxygenate Production, 1986–1997
Source: EIA 1998.

This trend of shifting alliances and reactive policymaking reflected the still emerging science of fuel oxygenates. Although more acceptable politically than alternative fuels such as methanol, reformulating gasoline faced a great deal of uncertainty from a technical standpoint. Although there had been some experience with oxygenating fuels to reduce CO emissions in Denver, Colorado, and other areas during the previous two winters,[7] little else was known about the impact of changing gasoline formulations outside of the laboratory. A study conducted by the Office of Technology Assessment (OTA) summed up the uncertainty associated with projecting the emissions reduction potential of RFG: "Reformulated gasoline is a concept, not a reality" (U.S. Congress, OTA 1990, 134). Compounding the challenge of reliable technical information for policymakers, a multiyear joint effort by the auto and oil industries designed to clarify these uncertainties did not even begin until the summer of 1989, with initial results not expected until late 1990 (U.S. GAO 1990). A more severe lack of knowledge existed regarding the negative environmental impacts of fuel oxygenates. Therefore, it was not surprising that regulators were caught off guard when the dramatic increase in MTBE use caused environmental consequences.

MTBE Emerges as an Environmental Threat

During the 1990 Clean Air Act debate, MTBE was recognized as an air pollutant, and the final bill included MTBE on the list of 189 hazardous air pollutants that EPA was required to regulate. It also was recognized that combustion of the methanol component of MTBE would increase emissions of formaldehyde—a known human carcinogen and reactive precursor to ozone formation (U.S. Senate 1990, 117). In general, regulators viewed this as an acceptable trade-off because the increased formaldehyde emissions were offset by the decrease in toxic emissions associated with compounds that MTBE was displacing. In California, for example, regulators explained, "any increase in formaldehyde emissions would be more than offset by the reduced toxic risk due to the entire package of Phase 2 gasoline specification" (CARB 1991b, 31).

These technical issues, buried as they were in the myriad details of the enormous new law, would soon pale in comparison to the impact of public perception. When the federal oxygenated fuels program began in 1992, the general public had little knowledge about the effects of fuel composition on emissions. What the public realized was that not only did the price of gasoline increase during the winter, but the new formula appeared to produce negative health effects in certain individuals. These complaints typically consisted of headaches, dizziness, nausea, and general malaise after exposure to fumes from gasoline containing MTBE. Concern first arose during the winter of 1992–1993 in cities required to use oxygenates to control CO emissions (Melnick and others 1997).

These unanticipated health complaints led to public protest against the new fuel in several cities and widespread media coverage of MTBE (Brown 1997). In New Jersey, concerned citizens formed the public interest group OxyBusters in 1993 and began a grassroots campaign to eliminate MTBE in their state. As complaints about MTBE spread, OxyBusters chapters appeared in other states. In California, an OxyBusters chapter was established in 1996 and soon had enlisted more than 100,000 people to sign a petition demanding a statewide ban of MTBE (Carlsen 1997).

Local governments responded to these complaints by calling for the removal of MTBE from gasoline pending additional studies. At the federal level, a program designed to assess the health effects of all fuel additives was already in place, but rules requiring comprehensive testing of more than 4,800 fuel additives (including MTBE) were not announced until 1994, and results were not expected until after 2000 (*Federal Register* 1994). EPA Administrator Browner, faced with the prospect of granting special exemptions to the far-reaching oxygenated fuel program, requested a special joint study by the EPA Office of Research and Development and Yale University in 1993 (Cone 1993). At the same time, the states of Alaska and Wisconsin undertook their own studies, and the Centers for Disease Control (CDC) initiated studies in Alaska, Connecticut, and New York (Melnick and others 1997).

The results from these initial studies were limited because they were mounted largely in an effort to respond expeditiously to public concerns

(Melnick and others 1997). EPA declared the program safe after concluding that MTBE most likely was not an acute health risk for most people at typical exposure levels and that the overall risk of oxygenated fuel was less than that of conventional gasoline (Ivanovich 1994). However, CDC testified before Congress that its results indicated a possible relationship between MTBE concentration in the blood and health complaints (U.S. Senate 1993b). Other researchers reported a connection between MTBE exposure and occurrence of symptoms such as headache, nausea, disorientation, allergic reactions, and even increased incidence of more chronic ailments such as asthma (Joseph 1995). The most controversial studies suggested a link between MTBE exposure and cancer in laboratory animals (Fadope 1995). Many of these researchers strongly recommended that additional investigations be conducted to better understand the health effects of exposure to MTBE before its continued use was allowed. The American Medical Association, for example, adopted a resolution calling for a moratorium on MTBE use in Alaska pending further study (Scanlon 1994). Comprehensive assessments of this research stopped short of establishing a link between MTBE and negative human health effects but concluded that acute symptoms associated with MTBE exposure could not be dismissed (Melnick and others 1997; National Research Council 1996).

Indications that MTBE posed a potential threat as an airborne contaminant were a precursor to problems that would arise when the chemical was found in water supplies. Because MTBE had been approved for use as an octane booster since 1979, it had circulated through the nation's fuel distribution infrastructure, and gasoline containing MTBE had spilled or leaked from this system on many occasions. As a result, groundwater specialists recognized MTBE as a particularly troublesome groundwater contaminant because of its high water solubility, extremely slow rate of natural biodegradation, and low odor and taste threshold when dissolved in water. Cases of MTBE contamination began to occur during the early 1980s, shortly after EPA approved its use as a gasoline additive (Garrett and Moreau 1986; Mckinnon and Dyksen 1984; McQuillan 1990). Although MTBE was recognized as less toxic than other gasoline components, the U.S. Interagency Testing Committee still included it on a list of chemicals recommended for priority consideration under the Toxic Substances Control Act in 1986 (Gilbert and Calabrese 1992).

The fact that MTBE was turning up as a problematic groundwater contaminant was not surprising, given that EPA had estimated during the mid-1980s that 11 million gallons of gasoline were released into the ground each year from leaking underground storage tanks (Tangley 1984). Comprehensive programs designed to replace and upgrade leaking tanks were undertaken at the state level beginning during the late 1980s, but approximately

170,000 contaminated sites associated with these tanks remained in 1995 (Zogorski and others 1997). Despite this widespread problem with the nation's gasoline delivery infrastructure, almost no concern was raised about the contamination of water by RFG during debate of the 1990 Clean Air Act Amendments and implementation of the resulting regulations.[8]

Regulators responsible for underground storage tanks responded to the problem of MTBE groundwater contamination by issuing *advisory levels.* This level of regulation is not legally enforceable, so local authorities are not required to monitor and report contaminant levels. An advisory generally establishes a level at which lifetime exposure is not expected to pose a health threat. Contamination discovered above the advisory level should be remediated or further monitored to assess the extent of the problem. In one of the first regulatory actions concerning MTBE as a groundwater contaminant, the Maine State Toxicologist "set a recommended maximum concentration level for MTBE in domestic water at 50 ppb [parts per billion]" in 1986 (Garrett and Moreau 1986, 235). At the federal level, EPA first issued a draft drinking water lifetime health advisory of 20–200 ppb in 1992 but did not offer any advice on the important elements of public risk perception associated with taste or odor thresholds (Squillace and others 1996). In California, the state Department of Health Services first became aware of MTBE water contamination in 1990 at two wells in San Francisco and adopted an interim action level of 35 ppb in 1991 (California State Auditor 1998). These actions were taken to address a problem that existed before the Clean Air Act Amendments of 1990 and under a completely different law.[9]

These scattered indications that MTBE posed a threat to groundwater were not enough to force changes in the regulation of reformulated fuels. This situation changed in rather dramatic fashion after an incident in Santa Monica, California. Beginning in late 1995, city officials became aware of a new contaminant affecting their drinking water wells. After extensive testing, it was found that MTBE was the unknown contaminant and had tainted 7 of the city's 11 drinking water wells, representing approximately 50% of the total supply (U.S. Senate 1997). The city was forced to close these wells and faced the challenge of providing new supplies while finding a way to clean up the contamination. Whereas isolated cases of MTBE contamination of drinking water wells had occurred in the past, this was the first instance of an MTBE contamination problem affecting large numbers of people in a major city. Santa Monica's dilemma was widely publicized and led to speculation over whether the problem could have been avoided.

Faced with this unexpected threat to water quality, regulators were caught unprepared and reacted slowly. At the time, state and federal regulators had not established an MTBE drinking water standard; the lengthy rulemaking process implemented under the Safe Drinking Water Act (SDWA) requires gathering sufficient scientific evidence to establish a safe level of contamination. As a result, local water distributors had never been required to test drinking water supplies for MTBE. Almost no information existed about the extent of the contamination or the risks associated with exposure to MTBE-contaminated drinking water. The first study designed to investigate this issue in California was released in 1998 and concluded that MTBE contamination affected at least 10,000 sites within the state (Happel and others 1998).

These unanticipated events delayed action as various regulatory agencies struggled to understand the scope of the problem. They took place in an atmosphere characterized by increasing publicity of MTBE that in turn raised public awareness and demand for action. As regulators gathered information and launched scientific studies, the legislature took action. California Senate Bill (SB) 521 was the most aggressive and required an outright ban of MTBE. Legislation was also introduced in both chambers of the U.S. Congress that would exempt California from the federal oxygenate requirements of the Clean Air Act.

Air regulators, environmentalists, and oil companies initially joined forces to shift focus to the air quality benefits of MTBE and opposed an outright ban. These groups advocated a policy that aggressively cleaned up contamination and prevented spills while still allowing MTBE use.[10] The oil industry, which had spent more than $1.8 million to lobby California legislators in 1996, was particularly effective in pressuring lawmakers to avoid an outright ban (Carlsen 1997). Opposition groups pushed for the elimination of MTBE as quickly as possible. Out of this charged political environment came an amended SB 521 that required a $500,000 study of the risks and benefits of MTBE be completed by the University of California before January 1, 1999. The bill also required the governor to take appropriate action, including an outright ban, if the study found MTBE to be a danger to human health.

The University of California study systematically evaluated all available information concerning MTBE and was completed in November 1998 (Keller and others 1998). It found that MTBE was placing the state's limited water resources at risk and that important data about the toxicity of MTBE had not yet been obtained. On the basis of these and other findings, the report recommended that MTBE be phased out over several years. This recommendation was not entirely unexpected but still sent shock waves through the refining industry. Using the University of California report as justification, Governor Davis declared that MTBE posed a significant risk to California's environment and issued an executive order in March 1999 directing state regulators to remove MTBE from the state's gasoline at the earliest possible date, but no later than December 31, 2002.

While debate raged in MTBE's largest market, actions in other states continued to erode its status as an air pollution control measure. One of the most important events took place in North Carolina, where MTBE had already been banned after classification as a probable human carcinogen. In August 1997, Conoco lost a lawsuit filed by residents of a North Carolina mobile home park who claimed that the company had been negligent, contaminating their water wells with MTBE and benzene (Wiley 1998). This $9.5-million judgment, in addition to an undisclosed punitive amount, was perhaps more troubling to the oil industry than pending legislation in California. Reflecting this concern, oil refiners began to announce that they would begin eliminating MTBE from their gasoline. Tosco Corporation, the nation's largest independent refiner and a key player in California, was the first to do so when it sent a letter to CARB recommending an immediate move away from MTBE (Bowman 1997).

EPA responded to these actions, perceived by some as threatening to unravel the federal oxygenate program, by appointing a Blue Ribbon Panel to investigate MTBE in late 1998. This panel concluded its work in the summer of 1999 and recommended a substantial reduction in the use of MTBE, removal of the requirement that RFG contain 2% oxygen by weight, clarification of federal and state authority to regulate gasoline additives that threaten drinking water, and comprehensive improvements in water protection programs (U.S. EPA 1999). After these findings were announced, EPA was forced to reverse its opposition to congressional proposals calling for an MTBE ban and develop a strategy to implement the panel's recommendations. Reflecting the complex nature of the MTBE problem, the agency developed a multidimensional approach that would be pursued under three different statutes.

The first was to regulate MTBE as a water contaminant under SDWA. Although regulations developed under this statute do not directly address the use of MTBE in gasoline, they provide the framework under which local water utilities can implement procedures to deliver drinking water that does not contain unsafe levels of a chemical. In August 1999, EPA followed up on earlier advisories regarding MTBE contamination of drinking water and issued a rule requiring public water systems to monitor for MTBE beginning in January 2001. This monitoring data, combined with ongoing health studies, is intended to provide the information necessary for EPA to develop a formal regulation for acceptable levels of MTBE in drinking water. Because of the amount of time required to gather this data and promulgate a rule, it is not anticipated that one will be in place before 2010 (McCarthy and Tiemann 2000).

A second statute offering an opportunity to regulate MTBE in a way that would reduce its use is the Toxic Substances Control Act (TSCA). Section 6 of this act grants EPA authority to ban or limit any chemical that poses an unreasonable risk to the public or environment. In March 2000, EPA issued an Advance Notice of Proposed Rulemaking announcing its intent to take action to regulate MTBE under TSCA and requesting comments on the issues surrounding such a regulation (U.S. EPA 2000a). However, reducing or eliminating the use of MTBE under TSCA is problematic because of the extensive amount of data that must be collected to show that the chemical is indeed a threat. In conjunction with gathering this data, EPA must also negotiate with affected parties as part of the implementation of any regulation that emerges from this process. Recognizing this time-consuming and uncertain process, EPA Administrator Browner admitted that negotiating the procedures required to regulate MTBE under TSCA could take as long as three years (Stout 2000).

Because of the amount of time required to develop regulations to limit MTBE use under TSCA, the only way to reduce its use in the near term is to modify the oxygenate requirements of the Clean Air Act. This option is a key problem because under current law, refiners must add oxygen to gasoline sold in ozone nonattainment areas, and the only viable alternatives are MTBE and ethanol. In an effort to draw congressional attention to this issue, EPA Administrator Browner appeared at a press conference with Secretary of Agriculture Dan Glickman on March 20, 2000, to propose a legislative framework that would allow the agency to implement the recommendations of the Blue Ribbon Panel. Specifically, EPA proposed an amendment to the Clean Air Act that would provide the agency with authority to reduce or eliminate MTBE, maintain air quality gains of the RFG program, and encourage the use of renewable fuels such as ethanol in place of the oxygenate requirement (U.S. EPA 2000b). The presence of Secretary Glickman, a former congressman from Kansas, at this press conference sent a strong signal that the Clinton administration supported a policy that would reduce MTBE use while encouraging ethanol production (Stout 2000).

Congress was already considering several legislative proposals to deal with the MTBE problem, and EPA's proposal outlined requirements for legislation that would be acceptable to the administration. Bills introduced to date encompass approaches ranging from exempting California from oxygenate requirements to a nationwide MTBE ban. One key issue underlying all of these legislative initiatives is the desire of ethanol backers to secure a place for their additive in the nation's gasoline supply. By 2000, approximately 6% of the nation's corn harvest was being used to produce ethanol, and farm state interests saw any effort to kill the oxygenate requirement as a direct threat to this market (McCarthy and Tiemann 2000). At the same time, these interests also saw an opportunity to dramatically increase demand for ethanol if they could successfully craft legislation that maintained the oxygenate requirement while banning MTBE. As was the case during the debate of the 1990 Clean Air Act Amendments, the goal of reducing air pollution through gasoline reformulation had become linked to encouraging energy independence and subsidizing agriculture.

Despite the momentum for some type of legislation dealing with MTBE, only one bill had successfully emerged from committee by the end of the 106th Congress. On September 7, 2000, the Senate Environment and Public Works Committee approved S.2962, authored by Committee Chairman Bob Smith of New Hampshire. Senator Smith attempted to strike a balance between interests that advocated a flexible approach through the elimination of the oxygenate requirement and ethanol backers who sought to retain a legislative mandate that would reserve market share for their additive. Specifically, the bill requires a nationwide phaseout of MTBE within four years and an increase in the use of "clean alternative fuels" such as ethanol to 1.5% of the nation's gasoline supply by 2011 (Eckert 2000).

While Congress and EPA struggle to address the known consequences of MTBE use, many uncertainties remain concerning its broader environmental consequences. Some studies completed to date have concluded that MTBE may be accumulating in the environment by way of mechanisms such as atmospheric deposition, storm water runoff, and discharge into surface waters from marine engines (Keller and others 1998; Squillace and others 1996; U.S. EPA 1999). The implications of these findings may reach beyond direct impacts to human health, because relatively little study has been undertaken to evaluate the impact of MTBE on ecosystems. One report concluded "to date there has been only minimal effort to investigate ecological impacts caused by exposure to … MTBE" and recommended that "research into the potential for MTBE to adversely affect ecological receptors is essential" (Carlsen, Hall, and Rice 1997, 1). These important unanswered questions about MTBE are ones that regulators will eventually be forced to consider in the attempt to find acceptable replacements.

Conclusions

What started as an attempt by President Bush and Congress to promote cleaner fuels resulted in an oxygenate mandate that introduced large amounts of MTBE into the environment, causing unanticipated and far-reaching consequences. Although Congress did not specifically mandate the widespread use of MTBE, regulations designed to implement the new law created a market decisively in favor of MTBE because of concerns about ethanol's higher volatility and potential to increase NO_x emissions. The consequences of these actions became apparent only after the reformulated fuel programs were well under way. The process that led to these unintended consequences illustrates three important challenges currently facing the environmental regulatory system.

One of the most important challenges is that successful environmental regulation must be integrated in its approach. The MTBE story dramatically shows the consequences of what Davies and Mazurek (1998) describe as the fragmented system of environmental policy that has evolved in the United States. Because of this fragmentation, regulators working under SDWA, the Resource Conservation and Recovery Act, TSCA, and the Clean Air Act failed to coordinate their efforts to meet the overall threat that MTBE posed until several years after its widespread use began. The complex web of regulations associated with these statutes eventually began to address MTBE use, but only after highly visible events generated pressure from the public and legislators. Several opportunities to apply a more integrated approach existed before MTBE became a widespread problem, but regulators at all levels of government never questioned placing gasoline with up to 15% MTBE into underground storage tanks, a significant percentage of which were known to be leaking. Had existing regulations developed under these statutes been coordinated with the RFG program, many of the problems associated with MTBE could have been avoided.

A second challenge facing environmental regulation that is particularly well illustrated by the MTBE story is the need for flexible implementation tools. This need arises to a large degree from the political dimension of environmental policymaking. As Adler (1992) concludes, the oxygenate requirements contained in the 1990 Clean Air Act Amendments were driven by special interest politics. The result was a complex law requiring specific amounts of gasoline additives to be introduced in a relatively short period of time. This trend has reemerged as Congress considers legislation to alter the Clean Air Act such that MTBE use can be reduced. It strongly suggests that future legislation concerning gasoline content will contain equally complex requirements, and regulatory agencies must find flexible methods of implementation if unintended consequences are to be minimized.

The MTBE experience also shows the consequences of a lack of timely and accessible scientific information. Regulators relied on the best available science in emissions control technology to design RFG regulations favorable to MTBE, yet failed to consider available evidence of its potential as a water contaminant. The regulatory process became, in large part, a quest for the oxygenated gasoline formula that would minimize tailpipe and evaporative emissions. As the science of cleaner gasoline evolved, regulators adjusted their specifications to the extent possible as they raced to meet the implementation deadlines set by Congress. In dealing with the initial consequences of MTBE use, agencies reacted at all levels by repeatedly taking action based on narrowly focused scientific studies. This narrow focus precluded consideration of published research available at the time which indicated that MTBE was a problematic groundwater contaminant.

This problem is still evident in recent attempts to phase out MTBE in favor of ethanol. Several months after EPA announced its desire to increase ethanol use in place of MTBE, the agency's Office of Research and Development changed the classification of acetaldehyde from "probable human carcinogen" to "likely human carcinogen" as part of a study of toxic emissions resulting from various gasoline formulas (*Risk Policy Report* 2000). Because acetaldehyde is a combustion by-product of ethanol, this action signals possible unintended consequences of dramatically increasing ethanol use and will further complicate efforts to reduce MTBE use within the confines of the oxygenate requirement. This announcement also illustrates how emerging science continues to play a critical role in the formation of policies concerning gasoline formulation.

The challenges illustrated by the MTBE experience point to the need for new ways of making environmental regulations. A more integrated approach is perhaps the most important area in need of more research. Achieving this goal suggests a broader consideration of all potential impacts to human health and the environment associated with new regulations. One approach to this problem would be to develop ways to involve a broader array of stakeholders in the existing rulemaking process. Regulators and independent groups concerned with underground tanks had little contact with air pollution specialists during formulation of the rules for implementing the sale of RFG. However, it is clear that at least some state regulators were aware of the particular challenges that MTBE posed as a water contaminant. Maine, New Mexico, and Connecticut, for example, had already experienced problems with MTBE water contamination before the Clean Air Act Amendments (Garrett and Moreau 1986; Gilbert and Calabrese 1992; McQuillan 1990). Had officials from these states been consulted during the rulemaking process, it is plausible that the process of evaluating MTBE as a drinking water contaminant could have started much earlier and been done in conjunction with the implementation of RFG. Whereas this step may not have prevented every instance of contamination, officials would have been better prepared to deal with problems such as those that arose in Santa Monica. This example of a missed opportunity is especially obvious because similar regulations for other components of gasoline had been in place for years.

A second way to achieve a more integrated process would be to conduct peer-reviewed environmental impact, risk, or benefit–cost assessments of the proposed RFG rules. The National Environmental Policy Act (NEPA) environmental impact statement process is potentially applicable to legislative actions such as the RFG program and would have required a comprehensive assessment of potential impacts to all media (Council on Environmental Quality 1986). This process is time-consuming, and it is not clear that the critical connection between MTBE-enriched gasoline and leaking underground storage tanks would have been made even if it had been conducted. Alternatively, and as possible even under NEPA, benefit–cost analysis that considers all impacts "to whomsoever they accrue" also may be a process to integrate concerns. Had this type of approach been used during the implementation of the RFG program, important questions about how oxygenates would be manufactured, delivered, and stored probably would have been asked. The rulemaking process actually used to implement the RFG program, in contrast, was focused on air pollution and did not provide this opportunity.

The development of more flexible tools to implement environmental regulations in an atmosphere of political pressure and changing public perceptions of risk is another important priority for future work. The advantages of a performance-based approach to fuel content regulations illustrate this point. The predictive model used by CARB for compliance with fuel regulations in California has proven to be flexible enough to address the MTBE problem without incurring unacceptable trade-offs in air quality. This model was originally promoted as a way for ethanol manufacturers to increase oxygen content and maintain their market share when CARB lowered oxygenate requirements below those specified in the Clean Air Act (CARB 1991a). The flexibility of this approach is evident in that some refiners now can use this model to eliminate oxygen (and thus MTBE) from gasoline during the summer months while still producing a fuel that complies with California's strict performance standards (U.S. EPA 1999).

The complex model developed by EPA is a similar tool and has been promoted as a way for ethanol producers to increase the use of their product in RFG without reducing air quality benefits (*Environment Reporter* 1992). Unfortunately, federal law still mandates minimum oxygen content in CO and ozone nonattainment areas, so regulators are not able to take full advantage of these tools to maintain air quality benefits while minimizing the environmental impacts of oxygenate use. However, the existence of these models illustrates the promise of developing similar regulatory compliance tools that can be used to react quickly and effectively to unanticipated problems.

More work is needed to find ways to incorporate science into the regulatory process in a broad, comprehensive fashion. As Franklin and others (2000, 3861) conclude, the scientific research used throughout the legislative and regulatory history of MTBE was "fragmented by reliance upon narrow disciplinary interests." Understandably, the years required to conduct a full evaluation of MTBE's impacts on the environment before implementing oxygenate requirements were not available given the strict implementation timetables mandated by Congress. However, published research that warned of problems associated with MTBE-contaminated water supplies was available as early as 1986 and was not incorporated into a regulatory decisionmaking process focused on air quality, suggesting flaws in the process by which science is communicated and used within the regulatory rulemaking process. Finding ways to take full advantage of existing and emerging science during the development of future environmental regulations in a comprehensive way will prove critical in avoiding unintended consequences.

It could be argued that the trade-offs of MTBE use are less severe than those of previous attempts to regulate automotive emissions, but the policy-making process still fails to anticipate them enough to proactively limit their impact or communicate the trade-offs. If the MTBE story teaches us anything about environmental policy, it is that there is no such thing as a "magic bullet" for reducing automobile emissions. MTBE turned out to be just another step in the long road toward mitigating the impact of automobiles on the environment.

Notes

1. These aides included C. Boyden Gray, President George Bush's chief counsel, and Bill Rosenberg, assistant administrator for clean air issues at the U.S. Environmental Protection Agency. Gray was so committed to alternative fuels that he drove a methanol-powered Chevrolet. Rosenberg was a businessman from Michigan and strongly advocated increased use of grain-based ethanol to stimulate the Midwest's economy. For a detailed discussion, see Cohen 1995.

2. A fuel's octane rating is a measure of its ability to resist knocking. Engine knocking occurs when the air–fuel mixture ignites prematurely under high compression and leads to reduced engine efficiency.

3. Methyl-*tert*-butyl ether (MTBE) was invented by the Atlantic Richfield Company (ARCO) during the 1960s and was first used in gasoline as a replacement for lead. A commercial market for MTBE was created in 1979 when the U.S. Environmental Protection Agency approved MTBE as a gasoline additive at a maximum concentration of 7% by volume (Piel and Thomas 1990). This action was based on two criteria established for approving gasoline additives in the 1977 Clean Air Act Amendments: it is substantially similar to gasoline, and it will not cause or contribute to the failure of an emission control device or system (CARB 1998). Using these criteria, MTBE was later approved for volumetric concentrations of 11% in 1981 and 15% in 1988.

4. The final Senate bill strongly favored ethanol by requiring 3.1% oxygen content in fuels sold in carbon monoxide nonattainment areas, stating that it was necessary to "encourage domestically produced ethanol to enter the fuel stream" (U.S. Senate 1990, 449).

5. The U.S. Environmental Protection Agency and the California Air Resources Board (CARB) also developed several other regulations concerning the properties of gasoline at this time. Most significant are restrictions on the content of benzene, aromatic hydrocarbons, olefins, and sulfur.

6. The study cited (Suflita and Mormile 1993) was funded by the American Petroleum Institute and found that no evidence for the anaerobic destruction of methyl-*tert*-butyl ether (MTBE) could be obtained for at least 249 days. Senator Grassley

emphasized this finding in his speech against MTBE. The authors warned of potential problems with oxygenates because of their recalcitrance in the environment.

7. Other cities instituted winter oxygenated fuels programs during the winter of 1989–1990, including Reno and Las Vegas, Nevada; Tucson and Phoenix, Arizona; and Albuquerque, New Mexico.

8. One of the few indications that some consideration was given to methyl-*tert*-butyl ether (MTBE) as a threat to water is summed up by the following statement from a California Air Resources Board staff report: "The potential risk associated with oxygenate transport includes additional traffic and the risk of spill. Oxygenates are not very toxic, probably less so than gasoline, although they can be more difficult to remove from contaminated water than the hydrocarbon portion of gasoline" (CARB 1991b, 79–81).

9. Under the Resource Conservation and Recovery Act, the U.S. Environmental Protection Agency regulates sites contaminated by leaking underground storage tanks through implementation programs developed by the states. In California in 1997, approximately 32,300 storage tanks subject to regulations under this program were known to be leaking (Happel and others 1998).

10. Specifically, environmental groups advocated (a) developing an immediate plan to investigate and clean up soil and groundwater contaminated with methyl-*tert*-butyl ether (MTBE), (b) repealing the "contaminant zone" policy that allowed some low-risk sites to degrade naturally without intervention, (c) rapidly developing drinking water standards for MTBE and requiring monitoring, (d) accelerating the introduction of non-petroleum-fueled vehicles, and (e) performing a multimedia risk assessment of MTBE and other oxygenates (Sierra Club 1997).

References

Adler, Jonathan H. 1992. Clean Fuels, Dirty Air. In *Environmental Politics: Public Costs, Private Rewards*, edited by Michael S. Greve and Fred L. Smith. New York: Praeger Publishers, 19–45.

Bowman, Chris. 1997. Oil Firm: Halt Use of Gas Additive. *Sacramento Bee*, October 29, A1.

Brown, William M. 1997. The Renewable Oxygenate Requirement: A Boon for the Environment or a Boondoggle for the Ethanol Industry? *New York Law School Review* 41(3–4): 1299–1328.

Bryner, Gary C. 1995. *Blue Skies, Green Politics: The Clean Air Act of 1990 and Its Implementation* (Second Edition). Washington, DC: Congressional Quarterly Press.

California EPA (Environmental Protection Agency). 1997. *MTBE Briefing Paper.* Sacramento, CA: California EPA. http//www.CARB.ca.gov/cbg/pub/pub.htm (accessed May 22, 1999).

California State Auditor. 1998. *California's Drinking Water: State and Local Agencies Need to Provide Leadership to Address Contamination of Groundwater by Gasoline Components and Additives.* Sacramento, CA: Bureau of State Audits. http//www.bsa.ca.gov/bsa (accessed June 27, 1999).

CARB (California Air Resources Board). 1991a. *Proposed Regulation for California Wintertime Oxygenates Program: Staff Report.* Sacramento, CA: CARB.

———. 1991b. *Proposed Regulations for California Phase 2 Reformulated Gasoline: Staff Report.* Sacramento, CA: CARB.

———. 1991c. *Public Hearing to Consider Adoption of and Amendments to Regula tions Regarding Reformulated Gasoline.* Hearings 91-11-1 and 91-12-1. November 21–22 and December 12. Sacramento, CA: CARB.

———. 1992. *Proposed Regulations for California Wintertime Oxygenates Program: Final Statement of Reasons.* Sacramento, CA: CARB.

———. 1998. *An Overview of the Use of Oxygenates in Gasoline.* Sacramento, CA: California Environmental Protection Agency. http//www.CARB.ca.gov (accessed May 17, 1999).

Carlsen, Tina, Linda Hall, and David Rice. 1997. *Ecological Hazards of MTBE Exposure: A Research Agenda.* Washington, DC: U.S. Department of Commerce, National Technical Information Service.

Carlsen, William. 1997. Plenty of Interest Groups Lobby Both Sides. *The San Francisco Chronicle,* September 16, A9.

Cohen, Richard E. 1995. *Washington at Work: Back Rooms and Clean Air* (Second Edition). Needham Heights, MA: Allyn and Bacon.

Cone, Marla. 1993. Questions of Health Risk Cloud Use of Special Fuel. *Los Angeles Times,* October 4.

Council on Environmental Quality. 1986. Executive Office of the President. *Regulations for Implementing the Procedural Provisions of the National Environmental Policy Act.* Washington, DC: Government Printing Office.

Davies, Clarence J., and Mazurek, Jan. 1998. *Pollution Control in the United States: Evaluating the System.* Washington, DC: Resources for the Future.

Eckert, Toby. 2000. State Ban on MTBE Not Quite Done Deal; Officials Fear New Mandate for Ethanol. *San Diego Union-Tribune,* September 8, A3.

EIA (Energy Information Administration). 1998. *Oxygenate Supply/Demand Balances in the Short-Term Energy Outlook.* February 6. Washington, DC: U.S. Department of Energy. ftp://ftp.eia.doe.gov/pub/forecasting/steo/special/rpt/oxy1.xls (accessed July 26, 1999).

Environment Reporter. 1991. Traditional Antagonists Agree on Makeup of Cleaner, Reformulated Fuel. 22(August 23): 1141–1142.

———. 1992. Environmentalists, Oil Industry, Others Team Up to Fight Break for Ethanol Makers. 23(September 18): 1429–1430.

Fadope, Cece Madupe. 1995. Washington Meeting Examines the Safety of MTBE. *Chemical Engineering* 102(5): 49.

Franklin, Pamela M., and others. 2000. Clearing the Air: Using Scientific Information to Regulate Reformulated Fuels. *Environmental Science and Technology* 34(18): 3857–3863.

Garrett, Peter, and Marcel Moreau. 1986. MTBE as a Ground Water Contaminant. In *Proceedings of the NWWA/API Conference on Petroleum Hydrocarbons and Organic Chemicals in Ground Water—Prevention, Detection, and Restoration, November 13–15, 1985, The Westin Galleria, Houston, TX.* Dublin, OH: National Water Well Association.

Gilbert, Charles E., and Edward J. Calabrese. 1992. Developing a Standard for Methyl Tertiary Butyl Ether in Drinking Water. In *Regulating Drinking Water Quality*, edited by Charles E. Gilbert and Edward J. Calabrese. Chelsea, MI: Lewis Publishers, 231–252.

Happel, Anne M., and others. 1998. *An Evaluation of MTBE Impacts to California Groundwater Resources.* June 11. Livermore, CA: Lawrence Livermore National Laboratory.

Ivanovich, David. 1994. Fuming over MTBE. *The Houston Chronicle*, June 26, Business 1.

Joseph, Peter M. 1995. Illness Due to Methyl Tertiary Butyl Ether. Letter to the Editor. *Archives of Environmental Health* 50(5): 395–396.

Keller, Arturo, and others. 1998. Health and Environmental Assessment of MTBE, Volume 1, Summary and Recommendations. In *Health and Environmental Assessment of MTBE: Report to the Governor and Legislature of the State of California as Sponsored by SB 521.* Berkeley, CA: University of California. http://tsrtp.ucdavis.edu/mtberpt/homepage.html (accessed 20 May 1999).

Kriz, Margaret, E. 1990. Politics at the Pump. *The National Journal* 22(22): 1328.

———. 1994. Fight over Clean Fuels Gets Dirty. *The National Journal* 26(16): 898.

McCarthy, J.E., and M. Tiemann. 2000. *MTBE in Gasoline: Clean Air and Drinking Water Issues.* Report for Congress No. 98-290 ENR (updated February 25, 2000). Washington, DC: Congressional Research Service.

Mckinnon, R.J., and J.E. Dyksen. 1984. Removing Organics from Groundwater through Aeration plus GAC. *Journal of the American Water Works Association* 76(5): 42–47.

McQuillan, Dennis M. 1990. *Ground-Water Quality Concerns of MTBE in Gasoline.* Santa Fe, NM: New Mexico Health and Environment Department, Environmental Improvement Division, Special Waste Bureau.

Melnick, R.L., and others. 1997. Potential Health Effects of Oxygenated Gasoline. In *Interagency Assessment of Oxygenated Fuels.* Washington, DC: Office of Science and Technology Policy, Executive Office of the President, Chapter 4.

National Research Council. 1996. *Toxicological and Performance Aspects of Oxygenated Motor Vehicle Fuels.* Washington, DC: National Academy Press.

Piel, W.J., and R.X. Thomas. 1990. Oxygenates for Reformulated Gasoline. *Hydrocarbon Processing* July: 68–73.

Risk Policy Report. 2000. EPA Study of Ethanol Emissions Prompts Questions about Safety. 7(9): 20.

Scanlon, Bill. 1994. Doctors Push Oxy Fuel Moratorium in Alaska. *Rocky Mountain News* 13(July): 20A.

Schneider, Keith. 1992. Bush Offers Plan for Wider Use of Ethanol in Fuel. *New York Times*, October 2, A15.

Seymour, Adam. 1992. *Refining and Reformulation: The Challenge of Green Motor Fuels.* Oxford, U.K.: Oxford Institute for Energy Studies.

Sierra Club. 1997. Environmental Groups Announce MTBE Strategy. *California Eco-Watch* 3(6). http://tamalpais.sierraclub.org/chapters/ca/ca%2Decowatch/1997/0004.htr (accessed August 3, 1999).

Squillace, P.J., J.S. Zogorski, W.G. Wilber, and C.V. Price. 1996. Preliminary Assessment of the Occurrence and Possible Sources of MTBE in Groundwater in the

United States, 1993–1994. *Environmental Science and Technology* 30(5): 1721–1730.

Stone, Peter H. 1994. The Big Harvest. *The National Journal* 26(31): 1790.

Stout, David. 2000. EPA Urges Substitution of an Additive to Gasoline. *New York Times*, March 21, A20.

Suflita, Joseph M., and Melanie R. Mormile. 1993. Anaerobic Biodegradation of Known and Potential Gasoline Oxygenates in the Terrestrial Subsurface. *Environmental Science and Technology* 27(5): 976–978.

Tangley, L. 1984. Groundwater Contamination: Local Problems Become National Issue. *BioScience* 34(3): 142–148.

Uhlman, Marian. 1990. Arco Chemical's Prize Product. *Philadelphia Inquirer*, May 13, F6.

U.S. Congress, OTA (Office of Technology Assessment). 1989. *Catching Our Breath: Next Steps for Reducing Urban Ozone.* OTA-O-412. Washington, DC: U.S. Government Printing Office.

———. 1990. *Replacing Gasoline: Alternative Fuels for Light Duty Vehicles.* OTA-E-364. Washington, DC: U.S. Government Printing Office.

U.S. EPA (Environmental Protection Agency). 1991. Establishment of Negotiated Rulemaking Committee on Clean Fuels Guidelines and Proposed Rules. *Federal Register* 56: 10522–10523, March 13.

———. 1994. Fuels and Fuel Additives Registration Regulations; Final Rule. *Federal Register* 59: 33042–33142, June 27.

———. 1999. *Achieving Clean Air and Clean Water: The Report of the Blue Ribbon Panel on Oxygenates in Gasoline.* EPA 420-R-99-021. September 15. Washington, DC: U.S. EPA. http://www.epa.gov/oms/consumer/fuels/oxypanel/blueribb.htm (accessed October 30, 1999).

———. 2000a. *Advanced Notice of Proposed Rulemaking to Control MTBE in Gasoline.* EPA420-F-00-012. March 20. Washington, DC: U.S. EPA.

———. 2000b. *Legislative Principles for Protecting Drinking Water Supplies, Preserving Clean Air Benefits, and Promoting Renewable Fuels.* EPA420-F-00-011. Washington, DC: U.S. EPA.

U.S. GAO (General Accounting Office). 1990. *Gasoline Marketing: Uncertainties Surround Reformulated Gasoline as a Motor Fuel.* RCED-90-153. Washington, DC: U.S. GAO.

U.S. Senate. 1990. Committee on Environment and Public Works. *Clean Air Act Amendments of 1989.* Report No. 101-228. October 2. Washington, DC: U.S. Government Printing Office.

———. 1993a. Committee on Environment and Public Works. *Congressional Record* 139(163). (Senator Grassley of Iowa speaking in support of congressional intent that ethanol should have a competitive role in the reformulated gasoline market.) 103rd Congress, 1st Session. November 20.

———. 1993b. Committee on Appropriations. *Health Risks Associated with Exposure to Gasoline Additives—Methyl Tertiary Butyl Ether (MTBE).* Hearing before a subcommittee. Senate Hearing No. 103-89. 103rd Congress, 1st Session. March 10.

———. 1994. Committee on Energy and Natural Resources. *EPA's Proposed Renewable Oxygenate Standard.* Hearing. 103rd Congress, 2nd Session. April 22.

———. 1997. Committee on Environment and Public Works. *Water Pollution Risks of Methyl Tertiary Butyl Ether (MTBE).* Field hearing held in Sacramento, California. 105th Congress, 1st Session. December 9.

Wald, Matthew L. 1989. ARCO Offers New Gasoline to Cut up to 15% of Old Cars' Pollution. *New York Times,* August 16, A1.

Weber, Edward P. 1998. *Pluralism by the Rules: Conflict and Cooperation in Environmental Regulation.* Washington, DC: Georgetown University Press

Wiley, Kip. 1998. *Clean Air vs. Clean Water: Does California Need MTBE?* Sacramento, CA: California Senate Office of Research.

Zogorski, J.S., and others. 1997. Fuel Oxygenates and Water Quality. In *Interagency Assessment of Oxygenated Fuels.* Washington, DC: Office of Science and Technology Policy, Executive Office of the President. Chapter 2.

Zyren, John, Charles Dale, and Charles Riner. 1996. 1995 Reformulated Gasoline Market Affected Refiners Differently. In *Petroleum Marketing Monthly,* January 1996. Washington, DC: U.S. Department of Energy.

5

The Barriers to Corporate Pollution Prevention

An Analysis of Three Cases

JAMES BOYD

The concept of pollution prevention (P2) is emblematic of a new, proactive environmental mind-set that promises more sustainable industrial management. By targeting the causes of polluting activities rather than the consequences, the P2 concept seeks to eliminate pollutants at their source and thereby avoid the need to treat or dispose of those pollutants later. It has given rise to talk of "win–win" opportunities in which innovation and new ways of thinking will lead to waste reduction while increasing profits to firms by reducing costs or stimulating new product development. Unfortunately, this vision is somewhat at odds with actual corporate experience. Although anecdotal evidence from several studies suggests that such opportunities exist and that many firms have pursued them, proponents say the pace of pollution prevention is too slow and that the private sector somehow fails to see opportunities in front of it.

Very little is known about how and why firms pursue or do not pursue pollution prevention goals. In this chapter, I report on a study that explored concrete P2-related decisions made at three different firms, all global chemical manufacturers headquartered in the United States. The details of that study are reported in a U.S. Environmental Protection Agency (EPA) publication (Boyd 1998).[1] I summarize those results and extend the discussion to broader issues regarding pollution prevention by the private sector.

What can be learned from three case studies of unsuccessful corporate pollution prevention investments? One answer is "very little," because drawing conclusions from such a limited sample is inherently risky. However, even a few case studies offer lessons for policy analysis and proponents of P2. First, much can be gained from any detailed, firm-level analyses of environmental investment decisionmaking. Policy and economic analysis is long on assertions regarding the linkages among environmental policy, private-sector profitability, and environmental investment and short on concrete knowledge about how the factors actually affect investment choices. These cases show how the technical engineering analyses of pollution prevention are brought together with financial and accounting methods and regulatory requirements to make or break P2 in the real world. Investment decisions about pollution prevention are highly complex and always idiosyncratic, depending on the technologies, facilities, and firms involved (a truth sometimes neglected by policy analysts). Moreover, the explanations for the failure of these investments are as different as the firms and technologies themselves.

The second broad conclusion of the study relates to the link between profitability and pollution prevention. Are firms somehow missing profitable P2 investments? For the three cases studied here, the answer is no. Investment in pollution prevention did not proceed, not because of some systematic organizational failure (as many proponents suspect) but for more prosaic reasons. The study found that the costs of pollution prevention were simply too high given their expected benefits to the firm. This result is likely to be frustrating to those who feel that more P2 should be easy to come by. Many hope that if the corporate sector would only "think differently," firms could simultaneously make more money and improve the environment. But the rewards to this kind of creative thinking may already be apparent to most corporations. On the basis of this limited sample, the inescapable costs of environmental innovation and the nature of regulation itself explain the failure to make P2 investments better than private-sector myopia or other forms of faulty decisionmaking.[2] This conclusion is somewhat disappointing, particularly if it holds more generally across projects, firms, and industries. Nevertheless, it is a possibility that deserves greater consideration.

As a final caveat, it is important to note that corporate pollution prevention is often successful. Most large corporations today are able to point to engineered reductions in input use and facility waste volumes. These successes speak to beneficial changes in law and corporate practice during the past 30 years of intensified environmental regulation. However, it is the failure to generate yet more P2 that is of inherently greater policy interest.

The Pollution Prevention Debate

The way that the private sector approaches pollution prevention is pertinent to a large set of ongoing environmental policy debates. Calls for P2 are at the center of a broad regulatory and corporate movement. Information-based programs (such as the Toxics Release Inventory) and experiments with flexible regulation (such as those associated with the Common Sense Initiative and Project XL) have in common the ultimate goal of preventing pollution.[3] Within the private sector, talk of "sustainable corporate environmental management" inevitably centers on firms' abilities to prevent rather than treat or dispose of waste.[4] A commonly expressed belief within this movement is that P2 is a win–win corporate strategy. In this view, innovation and new ways of thinking will reduce waste and, at the same time, make firms money by reducing costs or stimulating new products.[5]

Pollution prevention, cast as both a corporate and an environmental benefit, has ignited hope in less adversarial environmental regulation. It also has created optimism in the private sector's ability to come up with low-cost solutions to environmental problems. But whereas anecdotal evidence suggests that win–win opportunities exist and have been pursued by many firms, the slow pace of change is cause for much frustration.

The idea that firms neglect P2 opportunities, even though such opportunities can save them money, colors debate over regulatory reforms geared toward pollution prevention. First, it calls into question the desirability of regulatory reforms, such as regulatory flexibility, often associated with pollution prevention. If firms cannot be counted on to make environmental improvements that save them money, the argument goes, then only the blunt instrument of command-and-control regulation will get the job done.[6]

Another line of thinking purports that "organizational barriers" account for firms' failure to be aware of and pursue win–win investments. Organizational barriers may arise, for example, as a result of information barriers, accounting-based distortions, or inappropriate managerial incentive schemes.[7] The policy implications of this perspective point toward regulatory assistance or requirements focused on internal management and accounting processes.

For their part, corporate environmental managers tend to be skeptical of the profitability of pollution prevention.[8] They point to regulatory barriers that reduce the financial incentive to change production processes or introduce new products with uncertain regulatory mandates.

Much of the debate hinges on whether pollution prevention can actually save firms money. Consider the implications of evidence that companies fail to pursue P2 opportunities that would profit them. First, this evidence would lend credence to the argument that regulations should mandate pollution prevention.[9] Second, it would point to the need to reform private-sector capital budgeting, accounting, and environmental management techniques to overcome organizational barriers to P2. If, on the other hand, pollution prevention's financial benefits are overstated, then a different set of issues arises for regulators and firms. If firms do not pursue pollution prevention because it is simply not profitable to do so, then attention should be focused on factors that contribute to the difficulty, cost, and benefit of implementing P2 innovations. Perhaps environmental costs are not being adequately imposed on the firms that create them. Perhaps technical assistance and government R&D could be used to lower technical barriers. Or perhaps regulatory reforms should be used to lower regulatory barriers to P2 product and process changes.

These issues are of central importance to the future of environmental regulation. Unfortunately, very little is known about why firms pursue or do not pursue P2 opportunities. The three case studies reported on in this chapter analyze real-life investment decisions regarding pollution prevention. Two of the cases deal with investments that did not survive the firms' capital budgeting process. The third involved difficulties associated with a P2 product being marketed to a large group of commercial customers (who themselves had to decide whether to invest in pollution prevention).

Design of the Case Studies

The case studies center on pollution prevention initiatives at three different firms, all global chemical manufacturers headquartered in the United States. A particular type of business activity was sought for analysis.

First, the investment or product marketing effort had to involve a P2 opportunity. *Pollution prevention* was defined as a new product or process that allowed for pollutant source reductions or that involved in-process recycling. Environmental benefits had to come from these kinds of innovations, not from new disposal or treatment methods.

Second, the investment or marketing opportunity had to be promising enough to be evaluated by the firms themselves. More specifically, the opportunity had to involve not only technical but also financial analysis. The financial analysis is critical. Even if a P2 technology passes muster in engineering labs or environmental health and safety meetings, it will not succeed in a practical sense unless it survives a firm's strategic analysis and capital budgeting process. Strategic and financial analyses form the core of a corporation's decisionmaking. In the process of strategic and financial analysis, the widest variety of internal corporate expertise is brought to bear to evaluate costs, rewards, and risks. Because investment analysis is the principal information-processing function of a corporation, firms' investment analyses are the best place to look for answers.

Third, the investments or products had to be in some way "unsuccessful." That is, the firms chose to not invest in the product or process changes, or the investment was significantly delayed, pending the resolution of market, technical, or regulatory uncertainties. For the purposes of this project, unsuccessful P2 opportunities are of greatest interest because they allow a focus on the corporate rationale for not making P2 investments.

Fourth, projects with a capital, technical, or marketing "scale" sufficient to ensure a certain degree of complexity to the decision were sought. Small-scale P2 opportunities (for example, equipment purchases, certain chemical substitutions, general shop-floor housekeeping) are important and may be easier to achieve; however, they lack the complex range of factors that affect large-scale business decisions.

Finally, the analysis required the participating firms to provide detailed and often proprietary data on the investments considered. This last characteristic of the case study method represents the largest single hurdle for any researcher. An understanding of technical and financial decisionmaking requires information that is considered highly sensitive by most firms. Financial analysis methods themselves may be proprietary, and the possibility that an external researcher may turn up some kind of managerial error or environmental problem strikes fear in the heart of potential partners in this type of research. However, the greatest concern is that technical information will be revealed to competitors.[10]

The cases were meant to provide a deeper understanding of the relationship between P2 and corporate profitability. Of course, three nonrandomly selected cases cannot be used to draw broad policy or empirical conclusions. Instead, they should be viewed as a lesson on the practical challenges facing private-sector managers. Regulators, policymakers, and other corporate managers can presumably benefit from a better practical understanding of corporate P2 decisionmaking.

Brief Overview of the Cases

As background, a general description of each case follows. The cases are described in greater detail in Boyd (1998).

Dow Chemical

The Dow project involved process changes at a facility associated with polyurethane production. The facility manufactures methylene diamine diisocyanate, the primary input to polyurethane foam and thermoplastic products. Annually, the plant releases roughly 300,000 pounds of waste and incinerates approximately 2.5 million pounds of waste reported to the Toxic Release Inventory in a thermal oxidizing unit. The principal constituents of the incinerated waste stream are phosgene, methanol, and monochlorobenzene (MCB).

In 1995 and 1996, Dow participated in a study to identify P2 opportunities at the La Porte, Texas, plant. Assisted by an expert in P2 process redesign, the project identified an opportunity for in-process MCB recycling. The recycling of MCB is environmentally beneficial because MCB incineration creates "products of incomplete combustion." The quantity of emissions avoided by recycling was not quantified by the study but assumed to be environmentally significant.

This environmental "win" was accompanied by a potential financial "win" for Dow. The ability to recycle MCB, rather than incinerate it, created the possibility that the facility's incinerator could either be shut down or removed from its status as a regulated boiler. Retiring the incinerator would eliminate some operating costs. However, the primary benefit would be derived from avoiding costs associated with retrofitting the incinerator per new regulatory standards expected to be issued during 2000. Shutting down or converting the incinerator to an unregulated unit would allow Dow to avoid millions in upgrade costs associated with these new, tougher standards.

On close examination, this financial benefit was not sufficient to justify the process change. First, the MCB recycling option required a host of other collateral process changes associated with other waste streams. Because the incinerator disposed of these other waste streams, the savings from incinerator shutdown were offset by a new set of costs, namely, the need to find new disposal options for the other components of the plant's emissions. One problem was that several of these other disposal options were inhibited by regulations.[11] At the very least, costly repermitting would have been required.

Second, the Dow process change exhibited the characteristics of an investment in which a delay in the investment decision has a large option value (Dixit and Pindyck 1994). A delay has value (the option value) when future financial conditions are uncertain and investments made today are sunk, or unrecoverable. The Dow project possessed both characteristics. The benefit of retiring the incinerator would be primarily due to the avoidance of upgrade costs required by regulations that were uncertain in terms of both stringency and when they would take effect. Moreover, numerous technical and capital costs would be associated with the diversion of the non-MCB waste streams. Many if not most of these costs would be irreversible in that they could not be recovered if Dow decided to reopen the incinerator in the future. Thus, delaying investment in this P2 opportunity was financially rational.

Monsanto

The Monsanto case involves a process that, at the time, was associated with the firm's Rubber Chemicals business unit. The firm was producing a chemical (4-ADPA) used as an intermediate product in the production of antioxidants, primarily for use in tire production. The traditional process chemistry for 4-ADPA production used benzene, nitric acid, and chlorine to produce nitrocholorobenzene, which was then converted to 4-NDPA by reaction with aniline; 4-NDPA, in turn, was reacted with hydrogen to yield the desired 4-ADPA. The many process steps leading to creation of the 4-ADPA intermediate product required the disposal and handling of large quantities of chlorine. In addition, the process created carbon monoxide, xylene air releases, and an aqueous waste stream contaminated with inorganic salts. The process was the second-highest generator of waste in Monsanto's Chemical Group. These environmental characteristics made the process a prime target for process reinvention.

In 1991, Monsanto chemists and chemical engineers achieved a technical breakthrough with real promise: the possibility of a process with almost zero waste generation. The innovation, dubbed ADPA, eliminates the use of chlorine and several process steps in the production of the 4-ADPA intermediate product. The environmental benefits of the new process were significant: a 91% reduction in organic wastes, virtual elimination of inorganic wastes, and total elimination of chlorine. Another significant driver for the firm was the cost savings associated with the process change. The ADPA process required less than 50% of the old raw material inputs and required fewer process steps. For these reasons, ADPA was projected to reduce production costs by a magnitude almost guaranteed to have a positive effect on profits and market share.

Despite the potential benefits, Monsanto chose not to develop ADPA as part of its Rubber Chemicals business because of major effects on integrated business units. Although ADPA was ultimately developed, its transition from prototype process to the commercial market was neither immediate nor particularly smooth. Changes in Monsanto's strategic goals, relationships between the 4-ADPA product line and other business units, and a set of issues relating to divestiture and ownership of assets complicated the technology's development. For a host of reasons related to global markets, broad technical trends, and a cyclical downturn in its markets, Monsanto's Rubber Chemicals unit was significantly underperforming relative to new financial targets being pursued by Monsanto management. In effect, whereas the project was financially desirable in itself, the business unit of which it was a part was being starved of capital by corporate headquarters.

Strictly speaking, the firm's concern with returns in the Rubber Chemicals business should not have been relevant to the ADPA investment decision. An inviolable principle of financial project evaluation is that a new investment should be evaluated "incrementally," not on the basis of its effect on average returns across a larger business. In fact, Monsanto analysts sought to view ADPA independently but faced significant accounting challenges—created by an array of technical interdependencies—that made it difficult to divorce

ADPA from its parent business unit. Ultimately, these financial evaluation issues did not hinder the development of ADPA. The innovation's benefits were so obviously positive that development went forward in the context of a major business restructuring. But the case provides a lesson in the complexities and challenges that confront financial evaluation.

DuPont

The DuPont case relates to a product called the "DuCare" photochemical processing system, which is a package of products designed for sale to customers in the graphic arts (printing and publishing) industry. The industry, in the process of developing 1.5 billion square feet of silver halide photographic film annually, generates a huge amount of photochemical waste. The DuCare system was designed to reduce photochemical consumption by providing a recyclable chemical technology in lieu of traditional fixer and developer chemistries. The goal of the system is a zero-effluent photochemical process. Although chemicals must be transported off-site for recycling, it is technically a form of in-process recycling.

The environmental benefits of the DuCare system are, by design, significant. Traditional methods for dealing with photochemical waste include direct release to the drain for subsequent treatment at a publicly owned treatment works, transport of wastes for legal off-site disposal, or illegal dumping. DuCare, by recycling the photochemicals, would eliminate releases to the environment, whether those releases were destined for the drain, a treatment works, or a disposal facility. The process promised the elimination of releases that create problematic levels of chemical oxygen demand and vast reductions in silver, ammonia, and sulfates. DuPont estimated that the DuCare system (calculated on a base of 500 customers) would result in a 10-ton reduction in silver emissions and a 375,000-ton reduction in other chemical wastes annually.

From a private financial perspective, DuPont saw a commercial opportunity in its ability to sell this kind of pollution prevention. In contrast to the two cases discussed earlier (which relate to decisions regarding a firm's own investment in P2), this case concerns the demand for pollution prevention by DuPont's customers. It features financial and marketing analysis relating to a P2 technology that DuPont wishes to sell to other firms.

The DuCare system was envisioned as a way to reduce environmental compliance costs for its film and photochemical customers. DuPont expected to make money by selling P2 to its customers. The market opportunity became possible via the reality (and perception) of increasingly stringent environmental regulation. Product planning proposals spelled out clearly, "Customers want to minimize environmental risk and liability and want help in managing these issues." By eliminating emissions, DuCare would help its customers' regulatory compliance issues disappear. The possibility existed that the DuCare system might be viewed by regulators as a best-available control technology. Moreover, full recyclability meant that customers who used third-party disposal systems would eliminate any cradle-to-grave liability exposures associated with that transaction.

The product was introduced commercially by DuPont in 1994. Since then, the product has been successful from a technical perspective. Unfortunately, the product has failed to live up to its initial financial promise by a significant margin. The most direct explanation for this failure to meet expectations is that consumer demand for the product has been much weaker than predicted. In 1998, out of an estimated universe of more than 16,000 potential customers, the product was used by fewer than 600. A powerful explanation for the product's failure—after taking into account pricing and quality factors—is that regulation of photochemical processes has been relatively weak.

From the beginning, the project's financial and strategic analyses identified regulatory pressure as the primary driver of the market for the product. Many of DuPont's optimistic financial and marketing analyses were predicated on the assumption that regulatory compliance and standards would continuously tighten and affect more and more graphic arts facilities. The product is cheaper in a life-cycle sense only if customers are forced to internalize the costs of their waste disposal.

Marketing analysis at DuPont identifies insufficiently aggressive effluent monitoring and enforcement as the root cause of DuCare's difficulties. It is not entirely surprising that monitoring and enforcement would be problematic. Photochemical drain discharge occurs in relatively small amounts and by numerous, geographically diffuse, shallow-pocketed firms. Nevertheless, the case illustrates an obvious—but often forgotten—lesson for P2: pollution prevention most often pays because penalties and regulation impose costs on polluters. When those costs are not effectively imposed, markets for P2 are undermined.

Financial Methods and P2 Analysis

In this section, I present some of the broader questions that motivated the study. Tentative conclusions are offered based on what was learned in the cases.

Are Environment-Related Financial Benefits Being Captured Appropriately by the Decisionmaking Process?

One of the primary challenges to P2 is the need to define and quantify the benefits of such investments. An emerging literature emphasizes the desirability of methods such as environmental cost accounting as a means to improve corporate decisionmaking. With the identification and quantification of environment-related financial benefits, firms can be expected to make better private and environmental decisions.[12] For instance, environmental accounting can highlight how changes in a production process reduce future environmental compliance costs. It reveals a benefit to investment from a process change, a benefit that may not otherwise have been captured in a capital budgeting decision. The quality of environmental accounting is of clear importance, and not least to firms themselves.

As we evaluate firms' accounting of environment-related financial benefits and costs, it is important to distinguish between two types of questions. First, to what extent are financial benefits and costs monetarily quantified? Second, are environment-related financial benefits and costs, even if not quantified, being given sufficient weight relative to nonenvironmental benefits and costs? Note that the first question deals with the detail and numerical sophistication of quantitative estimation techniques—the way that the firm determines the bottom-line effect of activities that affect the environment. The second question deals with the impact of accounting techniques on the firm's decisions. Failure to quantify benefits and costs accurately could bias investment decisions against pollution prevention.

Consider first the issue of quantification. The financial analyses associated with the above cases revealed relatively little economic quantification of environment-related financial benefits and costs. In almost none of the documentation reviewed were dollar values attached to savings from reduced emissions or liabilities. In fact, one of the cases revealed an example of management spelling out its desire to explicitly avoid the use of certain quantified environmental benefits during economic evaluation.

How is this lack (or explicit avoidance) of environmental accounting data to be interpreted? Although economic values are not quantified, the technical quantification of environmental benefits (for example, pounds of a pollutant eliminated annually) is extensive. In all of the cases, emissions reductions are analyzed and quantified in various ways. The only step that is missing is the translation of these technical environmental benefits into financial benefits. Why is this step not taken? The best explanation is that it is simply too difficult to arrive at economic values with any precision. When a firm estimates conventional costs, such as the cost of a new piece of capital equipment, it can use something as simple and available as the market price of the item. No analogous list of prices, or costs, can be used to translate reduced environmental emissions into a dollar value. Methods chosen to do so may be somewhat ad hoc and thus of questionable accuracy.

Methods for estimating environment-related financial benefits are being improved. For instance, historical data can be used as a guide to future costs.

However, as technologies, consumer tastes, and regulatory standards change, the ability to quantify these benefits and costs accurately becomes difficult. It may be reasonable for upper management to discourage the use of quantification techniques that are untested and aimed at estimating values that are so inherently uncertain.

Evidence suggests that better forms of environmental accounting are needed; however, the collection and analysis of this kind of data pose significant challenges. For instance, when dealing with new products and processes, where can data on their environmental risks be found? To illuminate this problem, consider the distinction between *cost estimation* and *cost accounting*. Historical cost accounting data (that is, data that is certain) can be used for cost estimation, but only if past costs are a good guide to future costs. This may not always be so. Historical, actuarial environmental accounting data is not useful if it provides data on environmental risks, technologies, and legal situations that firms are not likely to encounter in the future. For instance, a firm's expected future liability costs probably should not be estimated by looking at its past Superfund liability costs. Firms dispose of wastes very differently than they did 20–30 years ago. Because of its highly uncertain technical and legal nature, environmental cost estimation may inherently be more speculative than we, or the firms themselves, would hope.

Given the lack of quantified financial benefits, do the cases suggest that potential environmental improvements are given inadequate weight by corporate decisionmakers? No. Although not monetarily quantified, environmental benefits were nevertheless "valued" qualitatively in the decisionmaking process. Evidence for qualitative weighting includes environmental benefits as key drivers in the strategic decisionmaking process. Environmental benefits—such as physical data on quantities of emissions reduced—were routinely featured in summary documents presented to upper management at the time of decisionmaking. The paper trail (agendas, strategic analyses, and decisionmaking summaries) followed to reconstruct the cases is permeated with a qualitative understanding of environmental benefits. Therefore, the cases suggest that the lack of quantified, monetary environmental benefits does not necessarily imply an inadequate weighting of environmental benefits in the decisionmaking process. Instead, lack of monetary quantification may simply indicate the inherent difficulty of establishing precise financial values.

Are Firms Missing Win–Win Pollution Prevention Opportunities?

Some observers believe that companies fail to pursue their own economic self-interest when it comes to P2. For instance, Porter (1995, 132) claims that companies exaggerate the risk of environmental investments, use inappropriate rate-of-return hurdles, and thus "leave ten-dollar bills on the ground." This type of assertion, while thought-provoking, is difficult to prove or refute without evidence related to actual corporate investment decisions. Even when an investment is explored in detail, as in a study of the same Dow investment I describe in this chapter (Greer and van Loben Sels 1997),[13] conclusions about an investment's economic desirability are often made without adequate analysis of business and financial considerations.

Study of the cases presented here was motivated in large part by the desire to better understand the corporate rationale for rejecting or delaying identifiable P2 opportunities. Basic concepts from business and financial theory suggest that the firms' investment decisions were financially rational, contrary to the view that firms suffer from a myopic inability to appreciate cost-saving P2 investments. Instead, significant unresolved technical difficulties, uncertain market conditions, and, in some cases, regulatory barriers or insufficient emissions enforcement rendered the investments financially unattractive. In many instances, the mystery of why firms do not pursue P2 opportunities can be explained by simply having a deeper understanding of the costs, benefits, and risks associated with those investments.

This conclusion implies nothing about the social desirability of the firms' decisions. Reasonable people's opinions differ regarding how much pollution prevention is the right amount. But the conclusion implies that there may be fewer high-return P2 opportunities than many observers believe. Those who favor mandated command-and-control regulations may wish to claim that many high-return, win–win opportunities exist but that firms ignore those opportunities. After all, if this were true, command-and-control regulations could be viewed as forcing firms to do what is in their own economic self-interest anyway. This line of thinking should be viewed with skepticism. The investments analyzed here were by no means clear-cut financial winners. However, they suggest that firms are quite capable of identifying the actions that are in their greatest financial self-interest.

Information Barriers and the Search for Clear Financial Benchmarks of Profitability

The cases challenge the belief that organizational barriers are to blame for missed or delayed P2 opportunities. Nevertheless, firms face significant informational problems when they evaluate new investment opportunities. Managers never have perfect information regarding the costs and benefits of a new investment. In fact, imperfect information explains a great deal about how firms analyze and make investments. However, it is important to distinguish between a firm's methods for dealing with imperfect information and evidence of organizational failure.

The term *organization failure* connotes the existence of a correctable management strategy, accounting procedure, or financial methodology. For instance, if the benefits of an environmental investment were analyzed only over a three-year horizon, the strategy could easily be labeled an organizational failure. The failure to account for benefits beyond three years could easily be corrected by a longer-horizon investment analysis, and better, more profitable decisions would be expected. Similarly, if financial managers rarely spoke with environmental managers or if they infrequently integrated regulatory expertise into business analysis, these problems could be easily and profitably corrected. However, the cases exhibit little in the way of these correctable kinds of failure. Instead, the cases depict managers struggling with much more formidable challenges to investment decisionmaking: challenges that are pervasive and not limited to the analysis of environmental investments.

Consider the concept of a "hurdle rate" for new investments. Most firms define a rate of return that new projects must exceed before capital is directed toward it. In general, firms will not make investments that fall short of the hurdle rate, even if the investments have a positive rate of return. This practice can be a source of frustration to advocates of P2, who see a positive rate of return as evidence of profitability; a positive rate of return seems to be a clear benchmark. When firms ignore that benchmark and focus on some higher hurdle rate, it is natural to suspect the firm's "decision rule." But a hurdle-rate decision rule should not necessarily be thought of as a correctable organizational failure. Hurdle rates and the rationing of capital serve an important, inescapable corporate function. They are means by which firms account for the risks of investment and inherent imperfection of information at their disposal.

Finally, a firm's cost of capital in general will not be the same as the cost of capital for a specific project. Because the cost of capital is a function of risk, capital costs will differ in different divisions, for different product lines, and across individual projects. If the project in question is riskier than the firm's business generally, the cost of capital will be higher—because riskier investments demand a higher return.

Moreover, businesses commonly depart from a pure positive net present value (NPV) decision rule. In other words, firms generally do not move forward with projects simply because they have positive NPVs. Instead, firms often set limits on the capital available to individual business units, placing a ceiling on funds available for new investment and forcing managers to prioritize across projects, all of which may have positive NPVs. This financial management technique, called *capital rationing*, is used commonly in the private sector.

Capital rationing was used in the Monsanto and Dow cases. The P2 projects had positive rates of return, rates that in fact may have exceeded the firms' costs of capital. Nevertheless, the supply of investment capital to the business decisionmaking units was limited by corporate headquarters. Thus, the capital constraint can be viewed as a culprit in the failure of the firms to move forward with the investments. After all, if the businesses had no such constraint, any investment with a positive NPV presumably would be financed. Should we therefore think of capital rationing as an organizational barrier to pollution prevention?

To answer this question, it is necessary to explore firms' rationale for limiting the availability of capital to their business units. The principal rationale is that capital rationing is necessary for internal financial control. In this view, rationing counters the tendency of managers to overstate the benefits of investment opportunities that they are involved with. That is, managers may be led to bias their forecasts of cash flows associated with a new investment (for example, overstate expected benefits or understate expected costs). Again, this concept emphasizes the inherently uncertain, and subjective, nature of financial data. Capital rationing, therefore, is a means to correct largely unavoidable problems associated with imperfect information, managerial monitoring, and investment incentives.[14] It should not be viewed as an easily correctable organizational failure.

When capital is rationed, investments are not judged on the basis of whether the rate of return exceeds the cost of capital (the positive NPV rule). Instead, projects are typically ranked, and capital is invested in those that promise the highest net return. In practice, the most rational investments need not be those that promise the highest rates of return or NPV. Instead, the most desirable projects are those that provide the highest NPV per dollar invested.[15] The implication is that capital rationing leads firms to prioritize their investments. Although certain P2 opportunities are passed over, in general, environmental investments are not disadvantaged.

Finally, the Dow case featured an investment that was best evaluated as an "option," that is, an investment that could be made in the future, pending the resolution of some uncertainty. In the presence of uncertainty and high irretrievable investment costs, simple NPV-based investment rules are inappropriate (Dixit and Pindyck 1994). Option valuation techniques typically yield a very different optimal investment rule.

Conclusions

For people concerned about environmental quality, it is natural to be frustrated with decisions that reject or delay P2 investments that could improve a corporation's environmental performance. Much of that frustration may be alleviated by understanding better the strategic and financial factors that affect the profitability of P2 investments.

The conclusions of my analysis should not be taken as either a defense or condemnation of the firms' environmental performance. The "right" level of pollution prevention to be undertaken by the corporate sector is a question left to a different study (however, regulations and the avoidance of regulatory costs were drivers motivating the firms' search for P2 in all the cases). Still, without a detailed accounting of social benefits and costs, little can be said about whether more or less stringent regulation was appropriate in these cases. Instead, the cases say much more about *how* firms are regulated.

The results of this analysis can be viewed largely as a defense of the soundness with which managers weighed the private benefits, costs, and risks of the P2 investments. Instead of revealing evidence of corporate organizational barriers or myopic decisionmaking, the cases demonstrate a set of complex but ultimately prosaic motivations for the decisions that were made. Appreciation of those motivations is important because it can help guide public- and private-sector efforts to improve corporate pollution prevention performance.

Although based on a very limited sample, the findings here run counter to the perception that firms are somehow failing to pursue win–win opportunities. Surely, some profitable P2 opportunities are going undetected by the private sector. But once unearthed, firms have both the motive and ability to evaluate profit opportunities. Convincing firms of the profitability of investments that they are uniquely capable of evaluating themselves is likely to be a relatively unconstructive route to P2 promotion. Instead, a more fruitful approach is to focus on barriers to the profitability of pollution prevention.

What policy changes are likely to enhance the profitability of pollution prevention? First, the cases reveal regulatory barriers of varying significance. The desire to experiment with P2 innovation often is thwarted by rigid media- and technology-specific regulations. The rigidity of many regulations is understandable, given the difficulties of environmental enforcement. Nevertheless, efforts to promote regulatory flexibility and innovations should be embraced as means to foster the corporate sector's ability to develop environmental innovations. The DuPont case featured a regulatory barrier associated with a material's permit classification. It also highlighted how lax regulatory enforcement can undermine the demand for P2 technologies.

Second, performance-based (as opposed to technology-forcing) regulation is likely to be a better way to promote private-sector P2 innovations.[16] Pollution prevention increasingly calls for firms to redesign complex products and processes in ever-changing product markets. Performance-based regulations, which allow greater latitude for technological experimentation and longer time horizons for compliance, enable firms to meet targets in the widest variety of ways.[17] In contrast, existing regulations feature substantial regulatory influence over the technologies used by firms. Not only are specific technologies often mandated, but technical constraints also arise because emission standards are applied to individual substances rather than broader categories of effluent. As a result, limits on the output of a single substance can significantly constrain the design (or redesign) of a production process. Performance-based environmental permitting should be explored as a means to lower these barriers and constraints.

Flexible, performance-based regulation has another important consequence: it enhances the private sector's demand for improved environmental accounting information. Rigid regulations do a particularly poor job of fostering the private sector's demand for and development of better environmental accounting information and methods. After all, end-of-pipe, single-media, and technology-forcing regulations leave firms with little reason to innovate and therefore even less reason to collect information that would reveal environment-driven financial opportunities. Better information helps firms only if they have the flexibility to act on—and benefit from—that knowledge. Regulatory flexibility, by expanding the technological options open to firms, increases the value of information relating to those options. In the end, regulation that allows for an array of innovative solutions is likely to be the best way to induce firms to invest in better environmental information and decisionmaking.

This study raises a host of issues for further examination. For instance, significant information barriers confront corporate financial decisionmakers. Their presence underscores the continued importance of improved environmental cost accounting methods to better estimate environment-related financial benefits. Improved data collection, estimation, and evaluation techniques can only improve corporate decisionmaking. Corporations know this, of course. The cost of better information is the best explanation for why there isn't more of it. Focusing on the costs of information itself might be a particularly fruitful approach to encouraging pollution prevention.

Another important question is, what forms of environment-related information are likely to be most valuable to the private sector? Perhaps the greatest challenge for firms is the initial identification of P2 opportunities. This study suggests that firms do a good job of evaluating such opportunities once they are identified. However, the technical and scientific process that identifies the innovation in the first place is much more difficult to evaluate.

The identification of P2 opportunities may be well served by greater efforts at basic R&D and firm-specific materials accounting. *Materials accounting* refers to efforts that track the physical flow of materials into, through, and out of a given facility. Knowledge of material flows is a prerequisite for P2 innovation and can help identify areas for process redesign.

Mandated accounting requirements (material or financial) are of questionable value given the idiosyncratic needs of specific firms and facilities. In addition, confidential business information issues undermine the practicality of mandatory, publicly disclosed P2 planning. However, government promotion of state-of-the-art accounting practices, actuarial analysis, and materials accounting tools are likely to concretely benefit firms that are increasingly concerned with environment-related costs and opportunities.

The firms in this study were all leaders in their industry. Smaller, less sophisticated firms will possibly, indeed likely, confront a different set of challenges in making their decisions about pollution prevention. Case analyses of this kind of firm will yield a different set of insights regarding corporate evaluation of P2 innovations. Finally, rigorous financial analysis of pollution prevention successes is highly desirable. Do investments in pollution prevention yield more or less financial benefit than the firm predicted? Understanding the reasons some P2 projects succeed will complement the analysis of why others fail.

Notes

1. The research was supported in part by a grant from the U.S. Environmental Protection Agency's Environmental Accounting Project.

2. I do not suggest that pollution prevention projects always fail; quite the contrary. Rather, the conclusion relates to why unsuccessful projects fail.

3. The Pollution Prevention Act of 1990 was the first federal legislation geared toward fostering this new way of thinking. It has been accompanied by state-led legislation and programs and calls for additional federal legislation. These statutory approaches are diverse, but they all attempt to motivate pollution prevention via a combination of information disclosure, mandatory prevention planning, and voluntary commitments. For an overview of the history and current status of the pollution prevention movement, see U.S. EPA 1997. Another good summary document is Freeman and others 1992.

4. Dupont claims that over the past 10 years, the company has shifted its environmental technology effort from 85% investment in pollution control to 60–70% investment in "green" products and pollution prevention (Pelley 1997, 138).

5. Barry Commoner (1994, 217) expresses the view as follows: "The strategy of prevention cures the conflict between environmental quality and economic development that is inherent in the control strategy .... Properly designed, the productive investments engendered by the strategy of prevention could trigger a much-needed economic renaissance."

6. One analysis of waste reduction at chemical plants concluded that waste reduction was brought about in some cases only when required by regulation to do so but that the plants found the measure to be "cost-effective once in practice" (Sarokin and others 1985). This type of result suggests that command-and-control regulations, in some cases, lead to lower private-sector costs and improved environmental quality and that private-sector managers were unable to identify the opportunity themselves. For a recent study with a similar overall perspective, see Greer and van Loben Sels 1997.

7. For example, Porter (1995, 131) suggests that assignment of environmental issues to corporate departments without full profit responsibility leads to excessively narrow and incremental decisions. He also suggests that firms use inappropriately high hurdle rates to screen environmental investments.

8. Evidence on the profitability of pollution prevention (P2) opportunities is decidedly mixed. For instance, a study by the U.S. Environmental Protection Agency (U.S. EPA 1992) evaluated a broad set of source reduction options at a large-scale petroleum refinery. Most of the options were found to have negative rates of return, and only one had a rate of return higher than the historical rate of return for projects at the refinery. This kind of evidence runs counter to the hope that plentiful, undiscovered win–win P2 opportunities exist. However, EPA also has been able to develop an extensive set of case studies depicting P2 successes; see the Environmental Accounting Project's web site (http://www.epa.gov/opptintr/acctg).

9. The state of New Jersey has pioneered mandatory pollution prevention planning. A study of New Jersey's experience suggests that mandatory planning leads to more ambitious corporate pollution prevention goals and can help reveal opportunities for cost savings (Natan and others 1996).

10. For these case studies, it was necessary to sign nondisclosure agreements with the participating firms. According to the terms of these agreements, the firms were allowed to review the reports for technical detail on completion of the studies and before publication. In the end, little in the way of technical detail was removed from the case write-ups.

11. Antibacksliding rules prohibit process changes that result in any (effluent-specific) reduction in environmental performance, even if aggregate performance improves.

12. See White, Becker, and Goldstein 1992 and White and others 1995.

13. Greer and van Loben Sels (1997) concluded that the monochlorobenzene recycling option was financially desirable, thus raising the question of why Dow did not pursue it.

14. For analyses that explore the theoretical justification for capital rationing see Holmstrom and Costa 1986, and Antle and Eppen 1985. They show how the private concerns of managers, combined with private information, can create managerial incentives to overinvest. Capital rationing is an effective corporate strategy to counter these incentives.

15. Brealey and Meyers (1991) give examples of how investment rules can violate the idealized net present value rule under capital rationing.

16. Instead of judging environmental compliance on the basis of specific technological inputs or narrowly defined emissions standards, performance-based regulation relies on more holistic measures of a facility's environmental performance. For instance, compliance with aggregate limits (on a larger "bundle" of chemicals), rather than chemical-specific limits, is consistent with a performance-based approach. The U.S. Environmental Protection Agency's Project XL—a voluntary industry–regulator initiative to promote innovative forms of environmental regulation—has fostered experimentation with performance-based permitting.

17. Care must be taken not to confuse *flexibility* with a lack of regulatory stringency. Flexible permitting, at the aggregate level, can be very stringent. But flexible

permitting allows firms to meet even stringent aggregate targets in the way they best see fit. Performance-based regulation is not without limitations of its own. Monitoring and enforcement issues, in particular, loom large. The uniformity (inflexibility) of standard command-and-control regulations is a virtue, because it is easier to monitor technology or emissions standards that are fixed and common to many firms.

References

Antle, Rick, and Gary Eppen. 1985. Capital Rationing and Organizational Slack in Capital Budgeting. *Management Science* 31(2): 163–174.

Boyd, James. 1998. *Searching for the Profit in Pollution Prevention: Case Studies in the Corporate Evaluation of Environmental Opportunities.* EPA 742-R-98-005. Washington, DC: U.S. EPA.

Brealey, Richard, and Stewart Meyers. 1991. *Principles of Corporate Finance.* New York: McGraw-Hill.

Commoner, Barry. 1994. Pollution Prevention: Putting Comparative Risk Assessment in Its Place. In *Worst Things First? The Debate over Risk-Based National Environmental Priorities,* edited by Adam Finkel and Dominic Golding. Washington, DC: Resources for the Future.

Dixit, Avinash, and Robert Pindyck. 1994. *Investment under Uncertainty.* Princeton, NJ: Princeton University Press.

Freeman, Harry, Teresa Harten, Johnny Springer, Paul Randall, Mary Ann Curran, and Kenneth Stone. 1992. Industrial Pollution Prevention: A Critical Review. *Journal of the Air and Waste Management Association* 42: 618–656.

Greer, Linda, and Christopher van Löben Sels. 1997. When Pollution Prevention Meets the Bottom Line. *Environmental Science and Technology* 31(9): 418A–422A.

Holmstrom, Bengt, and Joan Ricart I Costa. 1986. Managerial Incentives and Capital Management. *Quarterly Journal of Economics* CI(4): 835–860.

Natan, Thomas, Catherine Miller, Bonnie Scarborough, and Warren Muir. 1996. *Evaluating the Effectiveness of Pollution Prevention Planning in New Jersey: A Program-Based Evaluation.* Alexandria, VA: Hampshire Research Associates.

Pelley, Janet. 1997. Environmental R&D Shifts to Pollution Prevention. *Environmental Science and Technology* 31(3): 138A–141A.

Porter, Michael E., and Claas van de Linde. 1995. Green and Competitive: Ending the Stalemate. *Harvard Business Review* 73(5): 120–134.

Sarokin, David J., Warren R. Muir, Catherine G. Miller, and Sebastian R. Sperber. 1985. *Cutting Chemical Wastes: What 29 Organic Chemical Plants Are Doing to Reduce Hazardous Wastes.* New York: INFORM, Inc.

U.S. EPA (Environmental Protection Agency). 1992. *Amoco–U.S. EPA Pollution Prevention Project: Yorktown, Virginia,* Executive Summary. Washington, DC: U.S. EPA.

———. 1997. *Pollution Prevention 1997: A National Progress Report.* EPA 742-R-97-00. Washington, DC: U.S. EPA, Office of Pollution Prevention and Toxics.

White, Allen, Monica Becker, and James Goldstein. 1992. *Total Cost Assessment: Accelerating Industrial Pollution Prevention through Innovative Project Financial Analy-*

sis. EPA/741/R-92/002. Washington, DC: U.S. EPA, Office of Pollution Prevention and Toxics.

White, Allen, Deborah Savage, Julia Brady, Dmitri Cavander, and Lori Lach. 1995. *Environmental Cost Accounting for Capital Budgeting: A Benchmark Survey of Management Accountants.* EPA/742/R-95/005. Washington, DC: U.S. EPA, Office of Pollution Prevention and Toxics.

6

The Political Economy of Interstate Public Policy

Power-Sector Restructuring and Transboundary Air Pollution

ALEX FARRELL

Over the past 30 years, there has been a clear trend in the United States toward the economic deregulation of industry in pursuit of greater competition and innovation. However, over the same period, environmental regulation has moved in the opposite direction—it has increased, especially at the federal level. Meanwhile, intense debates over federalism have raged regarding which level of government (federal, state, or local) should be responsible for which functions (Andrews 1994; Gillespie and Schellhas 1994; DiIulio and Kettl 1995; Pagano and Bowman 1995; Revesz 1996; Engel 1997). In this chapter, I illustrate and analyze these issues by examining the dispute over the potential impacts of air pollution brought about by the deregulation of the generation sector of the electric power industry. Specifically, I look at the process by which the environmental impact statement (EIS) issued by the Federal Energy Regulatory Commission (FERC) in support of their Open Access Order was finalized (FERC 1995a, 1995b, 1996a).

A principal environmental issue in this debate was whether emissions of nitrogen oxides (NO_x) from large power plants should be controlled.[1] A pollutant that occurs mainly as a result of fuel combustion, NO_x contributes to the formation of photochemical smog (ground-level ozone) and other air pollution problems, acidifying deposition, and eutrophication (Smil 1990; Pitts 1993; Ayres, Schlesinger, and Socolow 1997). However, the debate about the EIS was carried on entirely in the context of smog, virtually ignoring other issues for reasons discussed later.[2]

Because smog formation and accumulation are highly sensitive to local conditions, different areas may have to adopt different requirements for emissions control to meet the same health-based standards for ambient air quality (National Research Council 1991). When emissions in one state affect another, or when different emissions control requirements have large economic consequences, the issue can become very contentious, as this case shows.

Economic Regulation

Economists often have argued for the deregulation of industries to improve economic efficiency; since the 1970s, it has been both politically and technically possible to do so in air and truck transportation as well as in long-distance telephone service. By the early 1990s, the electric power industry was also thought to be in need of deregulation. At the time, the sector contained many large and small firms, and it operated as a regulated monopoly franchise system.[3] The industry had demonstrated a lack of technological innovation, a failure to develop new services, and, most of all, rates that were perceived as too high. Under the local monopoly structure, significant differences in the power market existed across the country, with price differentials of more than 100% for relatively nearby states (Figure 6-1).[4] For instance, Indiana, Kentucky, and West Virginia had costs in the range of $0.04–0.05 per kilowatt-hour, whereas New York and New Jersey had costs of $0.10–0.11 per kilowatt-hour. The low price of electricity in the Ohio River valley states is largely due to the existence of many older coal-fired power plants whose capital costs have been paid off (Ellerman 1996; Lee and Verma 2000). Large consumers of electricity, often industrial customers, pay close attention to costs, and they believe that these differences affect the competitiveness between firms in different states (Brennan and others 1996).

The first elements of deregulation were introduced in the wake of the 1970s energy crises, and by the late 1980s the electric power industry already had incorporated some initial elements of competition, such as Independent Power Producers, wholesale power sales between large utility companies, and scattered moves toward restructuring in some states. However, the majority of electricity was generated by large vertically integrated firms, and most sales still occurred within the franchise monopoly system. Pressure for change increased so that by the mid-1990s the power sector was in the process of restructuring itself toward a poorly defined combination of regulated local distribution monopolies and transmission firms, plus a more competitive generation market (Bohi 1997; EIA 1997a, 1997b; Joskow 1997, 1998; Borenstein and Bushnell 2000). Perhaps needless to say, this process caused

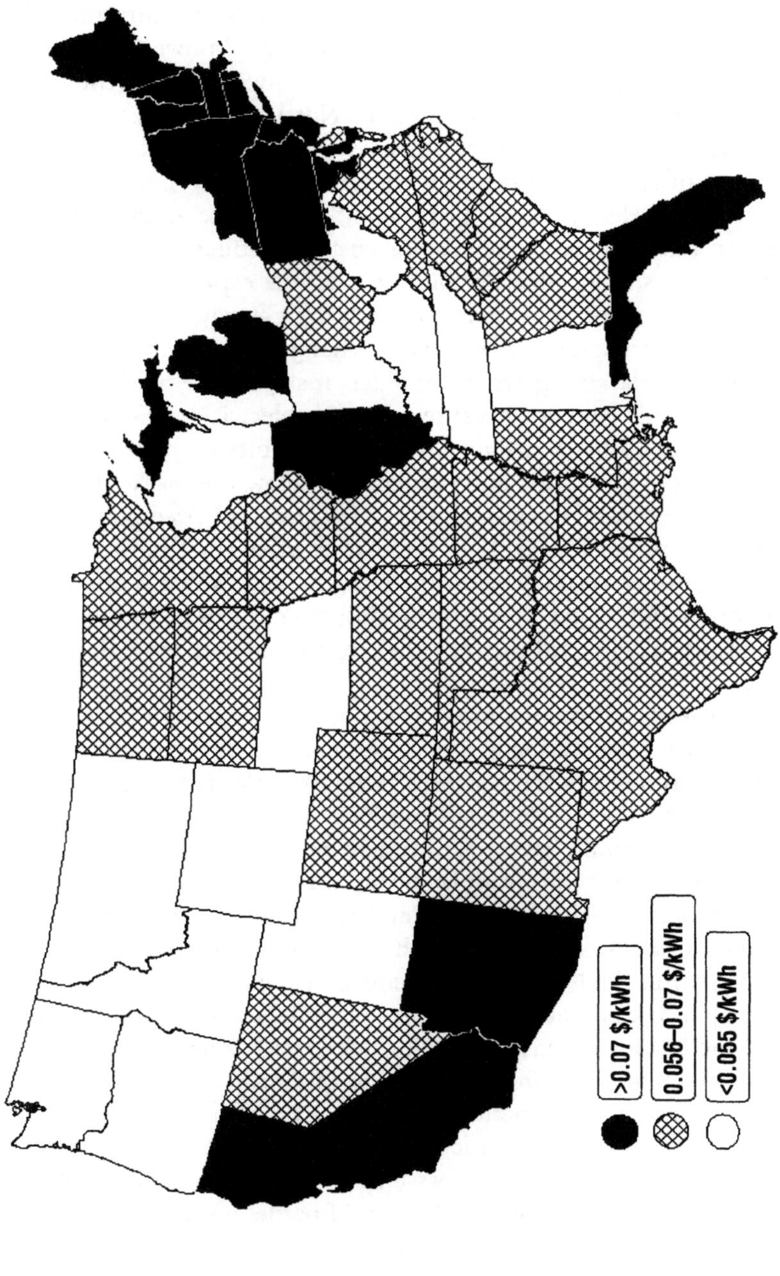

FIGURE 6-1. Average Costs of Electricity for All Retail Customers, 1995

Note: kWh = kilowatt-hour. *Source:* EIA 1996.

great uncertainty in an industry that had operated as a set of mostly non-competitive regulated monopolies for more than half a century. Firms in the states with higher costs were concerned about their continued existence in an increasingly competitive market. Part of this concern arose from the existence of considerable surplus generation capacity in the Midwest in the form of low-cost coal-fired plants, suggesting that Midwest power companies might be in a good position to compete in the Northeast market.[5]

Perhaps the most important part of restructuring was the elimination of traditional rate-of-return regulation for electricity generation, in which state officials determined electric power prices and thus the rate of return utilities would earn on allowable expenses.[6] Capital costs, including pollution control equipment expenses, were typically allowable, so the expenditures on emissions control equipment were borne entirely by ratepayers and provided extra revenue to shareholders. The scenario changes dramatically in a competitive generation market, where shareholders may bear the cost of environmental controls. This happens because wholesale electricity prices are typically set on an hourly basis at the value of the bid offered by the marginal generator, so cost increases for units generating below the market price will reduce net income instead of raising prices.[7] Because coal-fired power plants typically produce power for less than the market price (that is, their bids to the system operator are usually below the marginal bid, which is often set by the cost of low-polluting gas-fired generation), the costs of pollution control on coal-fired plants will tend to affect their operation and profitability, not power prices.

Power companies and mining interests (both firms and labor) in the Midwest naturally saw restructuring as a positive outcome, as did large commercial electricity consumers in the Northeast, who would be the main beneficiaries of lower-cost power. Northeast utilities, on the other hand, were concerned that they would have a hard time competing with Midwest firms, because the upwind emissions from Midwestern power plants would create a need for additional emissions controls (and therefore even higher costs) by sources in the Northeast. Regulators (both public utility commissions and air quality managers) in the Northeast were concerned about increases in transported ozone plus the potential burden on residential consumers as they were left holding the bill for expensive but clean electricity from local generators while industry bought power from the upwind Midwestern power plants.[8]

Environmentalists saw restructuring of the power sector as a negative outcome in general because the transition from a regulated monopoly structure left them and environmental regulators fewer means of influencing electric company decisionmaking. In several decades of battle in state electric utility regulatory hearings across the country, they had won victories on issues such as integrated resource planning, environmental cost adders, and demand side management. Restructuring threatened to reverse all these gains, and thus environmentalists tended to question the wisdom of restructuring at every step. More specifically, they believed (along with Northeastern environmental regulators) that restructuring would cause a shift in generation toward more coal-fired power in the Midwest and would raise barriers to the entry of cleaner and renewable energy sources.[9]

Conversely, the electric power industry was anything but unified on the issue of restructuring. Overall, there was great concern about the rapid and unpredictable changes that competition would bring, often over the expectation (soon realized) of mergers, takeovers, and layoffs. Moreover, firms in the industry differed greatly in their ability to provide low-cost power, to the point that some companies faced potential bankruptcy and others potential windfalls.

Finally, the turn toward a competitive industry accentuated the differences in interests between firms that had different emissions profiles and were already subject to varied environmental requirements (discussed in the next section); these differences didn't matter much under the monopoly franchise system. As a result, the once-monolithic power industry opposition to additional environmental regulation shattered during the mid-1990s as some firms began to realize that they would benefit from tighter national or regional NO_x control standards, whereas others saw they could substantially lose (based on interviews with Ozone Transport Assessment Group participants; see Keating and Farrell 1999, 27, 92, 188, and 190).

Environmental Regulation

From 1960 to 1990, the historical trend in air quality management had been toward greater centralization, exemplified in early struggles to create a federal role and the increasingly prescriptive amendments to the Clean Air Act (CAA) after 1970 (Jones 1974; Esty 1996; Bailey 1998). The system of air quality management that emerged is generally based on federally set standards for air quality and a mixture of federal and state programs for controlling emissions to meet these standards.[10] Supporters of centralization, then and now, argue that national policies are necessary because states and municipalities are fundamentally unable to appropriately control air pollution because their governments compete with each other for industrial development and are vulnerable to threats of industrial flight and job losses. In addition, supporters of centralization point to the transboundary nature of some air pollutants and to the need for national product standards in industries with transboundary markets, especially the auto industry (Stewart 1977; Esty 1996; Engel 1997).

In the early 1990s, the trend toward centralization in air quality management was countered by a larger trend in the opposite direction: a growing support for the decentralization of decisionmaking through the empowerment of state and local environmental agencies, a process variously called *devolution* or *civic environmentalism* (Gillespie and Schellhas 1994; John 1994; Walker 1996). There are several reasons for this trend. Most importantly, in the early 1990s, state governments were alienated in many ways from the federal government (Davies and Mazurek 1998). States had generally supported a stronger federal pollution control effort throughout the 1970s, partially because they always could rely on the ultimate threat of federal enforcement that former U.S. Environmental Protection Agency (EPA) Administrator William Ruckelshaus has referred to as "the 800 lb gorilla in the closet" to help secure compliance by recalcitrant firms (Davies and Davies 1975, 88; Standfield 1984; Portney 1990, 39).

However, after that early period, state agencies grew in competence and capabilities; federal funding became less important; and pollution control efforts turned from large, egregious sources to small, dispersed sources. States began to resent the federal presence in environmental regulation, often calling them "unfunded mandates." Contributing to this trend toward devolution was the development of more professional, more competent, and (according to some accounts) more honest state and local governments during the same time (Reinventing EPA and Environmental Policy Working Group 1997; Davies and Mazurek 1998). Finally, governors began to feel that increasingly tight environmental regulations were growing more costly in terms of competitiveness in attracting new development, especially industrial facilities with high-paying wages.

Despite the revolutionary rhetoric, devolution is really just a recent expression for the concept of federalism, an idea that has been part of the American political tradition from the beginning of the nation and is enshrined in the U.S. Constitution (Derthick 1996; Revesz 1996; Oates 1997). The rationale for devolution includes arguments about the fairness and efficacy of state and local policies compared with national regulations. Supporters of devolution argue that EPA, and the federal government in general, is not sufficiently representative of the interests in any particular state or municipality and that EPA does not, and perhaps cannot, know enough about heterogeneous local environmental and economic conditions to develop cost-effective policies. A particular complaint is that EPA often applies the same emissions control policy uniformly across the country and that this approach is inherently inefficient and unfair to almost every state and locality (Revesz 1996).

Although uniform federal environmental policies might be the norm in some cases, by the mid-1990s the regulation of power plant emissions was subject to a great deal of variation. At the federal level, the 1970 CAA and its 1977 amendments had created a significant disparity among different power plants by exempting existing and planned facilities from strict federal NO_x control requirements for new sources. As it turned out, this exemption covered a large share of the nation's generation capacity, because plants remained in service for longer than expected by Congress in 1977 (partially in response to these regulations) and because firms were able to successfully argue that in 1977, they had started planning many of the power plants that were actually built in the 1980s. At the same time, state-by-state variations have grown as well, the principal cause being the differences in the level of the ozone problem in different states. A sense of the spatial pattern in the stringency of air pollution regulations in 1995 can be gained from Figure 6-2, which shows the counties in the country that had the worst problems. The 1990 CAA Amendments included a set of increasingly stringent requirements for ozone nonattainment areas based on how badly they failed to attain the National Ambient Air Quality Standard. Figure 6-2 shows the three highest categories, Extreme, Severe, and Serious, in which federal requirements were particularly objectionable to state and local officials.

States with Extreme, Severe, or Serious Nonattainment Areas faced far more federal requirements for pollution control, a greater likelihood of fed-

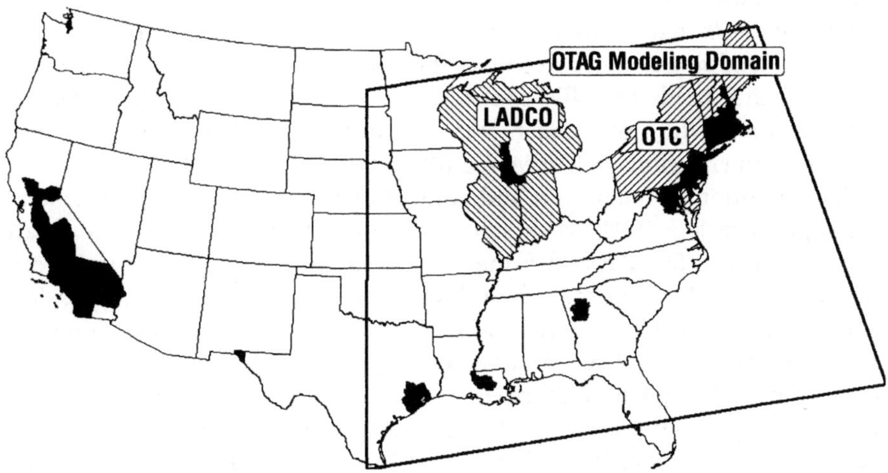

FIGURE 6-2. Air Quality Areas, 1995

Notes: Extreme, Severe, and Serious Nonattainment Areas for Ozone are shown in black. Multi-State Ozone Organizations are shown in the hatched area. LADCO = Lake Michigan Air Directors Consortium; OTC = Ozone Transport Commission.

Source: U.S. EPA 1997b and OTAG 1997.

eral sanctions for continued nonattainment, and considerable public pressure for improved air quality. Notably, outside of California, only six contiguous areas in 15 states received these designations, and only two of them are multistate metropolitan areas: one centered on Chicago, and the other on the Washington–Boston corridor. With the exception of Texas, by the mid-1990s, all states with severe ozone nonattainment areas had already taken steps to control ozone pollution that went beyond those taken by the rest of the nation, many of which were the mandatory requirements mentioned above. Important among these was the creation of multistate bodies for the study of ozone and the development of coordinated ozone policy, such as the Lake Michigan Air Directors Consortium (LADCO) and the Ozone Transport Commission (OTC) (hatched areas in Figure 6-2). Both of these organizations had been formed by the states under some pressure from the federal government, and they proved to be effective (Keating and Farrell 1999). One important effect was to increase the communication among the technical analysts and the decisionmakers in each group of states, greatly enhancing their understanding of the problem and their ability to trust each other and work together on regional air pollution control strategies.

The most important regional strategy was the OTC NO_x Budget, because it created an enormous difference between the levels of NO_x control placed on power plants in the Midwest and those in the Northeast (Farrell, Carter, and Raufer 1999; Farrell 2000). The OTC NO_x Budget requires emissions reductions of 75–85% from coal-fired power plants in the Northeast by 2003, whereas Midwestern facilities have to comply with only federal requirements for about a 40% reduction on some plants by the same time.[11] Marginal control costs tend to rise more steeply as the percentage reduction rises, a generalization that holds for NO_x controls, so the cost differential between the Northeast and Midwest is much greater than the doubling of the percentage reduction in emissions (STAPPA/ALAPCO 1994; NESCAUM 1998).

The question of federalism and control of tropospheric ozone reached a crisis point in late 1994, when state plans for attaining the ozone standard were due just a few days after the 1994 elections, which, as it turned out, brought some of the most strident devolutionists to power in Congress. The primary response to this crisis was the creation of the Ozone Transport Assessment Group (OTAG) process, a two-year effort led by the heads of the environmental regulatory agencies of 37 eastern states to study interstate flows of tropospheric ozone and its precursors, and to devise strategies for reducing them (OTAG 1997; Keating and Farrell 1999). Figure 6-2 shows the air quality modeling domain of OTAG; all the states at least partially inside this area participated in the process.

At the most basic level, during OTAG, the downwind states with the most longstanding and severe smog problems (the Northeast and Illinois) attempted to prove that upwind states (such as Missouri, North Carolina, Ohio, and West Virginia) had significant impacts on their ozone concentrations. If so, the upwind states needed to reduce their NO_x emissions to permit the downwind states to meet national air quality standards, although it was not clear how much reduction would be appropriate, because the impact of NO_x emissions on ozone concentration generally declines with distance. Coming immediately on the heels of the agreement to create the OTC NO_x Budget (which had been agreed to just a few weeks before the 1994 election), much of the discussion in the OTAG process essentially took the form of a debate about whether to extend those requirements to the Midwest.

The debate on ozone transport collided with the debate on restructuring, for obvious reasons. With a more competitive market for power, differences in environmental regulation came to have implications for market share and profitability, whereas previously they had not.

The Open Access EIS Debate

The general story is illuminated by the debate on FERC's EIS for the Open Access Order. For clarity, a basic chronology is shown in Table 6-1.

The Proposed Rule and Draft EIS

The National Environmental Policy Act (NEPA) of 1969 requires federal agencies to consider environmental impacts of their actions, principally through the development, publication, and public discussion of EISs (Carson 1992). Typically, an EIS is required when a project (for example, a dam) is supported by federal funds or when federal-owned land management decisions are made, but regulatory bodies (such as FERC) occasionally submit them as well.

Although many sharp criticisms of the EIS process have been leveled (Gregory, Keeney, and Winterfeldt 1992; Moffet 1994), the process is generally thought to be successful in cases where hiring environmental analysts to develop EISs has tended to institutionalize environmental values in organizations that previously did not have them (Taylor 1984). The EIS process also tends to make agency decisionmaking more transparent, because an EIS typically includes a comprehensive list of environmental impacts and possible options to the planned project, and it provides a forum for people from outside the bureaucracy to comment on the project's appropriateness. Usually, this "outsider" role is played by environmental groups that sue agencies for failing to conduct an EIS, or for doing so improperly. An important limitation of the process is that EISs are performed on a project-by-project basis

TABLE 6-1. Time line of the Air Pollution–Open Access Order Debate

Date	*Action*	*Reference*
Apr. 7, 1995	FERC Order 888 on "open access" and power sector restructuring proposed	FERC 1995b
Aug. 7, 1995	EPA comments on Order 888	
Nov. 1995	DEIS on FERC Open Access Order drafted	FERC 1995a
Dec. 20, 1995	OTAG holds a special session: FERC Forum	Elston and Shinn 1995
Jan. 10, 1996	OTAG Policy Group comments on FERC DEIS	Gade and others 1996
Feb. 1996	CCAP comments to FERC on DEIS	CCAP 1996
Feb. 21, 1996	EPA comments on DEIS	
Apr. 24, 1996	FERC issues Order 888 on open access	FERC 1996a
May 13, 1996	EPA referral letter to CEQ	Browner 1996a
May 14, 1996	FERC chair statement on EPA referral	Moler 1996
May 22, 1996	EPA letter to Kathleen A. McGinty noting readiness to implement cap-and-trade through FIP	Browner 1996b
May 29, 1996	FERC response to EPA referral	FERC 1996b
June 3, 1996	Letter from DOE on air pollution impacts of FERC's Order 888	O'Leary 1996
June 11, 1996	New England governors' recommendation	New England Governors' Council 1996
June 13, 1996	Letter from Council of Economic Advisors on air pollution impacts of FERC's Order 888	Stiglitz 1996
June 14, 1996	Letter from CEQ to EPA and FERC declaring that FERC Order 888 should go ahead and leaving environmental protection to the EPA, which might use the SIP or FIP processes; FERC and DOE commit to taking certain steps if "OTAG fails"	McGinty 1996
Sept. 1996	EIA publishes a service report on request by Senator Jim Jeffords showing that open access is small relative to the effect of increased demand but that emissions could go up by 2010 or so	EIA 1996

Notes: FERC = Federal Energy Regulatory Commission; EPA = U.S. Environmental Protection Agency; OTAG = Ozone Transport Assessment Group; DEIS = draft environmental impact statement; CCAP = Center for Clean Air Policy; CEQ = Council on Environmental Quality; DOE = U.S. Department of Energy; SIP = state implementation plan; FIP = federal implementation plan; EIA = Energy Information Administration.

and tend to ignore systemic problems, opening up the possibility of missing potentially important environmental impacts by choosing too small a unit of analysis. Finally, the EIS process does not require an agency to mitigate any environmental impacts it identifies but simply to report them.

The debate on the environmental aspects of power-sector restructuring formally started on March 29, 1995, when FERC proposed its major Open Access Order (FERC 1995b). This order contained important rules to promote competition in wholesale electricity trade, particularly by defining the rules for access to the nation's electric transmission system and for the treatment of "stranded assets." Stranded assets are generation units that were purchased under the traditional franchise monopoly system but are not financially viable under a competitive structure; they amount to between $100 billion and $200 billion nationally (Joskow 1997). Because changes in the economic regulation of wholesale power (not the actions of the owners of those assets) stranded these assets, power companies successfully argued that they should be compensated for these changes in the rules under which the investments were originally made. Most of the debate about the Open Access Order centered on access and stranded costs; FERC did not originally consider environmental issues at all.

However, environmentalists were very concerned about the potential for the order to cause considerable increases in Midwestern NO_x emissions, as noted above. Environmental regulators and governors in the Northeast had similar concerns. EPA was worried about the environmental implications of power-sector restructuring, which had been discussed during meetings between EPA and FERC, before the rule and EIS were published. FERC is an independent federal agency, not part of the administration, and so it can be at odds with EPA and White House positions. Nonetheless, EPA had the authority to intervene formally in the open access rulemaking (under Section 309 of the CAA and NEPA), and it did so.

Reluctantly, FERC developed a draft EIS, released in November 1995. It found that the Open Access Order would cause only minor environmental impacts compared with potential changes in fuel prices (which FERC has no control over) and that these impacts could be either beneficial or harmful (FERC 1995a). This position ran counter to the beliefs of Northeastern states and utility firms, which foresaw very significant environmental impacts from restructuring. This draft was rather limited, because it did not offer a scenario without open access for comparison with others. In addition, the draft EIS did not look at all at the issue of future emissions reduction requirements. Unfortunately, FERC effectively ignored the opportunity to fully evaluate the interaction between environmental and economic regulation of the electric power industry.

Initial Reactions

Unsurprisingly, the Open Access Order generated voluminous criticism; eventually, 430 comments were filed on the rule itself and 27 on the EIS. In additional to such traditional approaches, the existence of the OTAG process at that time created a unique opportunity for the interested parties to negotiate a solution to the problem, and a special meeting (the FERC Forum) was held by OTAG in Washington, District of Columbia, on December 20, 1995 (Elston and Shinn 1995). Much in the spirit of OTAG itself, the goal of this meeting was "to learn in an open process," whereas the meeting itself essentially consisted of presentations by OTAG participants and other interested parties (that is, stakeholders) to top state environmental regulators. Eight presentations were made. The primary criticism from environmental groups was that FERC's analysis assumed away the issue by its choice of parameter values for its model, specifically, by underestimating the capacity of the transmission system between the Midwest and Northeast and by assuming some form of state-by-state restructuring would occur without the Open Access Order. Upwind power companies and a lobbying group of upwind interests (the Midwest Ozone Group) supported the EIS but cautioned that because of the divergent interests of OTAG participants, it probably would be impossible to develop a meaningful consensus statement. In hindsight, this estimation seems to have been accurate.

In many ways, the presentations at the OTAG FERC Forum and the events that followed were a preview of the end of the OTAG process and EPA's response more than a year later. The individual presentations at the forum displayed widely divergent views and opinions, and OTAG's formal comments on the EIS (submitted shortly thereafter) say very little, indicative of the lack of agreement among the presenters at the meeting. The essence of the formal comments (Gade and others 1996) was that

> while there are varying opinions and data within OTAG regarding the consequences of the proposed rulemaking on NO_x emissions, and concomitantly on ozone formation, across the United States, there is consensus in OTAG that, to the extent there are impacts on emissions of ozone precursors as a result of FERC's proposed rule, as well as by the changes currently under way in the utilities industry reflected by the proposed rule, such environmental impacts must be both accounted for and addressed responsibly and appropriately ... OTAG believes that FERC should rely as much as possible on the most accurate information available in its EIS. Consequently, we urge FERC to utilize OTAG's emissions inventory and other data .... We believe the

most appropriate vehicle for addressing impacts on air quality is the state-led OTAG process.

EPA's analysis of the Open Access Order, published February 20, 1996, asked FERC to include additional specific analysis in the EIS. Other groups commissioned studies critical of the environmental impacts of the proposed rule and published reports with similar recommendations (CCAP 1996). For the next several months, FERC and EPA struggled to work out their differences. Thus, FERC conducted additional analyses and modified the EIS as requested, and it was made available in final form on April 19. Several days later, the final version of the Open Access Order was issued, in which FERC (1996a) concluded that any potential air pollution problem was far off in the future and, in any case, outside of its authority:

> The Commission believes, however, that there is time for federal and state air quality authorities to address any potential adverse impact as part of a comprehensive NO_x regulatory program under the Clean Air Act .... We share the view of most commenters that the preferred approach for mitigating increased NO_x emissions generally is a NO_x cap and trading regulatory program comparable to that developed by Congress to address sulfur dioxide emissions in the Clean Air Act Amendments of 1990 .... The Commission has concluded that a NO_x regulatory program could best be developed and administered under the Clean Air Act, in cooperation with interested states, and offers to lend Commission support to that effort should it become necessary.

Thus, FERC would not regulate power-sector NO_x emissions in this order, and probably not ever. With Order 888, FERC established that the economic regulation of the electric power industry was clearly separated from its environmental regulation, with wholly separate regulatory agencies. Many environmentalists saw this decision as a defeat, because it was clear by this point that the power sector would indeed be restructured, but it was not at all clear whether the air pollution controls they felt were needed would be required. Their attempt to link the two regulatory processes had failed, at least for the moment.

NEPA Referral

Several important events occurred after the final Open Access Order was issued. First, EPA remained unsatisfied with FERC's efforts and formally "referred" the Open Access Order to the Council on Environmental Quality (CEQ), an office in the White House (Browner 1996a). This action is authorized by CAA and NEPA and is used only when differences between the agencies are not otherwise resolved.

In the referral letter, EPA acknowledges the economic benefits of restructuring and recommends that the Open Access Order go forward, but expresses concern about the long-term consequences of open access. These concerns are almost entirely about NO_x emissions, but increased CO_2 emissions are also mentioned. Furthermore, EPA lays the groundwork for future action to control NO_x emissions in its referral, asking for a commitment from FERC to act if needed. Moreover, key portions of EPA's later comments signaled the agency's intentions about future actions (Browner 1996b):

> Among the measures under active study is a region-wide "cap-and-trade" program for NO_x emissions, with reductions from utility sources in the range of 55 to 85% below 1990 levels .... EPA believes that a region-wide cap-and-trade program of this kind is the best method for achieving the needed reductions .... EPA will use its authority under Title I of the Act to support successful completion of the OTAG process .... While the EPA has every expectation that all of the OTAG states will succeed in reaching agreement this year [1996] on the needed control strategy, EPA is prepared to establish a NO_x cap-and-trade program for the OTAG region through "Federal Implementation Plans" (FIPs) if some states are unable or unwilling to act in a timely manner.

Two things are particularly notable in this statement. For the first time, EPA officially announced it was ready to resolve the federalism question in the strongest means possible: direct regulation through an FIP, a powerful but rarely used tool. And EPA's definition of *success* for OTAG was announced: an agreement to sharply control NO_x emissions. However, this definition does not match OTAG's mission statement and goals, which were simply to study the problem and make recommendations. Moreover, OTAG was not in the end able to attain any such agreement, and under EPA's definition here, it must be judged a failure.

FERC responded the next day with a brief statement and followed up with formal comments in which it committed to initiating a Notice of Inquiry to examine NO_x mitigation under the Federal Power Act in the event that the OTAG process does not "succeed" (FERC 1996b; Moler 1996). FERC's definition for *success* appears to follow the definition used by EPA above—that is, a unanimous agreement by OTAG on a control strategy.

Moreover, in its order of May 29, FERC bluntly asserted that it was not subject to NEPA (and, in my opinion, misinterprets Congressional intent) (FERC 1996b):

> The rule violates no established national environmental law or policy …. The Commission objects to the decision by EPA to refer the rule and believes that there is not sufficient factual basis to justify that action. In addition, as an independent regulatory agency, the Commission cannot be bound by the decisions of the Council on Environmental Quality. Therefore, beyond this response, we do not intend to participate in the referral process.

Of course, the rationale for NEPA is not to prevent violations of the law, as FERC implies, but to illuminate the environmental impacts of government actions and foster informed debate about them. In the order, FERC volunteers to consult and exchange information with the Executive Branch but emphatically argues that it is obligated to maintain its status as an independent regulatory agency and thus cannot be bound by a NEPA referral. Nonetheless, the order of May 29 contains an extremely telling observation: FERC points out that EPA went ahead with the referral after admitting (in the referral letter itself!) that there would be no immediate and significant environmental impact due to the Open Access Order. The rationale for the referral is thus unclear.

The outcome of the debate on the environmental impact of power-sector restructuring was that FERC would go ahead with its Open Access Order, and that appropriate actions to control power-sector emissions would be left to future action by EPA. The two main agencies involved agreed, and additional support was offered by the U.S. Department of Energy and the Council of Economic Advisors (CEA) (O'Leary 1996; Stiglitz 1996). Throughout this debate, the administration was consistent on the appropriateness of potential EPA action to control NO_x emissions. For instance, a letter from CEA (Stiglitz 1996) read, in part,

> While CEA takes no position on the relative merits of FERC's and EPA projections regarding emissions, CEA believes that the economically efficiently way to address NO_x air pollution is through a regional cap-and-trade system that covers all sources of pollutants. Such a market-oriented approach recognizes that the problem is regional rather than site-specific, and permits the flexibility to comply with a regional pollution cap in the most cost-effective way. CEA therefore applauds the efforts of the EPA and the 37 states of OTAG to consider such a system to reduce emissions of NO_x from all sources …. In the unfortunate event that such efforts were to fail, CEA would support EPA using its authority under the Clean Air Act to address the problem.

On June 14, 1996, the CEQ effectively closed the issue with a letter concluding that the Open Access Order should go forward as proposed, reaffirming the split in responsibilities between economic and environmental regulations (McGinty 1996):

> EPA's concern about potential future increases in air pollution, which is the basis for the referral to CEQ, is an issue that deserves serious consideration. EPA outlined a set of sensible actions ... [which] include support for OTAG, EPA's use of its CAA authorities such as amendments to State Implementation Plans (SIPs) and issuance of Federal Implementation Plans (FIPs), and a commitment from FERC to take certain actions if OTAG fails. Furthermore, cap-and-trade proposals will be considered as part of the OTAG process, and EPA is initiating discussions about an interim cap-and-trade program to operate in the meanwhile. The Clinton Administration strongly supports the approach to ensuring clean air quality outlined by EPA.

All this suggests that the EIS referral for the Open Access Order was at least partially a method that the administration used to formalize and legitimate some of strategic decisions about the regulation of power plant NO_x emissions. EPA could now point to an independent regulatory agency (FERC) and to CEA for support in its contentions that NO_x emissions from power plants need to be regulated and that a nationally administered cap-and-trade program is the best means of doing so.

Seeking an Authoritative Decision

Although the CEQ letter effectively ended efforts to link environmental protection with FERC's Open Access Order, the book on power-sector restructuring and NO_x emissions was anything but closed. Wholesale access was only one part of deregulation; many other issues needed to be decided at the state and federal levels.

The issue continued to occupy the administration, and Congress got into the act as well. The first move came shortly after OTAG ended, when EPA formally proposed the regulations it had previously hinted at, doing all it could to encourage the states to create an NO_x cap-and-trade program on their own while proposing an FIP (U.S. EPA 1977a). In March 1998, the Clinton administration addressed the issue comprehensively in issuing its *Comprehensive Electricity Competition Plan* (Anonymous 1998). This plan proposed many new policies, most of which would require legislation, and recommended that EPA be given explicit authority to oversee an interstate NO_x emissions trading program. At the other end of Pennsylvania Avenue,

Congress was busy on the issue, too; 9 bills on restructuring were introduced in the House of Representatives and 7 in the Senate during the 105th Congress, and 18 and 13 bills, respectively, were introduced in the 106th Congress (EIA 1998, 2000). Most of these bills ignored air pollution issues, but some introduced by legislators from the downwind states would give explicit authority to EPA (or FERC, in one case) to regulate NO_x emissions from electricity generation, referring directly to emissions trading in some cases. Several dozen bills related to restructuring were introduced during the 106th Congress, a few of which address NO_x emissions—almost all of them prescribing a national cap-and-trade program (Solomon and Gorman 1998). In addition, several bills aimed more specifically at controlling the emissions of NO_x and several other pollutants from power plants have been introduced.

At the same time, EPA's proposed regulations were upheld by the D.C. Circuit Court and appeared to be moving ahead with some minor changes (Kelley 2000; U.S. Court of Appeals 2000). Although OTAG was not able to authoritatively resolve the debate on the regulation of NO_x emissions from coal-fired power plants, EPA and others subsequently relied on the analysis performed by OTAG. In particular, many of the fundamental issues that were resolved by OTAG (for example, baseline emission inventories and the cost of emissions controls) have not reemerged as contentious issues. In the consolidated lawsuit against EPA's proposed regulations, only one of the six arguments put forth by the petitioners involved technical issues at all. The remainder were about the legality of the decisionmaking process itself. And of that one technical argument, the court decision reads (in part): "Petitioners really do nothing more than quibble with the state-specific modeling .... Neither [of their] criticism affords ground for nonexpert judges to find a material likelihood of error" (section 1.B). Power-sector restructuring was not mentioned in this case, indicating that the court, too, separated the two issues.

Discussion

Restructuring changed the economics of the power sector significantly, but it may have changed the political economy of power-sector regulation just as much.

The most powerful effect of restructuring on the political economy of the power sector is that it fractured the once-monolithic industry opposition to air pollution regulation. Before competition, the power sector acted more or less unilaterally in opposing any new environmental regulations, but in OTAG some power companies promoted additional NO_x controls. This change resulted from the realization on the part of several Northeastern power companies that their interests diverged from those of the Midwestern and Southern firms over whether NO_x controls should be imposed on upwind sources. The clearest example of this new division was the departure of several longstanding members of the Utilities Air Resources Group (UARG), a nationwide lobbying firm organized and managed by Hunton & Williams, a powerful Washington law firm. These firms went on to form their own organization to promote their version of improving air quality and also acted independently with various environmental groups (NRDC and Public Service Electric and Gas Co. 1998).

An interesting phenomenon in the EIS debate was the widespread use of the term "a level playing field," which often was used to suggest that one policy option was inherently more equitable than others. However, the term is sufficiently vague to be used in contradictory ways. A firm in the Northeast might argue for a level playing field and mean uniform regulation for all power companies in the eastern United States so that environmental protection does not become a commercial issue. Conversely, a firm in the Midwest might argue for a level playing field in which firms should control emissions on the basis of their relative contribution to the ozone problem, which is presumably much less for them than for Northeastern power plants. Because of the vagueness of this language, it is often difficult to understand what was being said in the EIS debate without knowing who is saying it; terms such as "level playing field" lose any objective meaning.

In addition to the unsurprising upwind/downwind split, potential new entrants to the Northeastern electricity generation market created something of a third position within the power sector.[12] These firms were less concerned with where new NO_x control regulations would be applied as with the way in which new power plants would be treated. The main concern was that an emissions trading program would be created that would raise barriers to entry into the industry by distributing all of the allowances to existing firms (a common practice called grandfathering). Their idea of a level playing field was to make allowances available to all at a competitive price.

Environmentalist Opposition to Deregulation

Most environmentalists are opposed to (or at least wary of) the restructuring of the power sector, the deregulation of the generation business in particular. In contrast to the regulatory agencies, which see their role quite narrowly, environmentalists tend to see in deregulation a general loss of influence over the power companies.

Because environmentalists tend to seek to promote a public good of some sort (for example, reductions in air pollution or advances in energy efficiency) and unregulated markets are notoriously bad at delivering public goods, environmentalists generally wish to see more regulation rather than less. In particular, they have sought to maintain victories hard-won in rate hearings and lawsuits that they achieved in many different states over the course of the previous quarter-century. Environmentalists cast their role as much broader than the regulators who tended to separate economic and environmental regulation. For instance, environmentalists sought to defend subsidies for renewable energy and energy efficiency because of the improvements in air quality these policies would supposedly bring.

Differences between States and Power Company Interests

A very common perception during the OTAG process was that state governments tended to adopt the positions of the large power companies within their states. However, close analysis throws doubt on this assertion. Governors have constituents other than power companies to think about, and concerns other than economic growth. Restructuring has the potential to reduce energy costs for consumers in many states, both residential and (perhaps more important) industrial. Governors generally support reduced electricity prices because they could provide, or at least appear to provide, voters with lower costs and could improve the state's competitive position in attracting and holding on to industry.

In addition, some governors are interested in promoting environmental policy and thus want to see emissions reductions to improve air quality in their state. Of course, the heart of the problem is that downwind governors tend to want reductions in upwind states, whose governors have no incentive to do so because the beneficiaries are not potential electoral supporters. In addition, a clash between environmental agencies and electric power regulators often occurs at the state level as well as the federal level.

Agency Rigidity

Two distinctive features of the debate over FERC's EIS for the Open Access Order were that the regulatory bodies generally maintained their initial positions until the end of the process and that they construed their roles rather narrowly. Thus, FERC maintained that the environmental implications of the Open Access Order would be small and that they had no authority to regulate the environmental performance of the industry, whereas EPA insisted that emissions controls were needed and that a cap-and-trade program would be the best way to implement controls.

Although these features show some lack of flexibility, they also reflect the inherent structure of U.S. regulatory organization, which has created somewhat independent agencies with relatively narrow legislative mandates. Moreover, the impact of narrow legislative mandates is to make it exceedingly difficult for regulatory agencies to work out compromises on joint problems—hence the lack of an authoritative decision in this case and the probable need for additional legislation and adjudication.

The Geography of Interest

Interestingly, some of the divisions among power companies based on cost differences resemble those between upwind and downwind states. Thus, the economic geography of electricity restructuring in the debate over FERC's EIS for the Open Access Order was quite similar to the geography of the ozone transport debate in the OTAG process. In OTAG, the OTC states (and Illinois, to a lesser degree) were pitted against states along the Ohio River valley and in the south.

However, the differences in electricity costs (Figure 6-1) are too large to be attributed to differing emissions standards. Estimates of the costs of the level of emissions control contemplated even under the most severe cases in OTAG (85% reduction) were less than $0.01 per kilowatt-hour (some were less than one-tenth that value [Farrell, Carter, and Raufer 1999]), and the actual control levels in place when the data in Figure 6-1 were collected were significantly less than 85%. Thus, differences of several cents per kilowatt-hour between upwind and downwind states have to be caused by other factors, such as poor investments in nuclear power plants and different state reactions to earlier federal deregulation efforts (Borenstein and Bushnell 2000).

Emissions Trading as the Status Quo

One of the major changes in the political economy of power-sector emissions control is the near-universal agreement that emissions trading, with a cap-and-trade mechanism, is the preferred approach. A controversial and relatively unknown idea a decade ago, emissions trading is now so much the status quo that at least one state now requires regulators to explain why they are not using this approach when issuing new rules. All of the proposed congressional legislation that addresses NO_x controls does so through an emissions trading mechanism.

This approach has one potential problem, however, in that the proposals are all based on the assumption that NO_x is a "uniformly mixed pollutant," so that there is no need to account for differences in the location of sources engaged in emissions trading and no need to be concerned with the temporal pattern of emissions. Unfortunately, photochemical smog is a highly textured problem, with important patterns in both space and time, so the "uniformly mixed" assumption may be inappropriate and a more complex approach to emissions trading needed (Farrell, Carter, and Raufer 1999). This problem underlies many of the difficulties of the OTAG process, including the inability of the participants to agree on what kind of emission reductions were appropriate for Midwestern coal-fired power plants or what kind of emissions trading system could be used to implement it.

The Limits of Consensus

The outcome of the debate over FERC's EIS for the Open Access Order can best be described as inconclusive, particularly in regard to OTAG's participation. This lack of resolution highlights the limits of consensus, especially when the terms of debate are relatively narrow. By definition, states cannot be coerced into agreeing with the outcome of a consensus-based process, so the potential for such processes to create binding agreements that go against the interests of any state is small.

This is exactly what happened in the OTAG FERC Forum, and in fact, OTAG was able to deliver only a weak outcome, an agreement about the basic characteristics of ozone regionality and an essentially meaningless recommendation on emissions controls. Yet, as the outcome of the subsequent lawsuit indicates, OTAG was nonetheless quite important in that it eliminated confusion and conflict on many technical issues, allowing crucial judgments to stand on their own merits. In many ways, the EIS debate was a dress rehearsal for OTAG, in which the OTAG Policy Group played its own modest part. The debate drew in many OTAG participants and even OTAG itself, highlighting the tensions between upwind and downwind interests in OTAG.

Conclusions

Some important features of the political economy of air pollution and power-sector regulation have changed during the past decade, mostly because of changes in the incentive structure under competitive generation. They include the end of a unified position within the power industry regarding environmental regulations and the clear emergence of cap-and-trade programs as the preferred policy tools for use in controlling air pollution.

In the debate over the environmental implications of FERC's Open Access Order, EPA significantly improved its overall position in the debate about the need to control NO_x emissions from power plants. Several important features of this political landscape remain unchanged, however: many environmental groups continue to oppose the deregulation of industry, regulatory agencies seem to be resistant to changes in policy decisions, and the ability of consensus-based approaches to resolve deeply rooted controversies again remains rather limited. In addition, NEPA seems to have failed to have any appreciable impact on FERC. The commission has not only failed to institutionalize environmental values; it also denies that it needs to even take part in the procedural aspects of the EIS process.

Acknowledgements

The research presented in this chapter was made possible by support from several organizations, including the Center for Integrated Study of the Human Dimensions of Global Change. The center was created through a cooperative agreement between the National Science Foundation (SBR-9521914) and Carnegie Mellon University and has been supported by additional grants from the Electric Power Research Institute, the ExxonMobil Corporation, and the American Petroleum Institute.

This research also was supported by the Global Environmental Assessment Project at Harvard University, which is funded by the National Science Foundation (BCS-9521910), with supplemental support by other federal agencies.

The author thanks Terry J. Keating and Henry Lee for comments on earlier versions of this paper.

Notes

1. In addition, the future of energy efficiency and renewable energy programs were at issue.

2. Grumet (1998) provides a detailed description of the legal and political background of this debate.

3. In the most complete sense, the U.S. electric power sector contained about 3,200 "utility companies" during the mid-1990s, but only about 700 of them actually produced power. The remaining firms were exclusively distribution monopolies and are not directly affected by Order 888. In this chapter, I refer only to firms with generation capacity. Most of the traditional firms that fit this category are large, vertically integrated utilities that own generation, distribution, and transmission assets. For a good general description of the power sector and the changes it is undergoing, see EIA 1998, 2000.

4. Figure 6-1 contains data that were available in 1996 and thus are relevant to the debate described here. Since then, electricity pricing has changed substantially; see http://www.eia.doe.gov for more recent data.

5. However, it was not clear at the time whether sufficient transmission capacity was available to significantly change generation patterns, and the difficulty in getting new transmission lines sited means that it may be a long-term problem (EIA 1997a).

6. Regulated monopolies still exist in distribution, and interstate transmission of power remains a highly regulated industry. For a more complete discussion, see Brennan and others 1996; Bohi 1997; Joskow 1997, 1998.

7. In this model of wholesale power pricing, the system operator accepts "bids" from generators in ascending order, and the last bid that is accepted comes from the marginal generator. Thus, in any given hour, most generators are receiving higher payments than their bid.

8. Industrial and large commercial electricity consumers that can enter the wholesale power market directly could benefit from the lower prices that deregulation offered almost immediately, whereas households had to wait for some indefinite period for the introduction of retail competition.

9. Environmentalists eventually resigned themselves to the loss of leverage into power companies and switched to two new goals: establishing subsidies (or minimum production requirements) for renewable energy and eliminating the grandfathering provisions of the Clean Air Act.

10. See appendix D of Keating and Farrell 1999 for a more complete discussion of the U.S. air quality management system.

11. This requirement is part of the national Acid Rain Program in Title IV of the Clean Air Act Amendments of 1990.

12. Potential new entrants included power companies and their subsidiaries elsewhere in the United States (for example, PG&E National Energy Group or Enron) as well as companies trying to enter the generation business (such as Enron) and some large energy consumers who could potentially enter the market as co-generators.

References

Andrews, Clinton J. 1994. Electricity and Federalism: Understanding Regional Diversity. *Energy Policy* 22(7): 629–638.

Anonymous. 1998. *Comprehensive Electricity Competition Plan.* March. Washington, DC: U.S. Department of Energy.

Ayres, Robert U., William H. Schlesinger, and Robert H. Socolow. 1997. Human Impacts on the Carbon and Nitrogen Cycles. In *Industrial Ecology and Global Change,* edited by Robert Socolow, Clinton Andrews, Frans Berkhout, and Valerie Thomas. New York: Cambridge University Press, 121–155.

Bailey, Christopher J. 1998. *Congress and Air Pollution.* New York: Manchester University Press.

Bohi, Douglas. 1997. Introduction: Special Issue on Restructuring the Electricity Industry in the United States. *Resource and Energy Economics* 19(1): 1–2.

Borenstein, Severin, and James Bushnell. 2000. Electricity Restructuring: Deregulation or Reregulation? *Regulation* 23(2): 46–52.

Brennan, Timothy J., Karen L. Palmer, Raymond J. Kopp, Alan J. Krupnick, Vito Stagliano, and Dallas Burtraw. 1996. *A Shock to the System: Restructuring America's Electricity Industry.* Washington, DC: Resources for the Future.

Browner, Carol. 1996a. Letter from Carol Browner to Kathleen A. McGinty. May 13. Washington, DC: U.S. Environmental Protection Agency.

———. 1996b. Letter from Carol Browner to Kathleen McGinty. May 22. Washington, DC: U.S. Environmental Protection Agency.

Carson, J E. 1992. On the Preparation of Environmental Impact Statements in the United States of America. *Atmospheric Environment Part A: General Topics* 26(15): 2759–2768.

CCAP (Center for Clean Air Policy). 1996. *Comments of the Center for Clean Air Policy on the Draft Environmental Impact Statement.* February. Washington, DC: Federal Energy Regulatory Commission. RM95-8-000 and RM94-7-001. http://www.ccap.org/ (accessed February 23, 2001).

Davies, J. Clarence, and Jan Mazurek. 1998. *Pollution Control in the United States: Evaluating the System.* Washington, DC: Resources for the Future.

Davies, J. Clarence, and Barbara S. Davies. 1975. *The Politics of Pollution.* Indianapolis, IN: Bobbs-Merrill.

Derthick, Martha. 1996. Crossing Thresholds: Federalism in the 1960s. *Journal of Political History* 8(1): 65–80.

DiIulio, John, and Donald Kettl. 1995. *Fine Print: The Contract with America, Devolution, and the Administrative Realities of American Federalism.* CPM 95-1. Washington, DC: Brookings Institution.

EIA (Energy Information Administration). 1996. *An Analysis of FERC's Final Environmental Impact Statement for Electricity Open Access and Recovery of Stranded Costs.* SR/OIAF/96-03. Washington, DC: U.S. Department of Energy. http://www.eia.doe.gov/bookshelf/environ.html (accessed February 23, 2001).

———. 1997a. *The Changing Structure of the Electric Power Industry: An Update, December 1996.* DOE/EIA-0562(96). Washington, DC: U.S. Department of Energy.

———. 1997b. *Electric Power Annual 1996* (Volume 1). DOE/EIA-0348(96)/1. Washington, DC: U.S. Department of Energy.

———. 1998. *The Changing Structure of the Electric Power Industry: Selected Issues, 1998.* DOE/EIA-0562(98). Washington, DC: U.S. Department of Energy. http://tonto.eia.doe.gov/ftproot/electricity/056298.pdf (accessed May 4, 2001).

———. 2000. *The Changing Structure of the Electric Power Industry 2000: An Update.* DOE/EIA-0562(98). Washington, DC: U.S. Department of Energy. http://www.eia.doe.gov/cneaf/electricity/chg_stru_update/update2000.html (accessed May 4, 2001).

Ellerman, A. Denny. 1996. The Competition between Coal and Natural Gas: The Importance of Sunk Costs. *Resources Policy* 22(1/2): 33–42.

Elston, John, and Robert Shinn. 1995. *FERC Forum Presentation Summary and Proposed Comments.* December 22. http://www.epa.gov/ttn/rto/otag/docs/12-22min.txt (accessed February 23, 2001).

Engel, Kirsten H. 1997. State Environmental Standard-Setting: Is There a "Race" and Is It to the "Bottom"? *Hastings Law Journal* 48(January): 271–398.

Esty, Daniel C. 1996. Revitalizing Environmental Federalism. *Michigan Law Review* 95(December): 570–653.

Farrell, Alex. 2000. The NO_x Budget: A Look at the First Year. *Electricity Journal* March: 83–92.

Farrell, Alex, Robert Carter, and Roger Raufer. 1999. The NO_x Budget: Costs, Emissions, and Implementation Issues. *Resource and Energy Economics* 21(2): 103–124.

FERC (Federal Energy Regulatory Commission). 1995a. *Draft Environmental Impact Statement: Promoting Wholesale Competition through Open Access Non-discriminatory Transmission Services by Public Utilities (RM95-8-000); Recovery of Stranded

Costs by Public Utilities and Transmitting Utilities (RM94-7-001). Washington, DC: FERC.

———. 1995b. Proposed Rule: Promoting Wholesale Competition through Open Access Non-discriminatory Transmission Services by Public Utilities; Recovery of Stranded Costs by Public Utilities and Transmitting Utilities; Notice of Proposed Rulemaking. *Federal Register* 40: 17662, April 7.

———. 1996a. Final Rule: Promoting Wholesale Competition through Open Access Non-discriminatory Transmission Services by Public Utilities and Recovery of Stranded Costs by Public Utilities and Transmitting Utilities. *Federal Register* 61: 21540, April 24. http://www.ferc.fed.us/news1/rules/pages/order888.htm (accessed May 4, 2001).

———. 1996b. *Order Responding to Referral to Council on Environmental Quality.* May 29. Washington, DC: FERC. http://www.ferc.fed.us/news1/rules/pages/order888.htm (accessed February 23, 2001).

Gade, Mary, Robert C. Shinn, Donald Schregardus, and Harold Reheis. 1996. *Comments of the Ozone Transport Assessment Group on the Draft Environmental Impact Statement: Promoting Wholesale Competition through Open Access Non-discriminatory Transmission Services by Public Utilities and Recovery of Stranded Costs by Public Utilities and Transmitting Utilities.* January 8. http://www.epa.gov/ttn/rto/otag/docs/finaldoc.txt (accessed May 4, 2001).

Gillespie, Ed, and Bob Schellhas (eds.). 1994. *Contract with America: The Bold Plan by Representative Newt Gingrich, Representative Dick Armey, and the House Republicans to Change the Nation.* New York: Times Books.

Gregory, Robin, Ralph Keeney, and Detlof von Winterfeldt. 1992. Adapting the Environmental Impact Statement Process to Inform Decisionmakers. *Journal of Policy Analysis and Management* 11(1): 58–75.

Grumet, Jason. 1998. Old West Justice: Federalism and Clean Air Regulation 1970–1998. *Tulane Environmental Law Journal* 11(2): 375–413.

John, DeWitt. 1994. *Civic Environmentalism: Alternatives to Regulation in States and Communities.* Washington, DC: CQ Press.

Jones, Charles O. 1974. *Clean Air: The Politics of Pollution.* Pittsburgh, PA: Pittsburgh University Press.

Joskow, Paul L. 1997. Restructuring, Competition, and Regulatory Reform in the U.S. Electricity Sector. *Journal of Economic Perspectives* 11(3): 119–138.

———. 1998. Electricity Sectors in Transition. *The Energy Journal* 19(2): 25–52.

Keating, Terry J., and Alex Farrell. 1999. *Transboundary Environmental Assessment: Lessons from the Ozone Transport Assessment Group.* Knoxville, TN: National Center for Environmental Decision-Making Research. http://www.ncedr.org/casestudies/otag/index.htm (accessed May 4, 2001).

Kelley, Tina. 2000. Appeals Court Upholds EPA Rules to Reduce Smog in the Northeast. *The New York Times,* March 4, A11.

Lee, Henry, and Shashi Kant Verma. 2000. *Coal or Gas: The Cost of Cleaner Power in the Midwest.* E-2000-08. Cambridge, MA: Harvard University, Kennedy School of Government.

McGinty, Kathleen A. 1996. Letter from Kathleen A. McGinty to Carol Browner and Elizabeth Moler. June 14. Washington, DC: Council on Environmental Quality.

Moffet, J. 1994. Judicial Review and Environmental Policy: Lessons for Canada from the United States. *Canadian Public Administration* 37(1): 140–166.

Moler, Elizabeth A. 1996. Statement by Chair Elizabeth A. Moler to the Council on Environmental Quality on EPA Referral of Order No. 888, the Open Access Rule. May 14. Washington, DC: Federal Energy Regulatory Commission.

National Research Council. 1991. *Rethinking the Ozone Problem in Urban and Regional Air Pollution.* Washington, DC: National Academy Press, National Research Council, Committee on Tropospheric Ozone Formation and Measurement.

NESCAUM (Northeast States for Coordinated Air Use Management). 1998. *Status Report on NO_x: Control Technologies and Cost-Effectiveness for Utility Boilers.* SS-23. Boston, MA: NESCAUM.

New England Governors' Council. 1996. *Ensuring Cleaner Air From Electric Utility Restructuring: Recommendations to the Clinton Administration.* Boston, MA: New England Governors' Conference, Environment Committee.

NRDC (Natural Resources Defense Council) and Public Service Electric and Gas Co. 1998. *Benchmarking Air Emissions of Electric Utility Generators in the United States.* New York: NRDC.

Oates, Wallace E. 1997. On Environmental Federalism. *Virginia Law Review* 83: 1321–1329.

O'Leary, Hazel. 1996. Letter from Hazel O'Leary to Kathleen McGinty. June 3. Washington, DC: U.S. Department of Energy.

OTAG (Ozone Transport Assessment Group). 1997. *OTAG Executive Report.* December. Washington, DC: Environmental Council of the States.

Pagano, Michael A., and Ann Bowman. 1995. The State of American Federalism 1994–1995. *Publius: The Journal of Federalism* 25(3): 1–21.

Pitts, J.N. 1993. Anthropogenic Ozone, Acids, and Mutagens: Half a Century of Pandora's NO_x. *Research on Chemical Intermediates* 19(3): 251–298.

Portney, Paul (ed.). 1990. *Public Policies for Environmental Protection.* Washington, DC: Resources for the Future.

Reinventing EPA and Environmental Policy Working Group. 1997. *Environmental Goals and Priorities: Four Building Blocks for Change.* Washington, DC: National Environmental Policy Institute.

Revesz, Richard. 1996. Federalism and Interstate Environmental Externalities. *University of Pennsylvania Law Review* 144: 2341.

Smil, Vaclav. 1990. Nitrogen and Phosphorus. In *The Earth as Transformed by Human Action,* edited by B.L. Turner, William C. Clark, Robert W. Kates, and others. New York: Cambridge University Press, 423–436.

Solomon, Barry D., and H.S. Gorman. 1998. State-Level Air Emissions Trading: The Michigan and Illinois models. *Journal of the Air and Waste Management Association* 48(12): 1156–1165.

Standfield, Rochelle. 1984. Ruckelshaus Casts EPA as "Gorilla" in States' Enforcement Closet. *National Journal* May 25: 1034–1038.

STAPPA/ALAPCO (State and Territorial Air Pollution Program Administrators/Association of Local Air Pollution Control Officials). 1994. *Controlling Nitrogen Oxides under the Clean Air Act: A Menu of Options.* Washington, DC: STAPPA/ALAPCO.

Stewart, Richard B. 1977. Pyramids of Sacrifice? Problems of Federalism in Mandating State Implementation of National Environmental Policy. *Yale Law Journal* 86: 1196, 1210–1220.

Stiglitz, Joseph. 1996. Letter from Joseph Stiglitz to Kathleen A. McGinty. June 14. Washington, DC: Council of Economic Advisors.

Taylor, Serge. 1984. *Making Bureaucracies Think: The Environmental Impact Statement Strategy of Administrative Reform.* Stanford, CA: Stanford University Press.

U.S. Court of Appeals. 2000. State of Michigan and State of West Virginia vs. U.S. Environmental Protection Agency. March 3. Washington, DC: U.S. Court of Appeals.

U.S. EPA (Environmental Protection Agency). 1997a. Proposed Rule for Reducing Regional Transport of Ground-Level Ozone (Smog). *Federal Register* 62: 60317, October 10.

———. 1997b. *National Air Quality and Emissions Trends Report, 1996.* 454/R-97-013. Washington, DC: U.S. EPA.

Walker, D.B. 1996. The Advent of Ambiguous Federalism and the Emergence of New Federalism III. *Publius: The Journal of Federalism* 56(3): 271–280.

PART II

Behavior and Perception

The success of many regulations depends on understanding and acceptance by the public. Well-intentioned and scientifically justifiable regulations can fail if the public's concerns and reactions are not anticipated. Both the regulation design process and the regulations themselves can benefit from including the public. Current ideas on social risk management advocate the use of inclusive processes that engage the public and integrate both lay and expert perspectives. With an open and understandable process, social conflict and public mistrust might be reduced. Likewise, an information or warning label tailored to the needs and capabilities of the intended audience is going to be more effective than one that is hard to read and, although complete and correct, inefficient in communicating important facts. These insights have only recently been appreciated by large policy-setting organizations.

The first two chapters in this section discuss the use of two kinds of labels. In Chapter 7, Riley and others consider the design of warning labels on paint-stripping products. Understanding the details of a risk-exposure model is only one step in the process. To design an effective warning label, the knowledge and typical behavior of the product user must be assessed. In many cases, a product's actual use may differ significantly from its manufacturer's intended use. Using a warning label to correct a user's misconceptions can, in some cases, dramatically reduce risk. In Chapter 8, Nadaï discusses the use and impact of information or quality labels. The standards for earning a quality label will affect its value. Setting a high standard allows only a few products to qualify, perhaps frustrating consumers; setting a low standard lets many products qualify and has little meaning. Taking into account not only public but also industry concerns and regulatory goals makes the process far from simple.

The other two chapters look at the importance of public behavior and perceptions during the regulatory design process. In Chapter 9, Follin and Fischbeck use a conjoint analysis to explore the relationship between people's concerns and their actions as well as to determine how significant differences vary among stakeholder groups. Can the concerns of the public be adequately represented by vocal subgroups or by experts who calculate risks? In Chapter 10, DeKay and others explain the process of using lay risk-ranking groups to inform policymakers. Government and industry have been trying to open the risk assessment and risk mitigation processes to the public. How should it be done? A process that the public understands and trusts, that can be replicated with different groups at different times, and that policymakers can apply in a constructive manner conveys clear benefits.

These chapters raise numerous questions for research and discussion. While reading, you might consider questions such as these:

- Which people constitute the "public," and how do policymakers find them?
- How does one determine whether a representation of the public is broad enough? Does every subgroup have to have a voice? Should minority opinions be overrepresented to be "fair"?
- Can a mathematical equation that captures the concerns of most of the public be substituted for public hearings?
- Instructions and warning labels can hold only so much information. Where do you draw the line between one piece of information being important and another one not? Who is liable if a user is hurt by one of the risks deemed "unimportant"?
- Ranking risks is not the same as ranking risk mitigation alternatives. How should risk rankings be used by policymakers trying to establish regulations?
- How should risk mitigation alternatives be ranked? Who should be involved? What process should be used?
- How many different stakeholders should be included in a labeling process? If a risk label hurts sales, should the label be changed?
- How is the value of a risk information program measured?
- Suppose that people rank risks differently based on the display and type of information provided. What is the "correct" ranking?

7
Behaviorally Realistic Regulation

DONNA M. RILEY, BARUCH FISCHHOFF, MITCHELL J. SMALL, AND PAUL S. FISCHBECK

In the realms of environmental, occupational, and consumer product safety, the past three decades have seen extensive top-down regulation that has resulted in mandated processes, standard exposure levels, and outright bans on the most dangerous processes and products. Now that the "low-hanging fruit" has been "picked," it is necessary to refine our tools to address the remaining hazards. In the realm of chemicals and chemical products, we have eliminated some of the most hazardous substances, products, and processes and are left with those that offer considerable benefits but still pose substantial risks that we seek to minimize and control. For example, methylene chloride has survived as the "least hazardous" chlorinated solvent, while its chemical relatives (for example, carbon tetrachloride and chloroform) have been more strongly regulated because of their greater hazard level (Plaa and Larson 1965; Klaasen and Plaa 1966).

When risk levels depend on human actions, regulatory analyses must make behaviorally realistic assumptions. Otherwise, policymakers will not understand the impacts of their actions or identify the most effective ways to achieve regulatory goals. To that end, in this chapter we offer a general approach to behaviorally realistic regulation, illustrated with a case study of consumer use of paint-stripping products. The approach is grounded in an integrated assessment that characterizes the roles that human behavior plays in creating (and potentially controlling) risks. Its parameters are estimated by reference to published results or intensive interviews that examine potential users' "mental models" of the process. Our approach also assesses how well behavior is understood and what opportunities exist for reducing risk, through alternative product or process design.

Product labels are a potentially important risk communication tool, helping users of paint stripper make better informed decisions about product use, which often determines the user's total inhalation exposure concentration. Here we explore the promises and limitations of product labels and voluntary self-regulation for reducing consumers' risk. We present an approach for determining which information is most important to include in a label's limited space, and which risk mitigation strategies should be emphasized over others. First, consumers provide information about how they use a product and perceive its risks, which mitigation strategies they currently follow, and what barriers exist for their compliance with other risk-reducing behaviors. This stage allows for any local knowledge about controlling risks effectively and efficiently to be incorporated. Then, a quantitative exposure model links reported consumer behaviors with exposure outcomes to identify which actions are most effective in reducing risk, taking into account what we have learned about consumer beliefs and behavior. A similar approach evaluates labels for content and presentation, using quantitative modeling of exposure outcomes, incorporating our knowledge of how label reading, comprehension, and compliance relate to mitigatory behavior.

Estimating Behavior

Ideally, users' behavior will be estimated on the basis of observations collected in actual work environments, showing habits, time–activity patterns, amounts of hazardous substances used, typical operating routines, and the like. When these data are not available, user interviews can attempt to elicit this information. It is possible to ask directly for users to estimate the time, frequency, and intensity of their behavior. However, individuals may have limited awareness of their own action, limited ability to translate it into quantitative terms, and limited willingness to report it candidly. As a result, it is often useful to triangulate by eliciting users' mental models of their behavior—that is, the entire suite of beliefs a person has, regarding how the risks are created and controlled (Fischhoff 1999).

Triangulation is accomplished via semistructured interviews that direct users to task-relevant topics while allowing them full freedom of expression in the issues that they address and the language that they use. Doing so can also help users to achieve more realistic quantitative estimates and to identify potential interventions (in the sense of recommendations that users might realistically follow as well as the way to explain their rationale). When designing such communications, the users' mental models can be compared against the "expert model" of the integrated assessment to identify critical gaps in knowledge. Although such interviews are resource-intensive, with proper sampling, a relatively small set can identify the beliefs held with any great frequency in a group of users. If needed, the frequency of beliefs can be estimated with follow-up interviews, using structured surveys derived from the mental models research.

Mental Models

As a target audience for methylene chloride labeling, we chose home hobbyists. They might face similar risks and rely on similar combinations of intuition and product labeling, without any explicit training. Other relevant populations include professional refinishers, in workplaces suited to coordinated risk management, and professional remodelers, working on others' premises. The heterogeneity of beliefs and practices within these groups determines the feasibility of reaching all members with a single message.

Respondents were recruited at a home improvement store. They were asked how, where, and when they used paint stripper, in terms of a typical job (or their most recent one). The most structured of these questions explicitly targeted factors that influence exposure enough to be included in the integrated assessment exposure model: room size and type, position of windows and doors, time spent on different job tasks, location and duration of breaks, and amount of paint stripper used. Tables 7-1 and 7-2 show the results for the quantitative questions.

Every interviewee described taking at least one action to reduce inhalation exposure (although not all these measures were effective ones). Overall, consumers understand the importance of ventilation but not necessarily how best to achieve it. Few suggested using a fan. Only one-third of interviewees

TABLE 7-1. Respondents' Quantitative Estimates of Paint-Stripping Activities

Activity	*Mean*	*Median*	*SD*	*Minimum*	*Maximum*
Amount used (ounces)	82	32	93	12	384
Application (minutes)	18	10	20	NA[a]	60
Curing (minutes)	29	15–20	37	1	180
Scraping (minutes)	44	25	42	5	120
Cleaning (minutes)	24	30	16	10	60
Breaks (minutes)	19	15–20	13	5	30
Room volume (cubic meters)	51	61	31	4.5	92

Notes: SD = standard deviation; n = 20 respondents.

[a] Application took only seconds.

TABLE 7-2. Reported Precautionary Behaviors of Paint Stripper Users

	Respondents	
Behavior	*Number*	*Percentage*
Reading the label	20	100
Effective behaviors		
Taking breaks (outside of work area)	13	65
Working outside [exclusively]	12 [5]	60 [25]
Using a "safer" formulation	10	50
Wearing goggles	7	35
Working inside with open windows, no fan	7	35
Working inside with open windows and fan	4	20
Less effective and ineffective behaviors		
Wearing gloves	18	90
Taking breaks (inside work area)	4	20
Wearing dust mask	3	15
Wearing organic vapor cartridge respirator	1	5
Wearing army gas mask	1	5

Note: $n = 20$ respondents.

mentioned the use of goggles, and not all of them reported appropriate responses to the threat of eye injury. However, hospital admissions data show a high rate of eye injuries relative to other accidents caused by paint strippers (personal communication from R. Brown, Dow Chemical, March 1996).

Relating Behavior to the Creation and Control of Risk

Next, one must find a clear way of relating behavior to its effects. A good exposure-and-risk model will simulate the actual conditions of use as realistically as possible, will be validated against empirical data, and will use a level of complexity appropriate to available data and uncertainty. In some cases, existing models in the literature meet these criteria. For consumer paint stripping, ways to estimate users' peak exposures (which can cause heart attacks from carbon monoxide concentrations) and cumulative exposures (which can be related to the estimated carcinogenic potential of the solvent) are needed. No available model was satisfactory, so we developed our own (Riley, Small, and Fischhoff 2000).

In general, a risk model will have both fixed and user-determined parameters. In our model, parameters included physical constants (for example, volatilization rates), physical measurements (for example, room size, air exchange rates), and behaviors.

The primary hazards of paint strippers stem from inhalation of methylene chloride vapors. Therefore, an indoor air quality model is adapted from one developed by Van Veen (1996) to predict respiratory and dermal uptake of volatile chemicals. It is sensitive to several kinds of behavior that affect room concentrations of methylene chloride, including ventilation, amount used, and time–activity patterns. Riley (1998) describes the model in detail, including its validation with observational data collected by Girman, Hodgson, and Wind (1987).

Assessment of Current Practice

Interview data were incorporated into the exposure model to predict exposures for each respondent. Figure 7-1 shows the model's predictions of cumulative exposure for three interview participants, measured as a *potential inhalation dose*, the total mass of the chemical that enters the body through

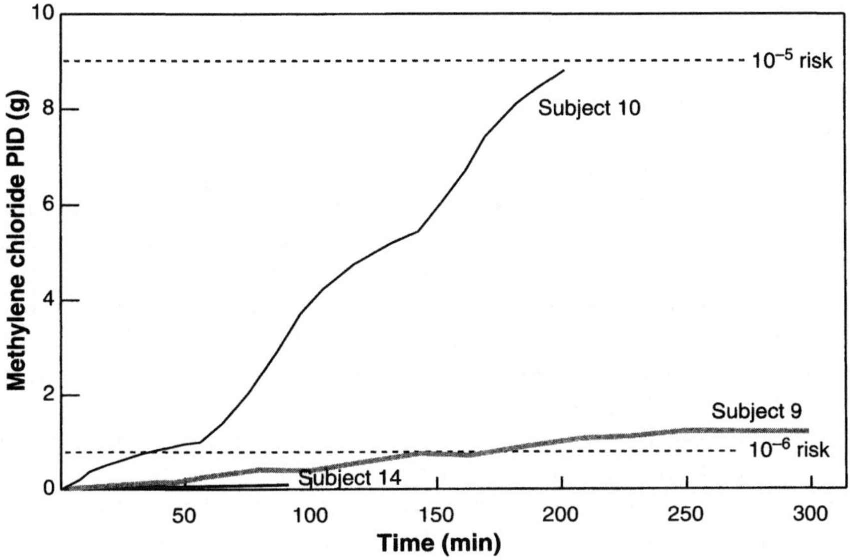

FIGURE 7-1. Modeled Predictions of Cumulative Inhalation Exposure for Three Interview Participants

Note: Risk is lifetime risk of cancer death for a person engaging in one stripping job per year for 45 years at this potential inhalation dose (PID).

breathing activity. According to the Occupational Safety and Health Administration (OSHA), a worker exposed to the eight-hour standard of 25 parts per million (ppm) of methylene chloride in a work day has a potential inhalation dose of 3.5 grams per day (about 900 grams per year), equivalent to an increased lifetime risk of death due to cancer of 10^{-3} (OSHA 1997). The horizontal lines in Figure 7-1 translate these OSHA estimates to the equivalents for these individuals repeating their project once a year for 45 years.

Subjects 10 and 14 represent low- and high-exposure single-session scenarios, respectively; Subject 9 represents an intermediate case.

- Subject 9 stripped an antique spinning wheel with six to eight coats of paint plus a stain on the bottom layer, working in a 26 × 14 × 7-foot garage with windows and doors open. The subject used more than one quart of paint stripper for the job over a five-hour period, taking breaks about once an hour.
- Subject 10 reported stripping window woodwork in an 8 × 13 × 9.5-foot kitchen with open windows and doors, using one quart of stripper in over three hours, taking breaks in the work area itself.
- Subject 14 reported stripping parts of a grandfather clock, working for 90 minutes each day with regular breaks for several weeks, 45 days in all. Although he used six to eight quarts of stripper for the entire job, he reported using only about six ounces on a given day. He also reported suffering from emphysema and therefore took 10-minute breaks about every 20 minutes, working in a garage with windows and doors open and a fan on.

Subject 10's exposure level is very high because a large amount of stripper was used in a small kitchen. Air exchange rates in an open garage, like the one where Subject 9 worked, are assumed to be about five times higher than those of a room in a house (Hodgson and Girman 1987; Schaeffer 1992). Therefore, the predicted exposures are much lower for Subject 9, who used the same amount of stripper over a longer period of time. Each of Subject 14's short sessions produce much lower exposures, punctuated by his long breaks. The maximum possible doses for the depicted sessions are dramatically different, with Subjects 9, 10, and 14 potentially receiving 8.8, 1.3, and 0.11 grams, respectively. Because Subject 14 distributed his work over 45 days, his total potential inhalation dose (total mass of methylene chloride inhaled over time) is about 5 grams; this total is higher than that for Subject 9, but still well below that for Subject 10, distributed over a period of over 4,000 minutes.

The absolute magnitude of these exposures also can be considered from a short-term perspective (not shown). The workplace short-term exposure limit (STEL) set by OSHA (1997) requires that the average concentration not exceed 125 ppm, as a 15-minute time-weighted average. Even Subject 9 could experience acute effects such as lightheadedness or drowsiness, observed in individuals exposed to concentrations as low as 100–300 ppm (ATSDR 1993). Paint stripper users with a pattern like Subject 10's may experience more severe acute effects, including cardiac arrest.

Evaluation of Regulation: Assessing the Impacts of Change

Estimating the effectiveness of a regulatory intervention is relatively straightforward when it specifies new equipment or workplace changes that can be easily reflected in the model (for example, the change in ventilation rate associated with a new ambient air standard). Being realistic about technology means considering such imperfections as maintenance and manufacturing problems. Being realistic about behavior (which is also a factor in maintenance and manufacturing) means considering how well consumers follow instructions. It, in turn, may depend on how well users understand how their actions affect their exposures, as captured in their mental models.

Figure 7-2 illustrates the value of linking behavior and exposure through an example. It considers the case of Subject 10, who reported stripping window woodwork in an 8 × 13 × 9.5-foot kitchen with open windows and

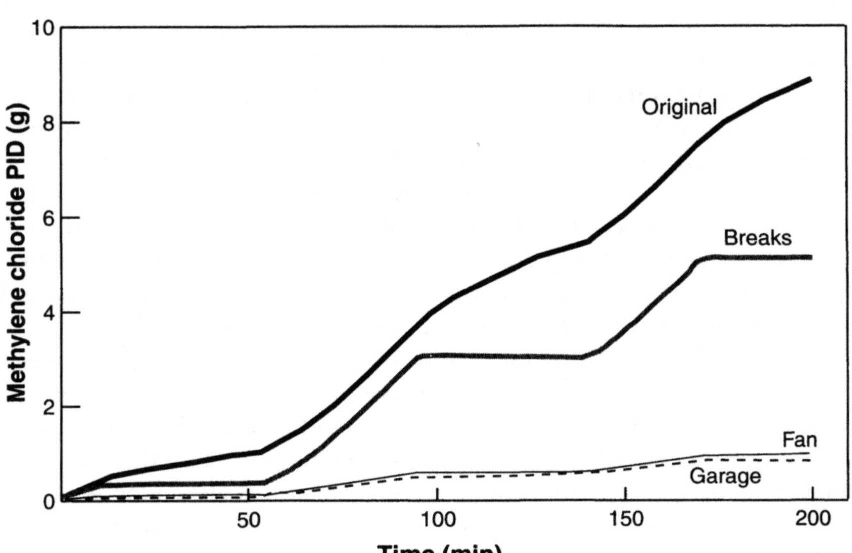

FIGURE 7-2. Exposure Reduction Options for Subject 10 (as PID)

doors, using one quart of stripper in more than three hours, with no outside breaks. Holding the amount of stripper and working time constant, the effects on potential inhalation dose of three alternative measures for reducing exposure are considered: taking breaks, using a fan, and changing the workspace to a two-car garage. Taking breaks reduces cumulative exposure nearly by half, and the new workspace or exhaust fan reduce it by almost an order of magnitude (these curves are superimposed in Figure 7-2).

Considering acute exposure (not shown), using a fan in the open kitchen or moving to a large open garage with no fan has similar, large effects throughout the job. Taking breaks keeps the same high peak exposures, which drop periodically to zero. For acute exposures, it is, of course, the peaks that matter.

The analyses depicted in Figure 7-2 assume that each mitigatory action has 100% compliance. Realistically, users may not perfectly understand or execute these courses of action (Wogalter, Allison, and McKenna 1989). For example, partial compliance with "open windows" may lead to opening only a single window or door, thereby failing to create a cross current. Partial compliance with "take breaks" can mean that users stop less often than they intend or that they fail to leave the room during breaks (for example, to clean up). A fan may be set up to blow inward, rather than outward, reducing its effectiveness.

Comfort level also can reduce the rate of compliance. For example, wearing gloves, goggles, or long sleeves can be uncomfortable, especially in warm weather. In cold weather, ventilation (or working outdoors) can create discomfort or incur additional heating costs. Reflecting these trade-offs, some interview subjects reported stripping paint only in summer; others reported practices such as ventilating with open doors and windows in warm weather, but working with a respirator in a closed basement in cold weather.

The model allows quantitatively analyzing the effect of partial compliance on exposure. As examples, scenarios were developed for two room types, with air exchange rates representing ventilation for rooms that were closed, open, or open with an exhaust fan. Partial compliance was assumed to be half as effective as full compliance. Break time was reduced by half.

As a way of summarizing the relative impact of advocating several alternative behavioral recommendations on exposure, we created a "marginal risk-reduction curve" (Figure 7-3). It shows the added benefit associated with each mitigatory action. These analyses identify the facts that should be emphasized in a communication, including the priorities among them. Producing general advice would require analyzing the impact of each possible measure on each member of the set of typical situations and then, considering the frequency of those situations, identifying the recommendations that

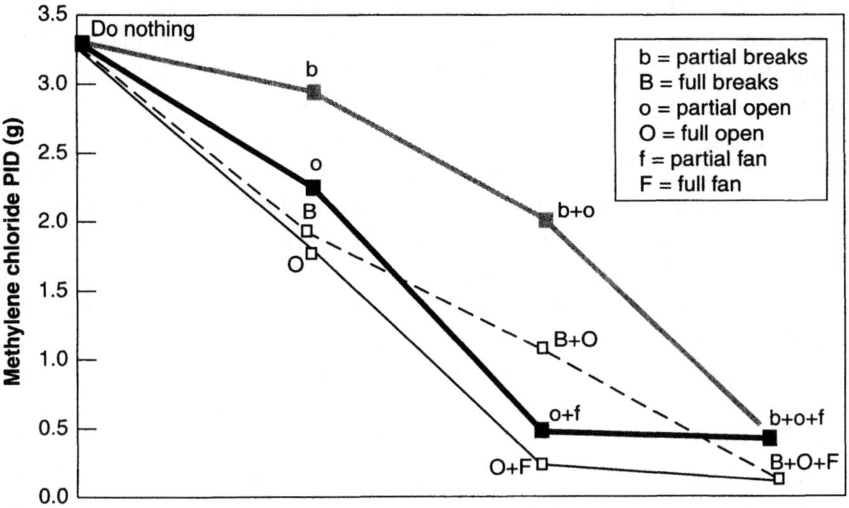

FIGURE 7-3. Exposure in Small Workroom for Full and Partial Compliance with Successive Precautions (as PID)

would bring the greatest expected benefits to the overall population of consumers. The information can be prioritized, assuming that consumers have no prior knowledge about the product or that they have the knowledge, gaps, and misconceptions revealed in the interviews.

Potential Risk-Reduction Messages

In our interviews, every subject reported having read the label on the can. This rate (like that for some other precautionary behaviors) is probably overreported, reflecting respondents' normative expectations (that is, what they thought we wanted to hear). Kovacs and others (1997) found that fewer than 5% of subjects even looked at the precautionary statement on the back label of a chemical cleaner in an experimental setting; nonetheless, in a questionnaire after the experiment, 18% reported having read the label during the experiment, and 76% reported that they "normally read" labels. Given the inefficiency of many reported work practices and precautionary measures, even if all consumers read the labels, they are unlikely to extract and implement what they need to know to avoid or reduce exposure.

Given consumers' level of concern about the risks of using paint stripper, a simple message about fans might find a receptive audience. Moreover, our modeling (Riley 1998) indicates that it could be one of the single most effective measures. However, the flammability of some formulations of paint removers means that this message would have to specify a nonsparking fan.

Another high-potential message emerging from the interviews is the use of goggles. The potential danger is easy enough to understand, once raised. Perhaps citing the high rate of hospital admissions for eye injury from paint stripper would help get the message across, as might asking consumers to examine the splatter patterns on their own work clothing. Goggles currently on the market could eliminate these risks effectively. Moreover, these goggles are suited for multiple uses, making them a much better investment than the gloves and breathing aids that most people use, which provide little protection; indeed, the solvent degrades many gloves. These measures might even increase exposures if they create a false sense of security, so that people remain in work areas longer than they otherwise would. Instructing consumers about the properties of different glove materials will be difficult. As a result, if the use of gloves is to be improved, an engineering or marketing solution may be needed. For example, appropriate gloves might be packaged together with stripper, while inappropriate gloves are explicitly labeled "not suitable for paint stripping." Note that behaviorally realistic strategies do not rely solely on individual users' behavior but use various risk-reduction approaches that fit improved understanding of behavior.

Evaluation of Information-Based Approaches

Information-based approaches are an increasingly popular item in the regulatory toolbox. If effective in reducing exposures, they can free individuals to continue to use a hazardous substance or process that brings them desired benefits along with its risks. In cases where individuals have a great deal of control over the risk, providing appropriate information can help them reduce their own risk and make better informed decisions about how to minimize their risk. Where the proper use of protective equipment or procedures will reduce risk, effective communication can go a long way toward proper implementation of risk-reduction policies.

However, to evaluate this approach, once must take into account the behavioral interactions that individuals have with the message content and the format of its delivery. One must understand the scientific and medical issues related to the product or process as well as the beliefs and behavior of typical users. Our approach to communication design begins with our formal model for estimating the risks associated with a product, including the key exposure routes and best-known approaches for preventing or mitigating the risk. Some key behaviors individuals can take to reduce risk may have already been identified and evaluated, as described in the last section. The key is to simulate the actual conditions of use as best as possible.

Does the communication provide users with the information that they need to manage the risks associated with a given product or process? To answer this question, we must consider the informational content of the message: does it offer consumers a basic understanding of the factors that influence the creation and magnitude of the risk, the potential health effects, and responses that can minimize risk? We also must consider whether the labels convey this information effectively, in terms of the order of information, the explicitness of instructions, the cost of compliance, and the proper use of symbols and emphasis (for visual communication). Our integrated assessment model affords a way of answering some of these questions.

Current Product Label Messages

Labels from 14 methylene chloride paint strippers from six formulators were reviewed. The analysis focused on ventilation instructions, given their critical role in reducing consumers' exposures. Table 7-3 lists the different health effects mentioned and precautionary actions advocated on the labels. Three formulators mentioned methylene chloride's tendency to be metabolized to carbon monoxide (which can lead to heart attacks) or recommended that people with heart or respiratory problems not use the product.

All but one of the products provided explicit instructions for ventilation. All included a discussion of where the product should be used. Eleven product labels noted that outside was the best place to use the product, and two products from Bix recommended (on the side panel) working outside only if the user observed negative health effects. Six product labels additionally recommended avoiding basements and other enclosed areas; five products from one formulator recommended not using the product anywhere that "vapors can accumulate and concentrate, such as basements, bathrooms, or small enclosed areas." Seven products (from three formulators) warned that dust masks do not provide adequate protection from vapors. In our interviews, several respondents reported using such masks. Six product labels (from three formulators) suggested wearing protective clothing on exposed skin, and another six (from two formulators) recommended avoiding contact with skin. Only two product labels (from one formulator) suggested leaving the area for fresh air while the stripper works.

All 14 methylene chloride strippers included some precautionary information in the directions section of the label (ventilation, working outside, and breaks). However, the ventilation instructions were always nonexplicit, such as

TABLE 7-3. Selected Health Effects and Precautionary Behaviors Mentioned on Paint Stripper Labels

	Health effect				*Precautionary information or instructions*							
	Skin, eye irritant	*Cancer*	*Headache*	*Cardiac arrest*	*Gloves*	*Explicit ventilation*	*Location*	*Work outside*	*Goggles*	*Skin cover*	*Dust mask*	*Break*
Klean Kutter	F, B	B	B		S	B	•••	B	B	•	B	
K-3 Premium	F, B	B	B		S	B	•••	B	B	•	B	B
KS Sprayable	F, B	B	B		S	B	•••	B	B	•	B	B
KS Aerosol	F, B	B	B		B	B	•••	B	B	•	B	
KS Graffiti Rmvr	F, B	B	B		B	B	•••	B	B	•	B	
Savogran Strypeeze	F	B	B		B	B	••	B	B	B		
Savogran Sprayable	F	B	B		B	B	••	B	B	B		
Savogran Superstrip	F	B	B		B	B	••	B	B	B		
Savogran Kutzit	F	B	B		B	B	••	B	B	B		
Bix Stripper	F	S	S	S	S	S	(B)	(S)	B			
Bix Spray-on	F	S	S	S	S	S	(B)	(S)	B	B		
Behr Stripper	B	B	B	B		B	••	(B)			B	
Parks Pro Stripper	F, B	B	B	B	S	B	••	B	B	•	B	
Nasco	F	B			B		B	B		B		

Notes: Precautions were given on the front (F), back (B), or side (S) of the container; parentheses around location indicate that contradictory messages were given (for example, "good for indoor use" and "work outside" on the same label). • = avoid skin contact; •• = avoid basements and other enclosed areas, ••• = avoid basements, bathrooms, and other small enclosed areas.

"Use with adequate ventilation." This directive can serve at best as a reminder or a pointer to another part of the label that offers more explicit information.

The quantitative model was used to estimate the effects of these label designs on users' exposure, making different assumptions about users' label-reading behavior and compliance. As mentioned, although most brands contain explicit ventilation instructions, some are more accessible to readers than others. Savogran puts its instructions for ventilation in the first paragraph at the top of the label, whereas Klean-Strip puts it in the middle of a long column, after the discussion of hazards, effects, and the need for engineered air control systems for prolonged use. Only determined readers would extract the ventilation information from the Klean-Strip label.

Consumers' interactions with warning labels are complex (Laughery and others 1993). They must decide which parts to read, then how extensively to read them. How effective will a label's content be, assuming that consumers choose to look at all? Specifically, we considered combinations of four reading approaches and three room types.

The reading scenarios are described as follows:

- Scenario 1: User reads only the first five statements on the label.
- Scenario 2: User reads only the items in boldface.
- Scenario 3: User reads only the product use directions.
- Scenario 4: User reads (and complies with) all information found anywhere on the label.

When information is less explicit, partial compliance is assumed. Because participants in our interviews reported such a variety of beliefs and practices, we assumed that consumers as a whole have no common prior knowledge about product safety.

Studies have found that consumers are most likely to read the first few sentences at the top of the back panel (Scenario 1; Friedmann 1988) or in the directions section (Scenario 3; Frantz 1994). Boldfaced words add emphasis and draw the reader's attention (Scenario 2; Edworthy and Adams 1997). Explicitness and understandability of instructions increase the likelihood of compliance (Laughery and others 1993; Frantz 1994). Because formulators tend to treat safety information similarly on all their products, these reading scenarios were examined for only one product from each formulator.

The three room types were described as follows:

- Room 1: A worst-case basement, small and tight (base air exchange of 0.5 air changes per hour), with only one small window, limiting the potential for increasing ventilation. The air exchange rate increases to only 1.0 ACH when the window is open.

- Room 2: A small workroom, the same size as the basement and similarly tight (0.5 ACH), but with more windows. Its air exchange rate increases to 3.0 ACH when doors and windows are open.
- Room 3: A large garage that is leaky to begin with (2.1 ACH) and has great potential for increased ventilation when the garage door and windows are open (10.6 ACH).

The room evaluated for each label and reading scenario reflects a preliminary screening of the label content. Room 1 is used when an individual pursuing that reading scenario would encounter no room choice information; Room 2 when the reader would encounter a warning against basements, but not against small, enclosed areas; and Room 3 when the reader would see (and follow) advice for working in an airy space. No label recommended using an exhaust fan to increase ventilation; as a result, no scenario included the use of a fan.

Table 7-4 shows peak exposures for the different paint strippers under each scenario, as an indicator of acute health effects resulting from short-term exposure. Above 500 ppm, mild effects such as central nervous system depression, headaches, and lightheadedness have been observed; at these levels, sensitive individuals may be at risk for cardiac arrest (ATSDR 1993). As can be seen, users who read just the boldfaced information (Scenario 2) or just the usage directions (Scenario 3) will face peak exposures well above those associated with acute health effects, because these labels use boldfaced text primarily for section headings and because their directions lack explicit precautionary instructions. For readers who naturally follow these reading strategies, effective and explicit instructions for exposure reduction would have to be in boldface and included in product use directions.

TABLE 7-4. Peak Inhalation Exposure to Methylene Chloride Based on How Much of the Precautions on the Label the User Reads (and Follows) (ppm)

	Scenario			
Paint stripper	*1* *First five*	*2* *Bold*	*3* *Directions*	*4* *Entire label*
Klean-Strip	1,860	1,600	1,600	270
Savogran	710	1,600	1,400	710
Bix	1,860	1,860	1,860	1,860
Behr	1,600	1,600	1,600	710
Parks	1,860	1,600	1,860	710
Nasco	1,600	1,600	1,600	1,600

Except for Savogran, the same would be true for those reading only the first five label items (Scenario 1). Savogran gives prominence to specific and explicit ventilation instructions (for example, the product should be used outdoors when possible, or with open windows and moving fresh air across the work area). However, readers who read the entire Savogran label would not learn anything else of value in reducing exposures. Those who read all of the labels for Behr and Parks would reach equivalent protection, while those reading all of Klean-Strip would achieve a higher level (Scenario 4). At the other extreme, the labels for Bix and Nasco contained little useful information for reducing exposures.

Figures 7-4 through 7-7 show the cumulative exposure for each reading strategy, measured as potential inhalation dose. Generally speaking, the results parallel those for peak exposure, although the magnitude of the effects requires explicit modeling. Figure 7-4 shows that Savogran's label brings the greatest exposure reduction to readers of the first five statements because more critical information is placed at the top. Figures 7-5 and 7-6 show small differences among brands for users who read only emphasized text or directions, because these labels use bold text (Figure 7-5) primarily for section headings and because their directions (Figure 7-6) lack explicit precautionary

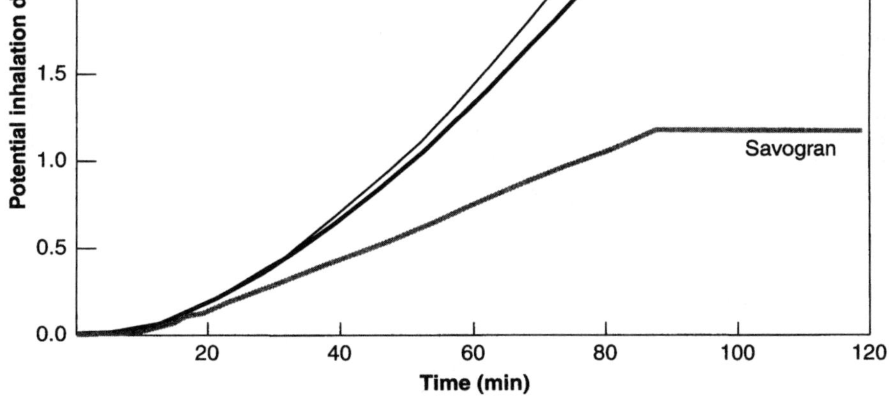

FIGURE 7-4. Exposure (PID) for Users Reading the First Five Points (Scenario 1)

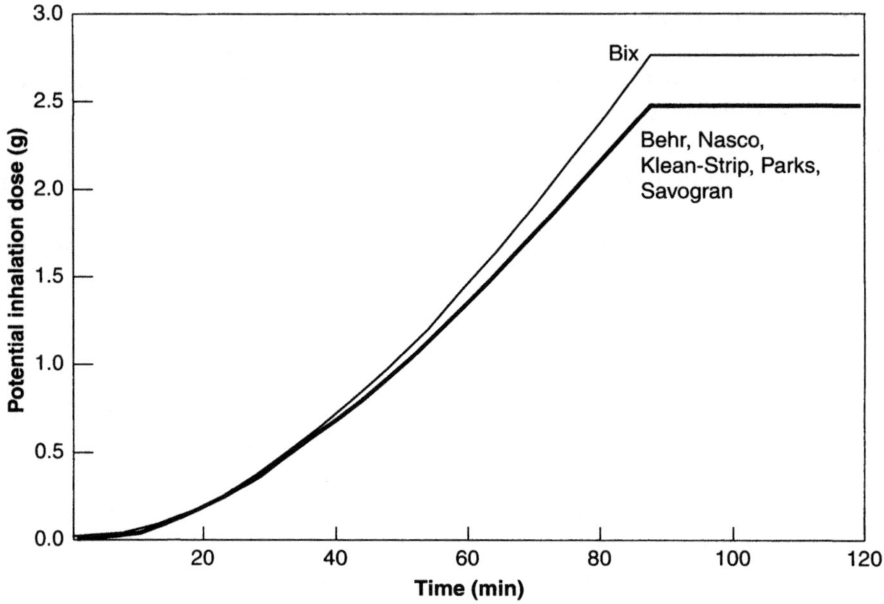

FIGURE 7-5. Exposure (PID) for Users Reading the Emphasized Text (Scenario 2)

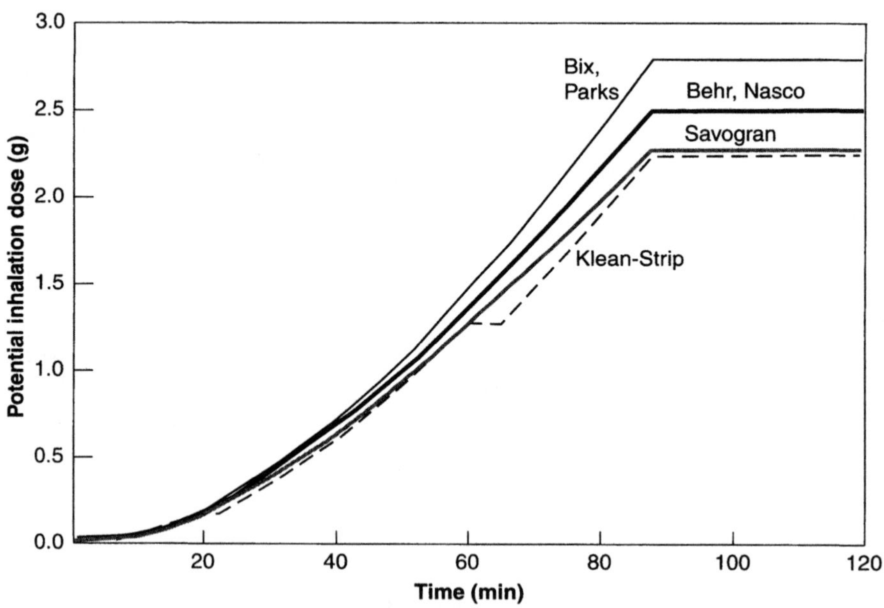

FIGURE 7-6. Exposure (PID) for Users Reading the Directions Only (Scenario 3)

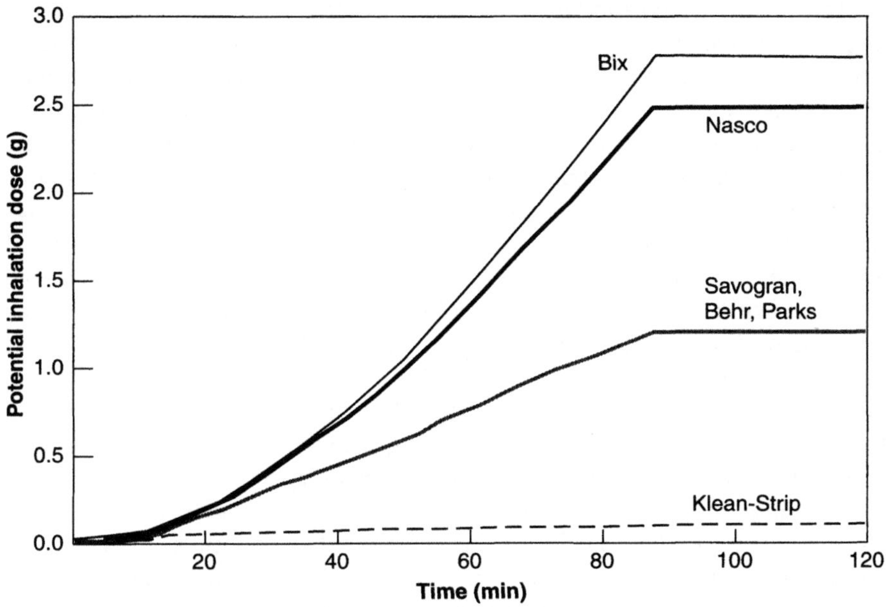

FIGURE 7-7. Exposure (PID) for Users Reading the Full Label (Scenario 4)

instructions. For readers who naturally follow these reading strategies, effective, explicit instructions for exposure reduction would have to be in boldface and included in product use directions. Figure 7-7 shows Klean-Strip's superiority—for people who read and process everything found anywhere on a label. Savogran, Behr, and Parks all have good information, but Behr's and Parks' information is not placed optimally for less thorough readers. Again, Bix and Nasco lacked key information that could help consumers reduce their exposures, however much they read.

Recommendations and Conclusions

The research presented here demonstrates an integrated approach for assessing the effectiveness of regulations that also combines knowledge of behavior with quantitative estimates of its resultant risk. The approach can be used to characterize individual exposures, to assess the relative effectiveness of existing regulatory strategies, and to design better interventions. It is flexible enough to consider alternative behavior patterns and different levels of compliance with risk-reduction strategies.

It would be straightforward to extend the approach to include factors such as prior knowledge and experience, attitudes toward the risks, and costs of compliance. Doing so would require more systematic sampling to estimate the prevalence of various beliefs and attitudes. It would allow more precise estimation of the effects of specific messages in information-based strategies.

If our sample of paint stripper users was any indication, most users do not know enough about risks to take the associated precautions without receiving explicit instructions and information. As a result, information-based approaches should focus on users' choices of workspace and ventilation. For example, explicit ventilation instructions (including the suggestion of a non-sparking fan) could significantly reduce exposure for many users. The labels analyzed in our study varied considerably in usefulness of information for consumers who need to make decisions about product use. Some formulators could enhance their labels by adding information; all could improve their effectiveness by arranging information more appropriately.

Providing information accomplishes only so much. Goggles and gloves are usually recommended together on labels. Nonetheless, reported use of goggles is much lower than reported use of gloves, even though goggles are typically much more effective. Possible reasons are that goggles are more expensive than gloves, goggles are less readily available than gloves, and goggles are less comfortable than gloves. At one of the home improvement centers where our label data were obtained, solvent-resistant gloves were available next to the paint strippers, but goggles were not. Packaging and sales strategies might encourage the use of goggles, even without explicit instruction.

The integrated approach to behaviorally realistic regulation presented here cuts across the current debate over regulatory reform. It fits the calls of proponents of regulatory reform for stricter adherence to cost-effectiveness criteria by facilitating the identification of such solutions and by empowering users to make their own risk decisions, given more complete information. This approach also should appeal to consumer and workers' rights movements, which advocate communicating information in ways that individuals on the "front lines" can understand and offering realistic alternatives. People are encouraged to make decisions based on scientific reality rather than overblown fear or blissful ignorance.

Acknowledgements

Michael Wogalter, Dan Kovacs, and Gordon Butte offered helpful suggestions on this work.

This work was supported by a grant from the Center for Emissions Control, a Dolores Zohrab Liebman Fellowship, and a grant from the National

Science Foundation to Carnegie Mellon University's Center for the Integrated Assessment of Human Dimensions of Global Change.

The views expressed are those of the authors.

References

ATSDR (Agency for Toxic Substances and Disease Registry). 1993. Methylene Chloride Toxicity. *American Family Physician* 47(5): 1159–1166.

Edworthy, J., and Adams, A.S. 1997. *Warning Design: A Research Prospective.* London, U.K.: Taylor and Francis.

Fischhoff, B. 1999. Why (Cancer) Risk Communication Can Be Hard. *Journal of National Cancer Institute Monographs* 25: 7–13.

Frantz, J.P. 1994. Effect of Location and Procedural Explicitness on User Processing of and Compliance with Product Warnings. *Human Factors* 36(3): 532–546.

Friedmann, K. 1988. The Effect of Adding Symbols to Written Warning Labels on User Behavior and Recall. *Human Factors* 30(4): 507–515.

Girman, J.R., Hodgson, A.T., and Wind, M.L. 1987. Considerations in Evaluating Emissions from Consumer Products. *Atmospheric Environment* 21(2): 315–320.

Hodgson, A.T., and Girman, J.R. 1987. Exposure to Methylene Chloride from Controlled Use of the Paint Remover in Residences. In *Proceedings of the 80th Annual Meeting of the Air Pollution Control Association.* Pittsburgh, PA: Air Pollution Control Association.

Klaasen, C.D., and Plaa, G.L. 1966. Relative Effects of Various Chlorinated Hydrocarbons and Liver and Kidney Function in Mice. *Toxicology and Applied Pharmacology* 9: 139–151.

Kovacs, D.C., Small, M.J., Davidson, C.I., and Fischhoff, B. 1997. Behavioral Factors Affecting Exposure Potential for Household Cleaning Products. *Journal of Exposure Analysis and Environmental Epidemiology* 7(4): 505–520.

Laughery, K.R., Vaubel, K.P., Young, S.L., Brelsford, J.W., and Rowe, A.L. 1993. Explicitness of Consequence Information in Warnings. *Safety Science* 15(5/6): 597–614.

OSHA (Occupational Safety and Health Administration). 1997. Occupational Exposure to Methylene Chloride: Final Rule. *Federal Register* 62: 1494.

Plaa, G.L., and Larson, R.E. 1965. Relative Nephrotoxic Properties of Chlorinated Methane, Ethane, and Ethylene Derivatives in Mice. *Toxicology and Applied Pharmacology* 7: 37–44.

Riley, D.M. 1998. Human Factors in Exposure Analysis for Consumer Paint Stripper Use. Doctoral Dissertation, Engineering and Public Policy. Pittsburgh, PA: Carnegie Mellon University.

Riley, D.M., Small, M.J., and Fischhoff, B. 2000. Modeling Methylene Chloride Exposure-Reduction Options for Home Paint-Stripper Users. *Journal of Exposure Analysis and Environmental Epidemiology* 10(3): 1–11.

Schaeffer, Valentine. 1992. Updated Exposure and Risk Assessments for Selected Methylene Chloride-Containing Consumer Products. Memo from Valentine

Schaeffer, Ph.D., Pharmacologist, Division of Health Effects, Directorate for Health Services, Consumer Product Safety Commission (CPSC), to Charles Nichols, Methylene Chloride Project Officer, Directorate for Epidemiology, CPSC. Washington, DC: CSPC. February.

Van Veen, M.P. 1996. A General Model for Exposure and Uptake from Consumer Products. *Risk Analysis* 16(3): 331–338.

Wogalter, M.S., Allison, S.P., and McKenna, N.A. 1989. Effects of Cost and Social Influence on Warning Compliance. *Human Factors* 31(2): 133–140.

8

The Impact of Industrial Strategy and Expert Information on Eco-Labels

ALAIN NADAÏ

Consumers' limited capacity for processing the information displayed on a label is a major issue for understanding the effectiveness of labels. (For example, Chapter 7 deals with consumers' perception of warning labels and their behaviors.) Another element of this complex issue consists of sorting out and ranking the most relevant risks to consider when developing labels. For instance, should the potential effects on human health, global warming, or water pollution be emphasized? I present a case study of paint products in this chapter. Because the paint product environmental label (eco-label) requires a high degree of scientific expertise to interpret, the discussion reaches beyond this particular case.

Product labeling is increasingly based on scientific expertise. A currently debated instance of such an "expert label" at the time of this writing is the "GM-free" (no genetically modified organisms) label used in the European Union and under consideration in the United States. Another instance is the "organic/bio" label for food products (that is, the U.S. National Organic Program), the underlying requirements of which were updated by the U.S. Department of Agriculture to exclude genetically modified organisms from the definition of organic products. Standardized medical claims on food products (in the United States, in China, and under negotiation in the European Union) and various eco-labels are other examples.

Eco-labels have become the most common type of expert labels since their initial diffusion in the early 1990s. Approximately 20 programs are in effect within countries of the Organisation for Economic Co-operation and Development; Brazil, China, India, Malaysia, Singapore, South Korea, Taiwan, and Thailand also have adopted national programs.

The criteria for an eco-label typically are determined through formal or informal negotiations between industry and governmental regulators. Both the input of expert information and the negotiation process affect the criteria underlying the label. With such products, the presence (or absence) of the label itself connects the interests of consumers and the objectives of industry and regulators. Consumers may alter their purchases and hence the market shares of competing firms. In dynamic response, firms also may innovate so that their products receive a label.

In this chapter, I briefly examine the theory of industrial strategy behind eco-labeling and then review the negotiations for a successful eco-label: the European eco-label for indoor paints and varnishes. The role of expert information in the regulatory process, particularly life-cycle analysis (LCA) and market data, is shown to play an important, albeit limited, role. LCA attempts to answer questions about environmental impact from all stages of product manufacture, including materials production, processing, distribution, recycling, and disposal. It can clarify the issues for various constituencies in the process. It also can identify the extent to which technological progress can help mitigate the environmental risks associated with the product. By doing so, expert information could enable the government to understand which strategy industry is pursuing in a regulatory negotiation. However, it cannot reduce the divergence of interests between the protagonists.

Because expert information is costly to produce, regulators might want to rely on some proxies, suggested by the economic theory, to produce this information only in the cases where the chances of a regulatory success are significant.

Economics and Strategy of Eco-Labeling

When analyzing labeling programs, economists tend to reduce them to market mechanisms such as signaling, reputation, and market differentiation strategies (for example, Crampes and Ibanez 1996; Linnemer and Perrot 1997; Mattoo and Singh 1994; Sedjo and Swallow 2000). This approach assumes that consumers either know the product quality beforehand or learn about its quality shortly after purchase. Given this information, consumers can sanction fraudulent sellers by no longer purchasing their products or by publicly criticizing them, and thus drive the market to an efficient equilibrium. Although these models generate insight into the impact of some labels on the market and competition, they overlook the informational barriers to consumers for some products and the resulting opportunity for business and regulatory strategy.

As a matter of fact, any expert labeling program relies on requirements that determine a product's eligibility for the label and signal a minimum level of quality to the consumer. Economists interested in product information have opposing stands regarding the need for transparency on these requirements. According to Viscusi (1994), a scientific assessment may not be within the reach of consumers, and it also might not be possible to reduce the discrepancy between the risks consumers perceive and the actual risks or benefits associated with a product. Accordingly, the policy should not necessarily be transparency on these requirements. It should be the consumer's expected welfare based on the actual risks or benefits, not the perceived ones; that is, the requirements should be based on science, not on consumers' perception. Alternatively, Haldfield and Thompson (1998) suggest that informational regulation can bring consumers' assessment of risk in line with scientific assessment. Labels point out to consumers the products about which they may choose to get more information. By raising the expected value of product-related information, warning labels prompt consumers into a learning process.

These views still do not take standard process into account. The development phase of an eco-label consists of negotiating the requirements, and the diverging interests of the stakeholders might be as influential as scientific information in shaping these requirements. The importance of this negotiation has been emphasized in the case of the European product eco-label. Focusing on the role of industry, Nadaï (1999) points to three strategies during the eco-label negotiation: outright opposition of the whole industry to the regulator and the development of the label, industrial cooperation in the development of the label, and division of the industry into different strategic groups fighting about the content of the criteria. Theoretical work (Nadaï and Morel 2000a,b) confirms this point and demonstrates that the outcome of this negotiation—that is, the list of environmental criteria that eco-labeled products will have to meet—also determines how sales of labeled and unlabeled products will be affected and, in turn, the eventual impact on the environment.

Nadaï and Morel also show that industrial strategies are influenced by different variables. In short, high demand for environmental quality and low innovation cost to change products bring a favorable context for eco-labeling. The higher the demand for green products, the more likely firms are to increase their market share by eco-labeling their products and, hence, regard the eco-label as a profitable opportunity. However, this incentive might not be sufficient if the cost of complying with the eco-labeling requirements is very high (for example, when requirements are stringent or innovation costs are high in the concerned industry).

In any case, the degree of technological heterogeneity of the industry (that is, the difference in the environmental profiles of the products sold by the different firms) is the third variable that determines the strategy and outcome of the negotiation phase. Consider a homogeneous industry in which all firms are selling the same range of products. They all have access to the technologies needed to meet the eco-label criteria under the same conditions, regardless of the content of the criteria. Their sole concern is to determine which type of criteria will be the most profitable for them all. The cooperation of the whole industry may be expected if the eco-label generates a gain for the firms (that is, significant increases in final demand for green products, low innovation costs). Alternatively, consider an industry that is technically heterogeneous. It is made up of firms that each sell one product that has a different environmental profile from the products of the other firms. The exclusion of a product from the eco-label is equivalent to the exclusion of a firm from the program. Rivalry appears in the negotiation of the eco-labeling criteria, because satisfying one firm's interest will go against the interests of the other firms. If the gain at stake is significant, the industry will split into strategic groups that will disagree about the eco-labeling criteria.

Whereas this kind of analysis focuses on the behavioral and design dimensions of the eco-labeling process, other authors have focused on the performance of firms in a market with an eco-label. Arora and Gangopadhyay (1995) and Kirchhoff (1998) show that the demand for green products might provide firms with an incentive to comply beyond regulatory requirements. Mattoo and Singh (1994) and Sedjo and Swallow (1999) demonstrate that "green" (eco-labeled) firms always gain from the implementation of an eco-label, but the impact of this implementation on "brown" (non-eco-labeled) firms depends on the eco-label selectivity, that is, the percentage of products granted eligibility for the eco-label without any change in their existing environmental performance.

Nadaï and Morel (1999, 2000) bridge the behavioral and performance approaches. They analyze how the selectivity depends on industry behavior, as the criteria are negotiated, and determines the environmental performance of the eco-label. Environmental effectiveness is measured by either green innovation or, if firms are not enticed to innovate with their product, the relative increase in the green product market share after eco-labeling (that is, the increase in the ratio of green market shares over brown market shares).

This approach has been applied to both a heterogeneous industry (single-product firms, each of which sells a product that has a different environmental impact) and a homogeneous industry (multiple-product firms, each of which sells the whole range of products). In both cases, it is assumed that firms are maximizing profits by choosing the volume of product(s) they supply. Depending on the market equilibrium, some firms might earn positive profits and undertake green innovations to eco-label their products. The authors examine profits, green innovation, product market shares, and the possibility of compromise on the criteria between the regulator (advocating for environmentally effective criteria) and firms (advocating for profit-maximizing criteria) during the negotiation phase.

Results show that in a homogeneous industry, all firms gain from the implementation of an eco-label. All firms will agree on profit-maximizing criteria even if some of them might increase their profits less than others. When firms' interests are not aligned with the regulator's goal, firms are still better off with suboptimal criteria than with no eco-label. Hence, there is always room for a compromise that will be beneficial to both private firms and the environment. The more significant the demand for green (that is, eco-labeled) products, the wider the window for "win–win" criteria. However, the model shows that, with the current low demand for green products in most developed countries, if innovation costs are significant compared to production costs, then it is unlikely that the eco-label will induce green innovation. Stringent criteria might increase the relative market share of green products but not lead to innovation.

More significantly, when innovation costs are negligible compared with production costs, firms from a homogeneous industry might find it profitable to innovate under permissive criteria. They do so because they expect to increase profit on brown products by turning brown products into green ones, thus reducing brown market share.

In contrast, firms in a heterogeneous industry are neither uniformly enticed to innovate by permissive criteria nor willing to compromise on stringent ones, although some of them might do so. Other firms will oppose stringent criteria, seek to obtain permissive ones, or try to nip the eco-label in the bud because selective criteria will exclude them from the eco-labeling program. Low innovation costs might allow more firms to find innovation profitable and to benefit from the program, but opposition to negotiation and tension between firms during the negotiation are structural features in this case.

LCA, which attempts to describe the direct and indirect environmental impact of a product, is viewed as an important tool for interpreting this impact. Most existing worldwide labeling programs rely on some sort of LCA approach, whether a general LCA matrix (European Union), a streamlined LCA (United States' Green Seal), or a detailed LCA (France, Nordic countries).[1] Even if the International Organization for Standardization standard for LCA (ISO 14040) is widely recognized, methodological issues are still unresolved. These issues have been discussed in the specific case of some existing programs, but no general approach has been undertaken (Karl and Orwat 2000).

For instance, what role might be played by an LCA in light of strategic behavior by firms? For situations in which tensions are likely to occur, producing technical and scientific information about products might reveal the heterogeneity in their environmental impacts. This information would seldom reduce the divergence in the firms' interests or relax the existing tension in the negotiation of the criteria. Hence, the regulator might allocate resources to situations with realistic hope that industry will cooperate in the development of the eco-label. These theoretical predictions on industry behavior and the role of life cycle information are investigated below in the case of the European eco-label for indoor paints and varnishes.

European Eco-Label for Indoor Paints and Varnishes[2]

In 1992, the E.U. Commission adopted eco-labeling regulation 880/92. It imposes rules that set up a consultative process to develop product eco-labels (CEE 1992, 1994) (left side of Figure 8-1). Each member state has to designate a national "competent body" in charge of participating in the negotiation and managing the European eco-labels at the national level.

Any firm, interest group, or public authority can ask for the development of a European eco-label on a new group of products by addressing a request to the country's competent body. When such a request is made, the E.U. Commission designates a member state whose competent body will lead eco-label development. This leading state consults an ad hoc group made up of representatives from the member states and interest groups. The proposed criteria that result from this preparatory phase are submitted to the E.U. Commission, a forum of interest groups (trade unions, industry, trading sector, consumers, and green organizations), and a regulatory committee made up of representatives of the member states. If the proposal receives adequate support from these parties, the E.U. Commission can either adopt the criteria or, if it does not agree with its content, forward it to the Council of Ministers for a final decision.

Once adopted, eco-labeling criteria are legally valid for three years (CEE 1992), but renegotiation may be scheduled periodically. Unofficially, the commission is also worried about the percentage of products receiving the eco-label—that is, the label's selectivity. Acknowledging this unwritten policy, the commission aims for an initial selectivity of between 10% and 15%, up to 35% at the time of renegotiation. Although this target rate is not registered in the regulatory text of 880/92, all the stakeholders involved in the negotiation of the criteria are aware of this objective.

As of the late 1990s, development of the E.U. eco-label program was still limited; 10 groups of products were covered by eco-labeling criteria, and 200

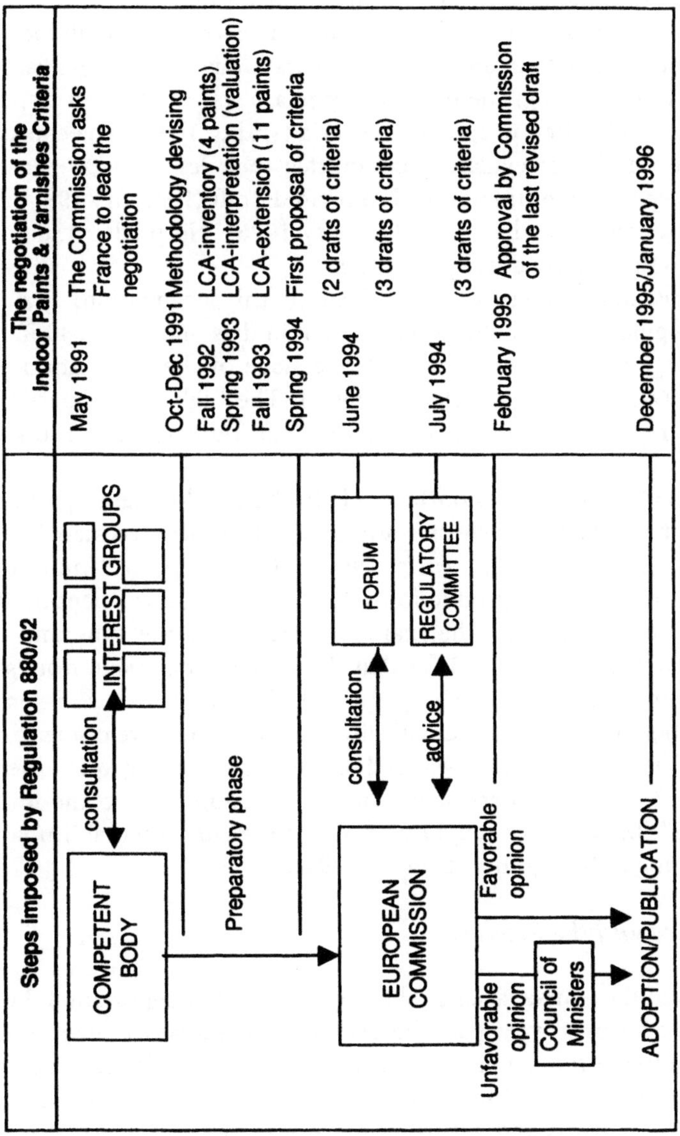

FIGURE 8-1. Negotiation of Paints and Varnishes Criteria

products were eco-labeled at the E.U. level. Although it was a major success in some product groups, it failed to be a significant factor in other markets. For instance, an eco-label was present on only one detergent product three years after the adoption of the criteria for this group of products (Nadaï 1998), whereas indoor paints and varnishes made up about 50% of all E.U. eco-labeled products. Negotiation over the label criteria for indoor paints and varnishes lasted four and one-half years (May 1991 to January 1996; right side of Figure 8-1). The preparatory phase was the longest step, taking two years to reach a first proposal of criteria (spring 1994). The consultation of the forum and the regulatory committee took less than a year. A year passed between the adoption of the last revised draft (February 1995) and the publication of the final document (January 1996). No further modification was made to the criteria during this time.

Numerous stakeholders were involved in this negotiation. Besides the European regulators and national competent bodies, more than 40 industrial stakeholders were involved in at least one stage of the negotiation process (AFNOR 1992, 1993, 1994a, 1994b, 1994c). The high rate of participation from industry attests to the strong interest that industry had in this regulatory process.

Product LCA played an important role in the development of a European eco-label for paints and varnishes and in the European eco-label program more generally. During the early 1990s, the European Union debated whether to invest in the production of publicly available scientific data to inform the policy process. The main concerns were the time and money involved as well as the kind of LCA that should be undertaken, from streamlined to comprehensive and fully quantified. The debate was so intense that some member states were systematically opposing the development of E.U. eco-labels that had not been developed on the basis of a fully quantified LCA. Since then, E.U. stakeholders have reached a compromise in the form of a "matrix-use" LCA, which explores only the environmental impact that experts assume to be significant (Pesso 1994).

The Industry and the Product

To understand the course taken by the negotiation, it is important to be familiar with some features of the paint industry and its regulatory environment.

Although the European paints and varnishes industry is made up of a few hundred producers, the nine leading firms made up the following percentages of the E.U. paints and varnishes market as of 1992: Akzo (the Netherlands, 6%), ICI (United Kingdom, 6%), Casco-Nobel (Sweden, 5.8%), BASF (Germany, 5%), Hoechst (Germany, 4.7%), Total (France, 4.4%), Becker (Sweden, 4%), Courtaulds (United Kingdom, 3.8%), and PPG (United

States, 3.6%).[3] Together, these firms serve about one-half of the European market. They account for most of the supply of indoor paints and varnishes, which is the only market targeted by the E.U. eco-label. Technological innovations are rather quickly imitated among these firms and do not provide durable competitive advantage. Each of the leading producers is actually selling a range of products that includes all the technologies available.

As for the final demand, consumers differentiate the products according to the kind of finish desired (for example, glossy or flat) and their functional needs (water base versus solvent base, washability, durability, and odor). Because most of these characteristics are typical of experience goods,[4] brand and reputation play key roles in sustaining consumers' choices on this market. Currently, product characteristics are strongly correlated with the type of solvent on which the product is based. The traditional distinction between glycerophthalic (organic or oil solvent–based) and acrylic (water-based) products shapes the overall market differentiation. Until now, firms have mostly competed through brand and reputation, managing their product range so as to serve all market segments.

Paints and varnishes contain several components, among which pigments and solvents generate the most environmental concern. Regarding pigments, the LCA highlighted the high amounts of energy consumption, emissions into the atmosphere (sulfur dioxide, carbon dioxide, nitrogen dioxide, and nitrogen oxides) and into water (manganese, chromium, and chlorine) as a result of the production of titanium dioxide (TiO_2), a pigment commonly used to achieve whiteness and opacity. Yet, whereas only one of the two processes leading to TiO_2 production had been studied with LCA, the by-products of these different processes are not the same. A process with sulfates, well-developed in Europe, leads to tremendous amounts of solid waste (titanogypsum or calcium sulfate). A process with chlorides, more developed in the United States, leads to chlorides that are difficult to precipitate and currently are injected into groundwater, where the metal impurities they contain accumulate. LCA has not been used to analyze this process. An alternative white pigment, lithopone, is used in paints recently marketed as "natural paints" by the German company Biofa. The environmental impact of this pigment is not known because the firm refused to undertake an LCA.

Solvents are a source of volatile organic compounds (VOCs), considered to be contributors to global warming and respiratory irritants. Whereas indoor paint and varnishes are estimated to contribute 10% or more of the national VOC emissions in European countries,[5] this amount is obviously related to the consumption of glycerophthalic products. Indeed, paint is commonly distinguished according to the kind of solvent it contains (Table 8-1). Acrylic products are water-based; glycerophthalic products are solvent-based. Two new technologies might allow reducing the solvent con-

TABLE 8-1. Primary Types of Paints

Appellation	*Thinner contained in the product*	*Thinner during usage*
Water-based or acrylic products		
Dispersion in water	Water and cosolvents	Water
Solution in water	Cosolvents	Water
Solvent-based products		
Glycerophthalic products	35–55% solvents	Solvent
High-dry-extract paints	20–30% solvents	Solvent
Powders	None	None (electrostatic application)

Sources: Ecobilan 1991a; interviews with the Federation Française des Fabricants de Peintures (FIPEC; French Paint Industry Federation).

TABLE 8-2. Volatile Organic Compound Content of Paints and Varnishes (grams/liter)

	Acrylic	*Glycerophthalic*	*High-dry extract*
Average	250	450	300
Amplitude	0–450	400–600	250–350
Satin/flat	≤200		
Shiny	≤450		

Sources: Ecobilan 1991a; interviews with FIPEC.

tent of paints. High-dry-extract paint,[6] a solvent-based innovation under development, might allow paint producers to almost halve the solvent content (Table 8-2). The second innovation is solid powder. It contains no water, solvent, or cosolvent. Solid powder is applied with an electrostatic field, which necessarily restricts its use to industrial applications. Paint producers are trying to formulate powders that could be dissolved in water for standard application.

In the European Union, glycerophthalic products traditionally have been marketed as top-of-the-line items, and paint manufacturers still are making higher profits from them. They are commonly considered more resistant to extreme climatic conditions (humidity, temperature, and so forth) and as having better durability, washability, and gloss. Acrylic products are a more recent invention and have benefited from significant technological improvements over the past 15 years. The respective market shares of these two kinds of products are evolving very quickly. Northern E.U. countries have almost entirely turned to acrylic products, whereas countries in the south tend to use more glycerophthalic paints. Several reasons support the movement toward acrylic products in the European Union. Acrylic products are easier to use, their performance (for example, shine and durability) has been constantly improved, and there is a regulatory trend toward a reduction in VOC emissions (European Commission 1993).

In the current state of technology, a straight extrapolation of this environmental pressure supports the development of acrylic products. Such a change could erase the current differentiation between top- and bottom-of-the-line products (that is, between glycerophthalic and acrylic paints) and reduce firms' profits. Because producers would prefer a smooth transition that preserves a top-of-the-line segment, two paths are possible. Firms can progressively market the best-performing acrylic products as the top of the line or introduce innovative low-VOC solvent-based products (for example, high-dry-extract paints) to the top-of-the-line segment. The new market differentiation would match, rather than oppose, the status of green and top-of-the-line qualities, allowing paint producers to focus on green innovations and better reward them. It also would reduce market differentiation throughout the European Union by making southern and northern markets more similar. It might allow paint producers to create and take advantage of scale economies in production and marketing.

The leading producers exhibit no significant technological asymmetries, and the whole core of the paint industry may presume to benefit from the ensuing gains. Paint producers are not divided by conflicting interests. They have been proactive in this area. In fact, they devised a self-proclaimed environmental standard for their products during the late 1980s. In 1989, the Confédération Européenne des Fabricants de Peintures et Vernis (CEPE; European Confederation of Paint Manufacturers) was on the verge of registering this standard as a trademark when the European Commission proposed that the paint industry participate in the eco-labeling program. The French and the European trade associations (Federation Française des Fabricants de Peintures [FIPEC; French Paint Industry Federation] and CEPE, respectively) accepted and abandoned registration of their own environmental trademark for the paint industry.

CEPE's trademark was going to be based on composition criteria (such as VOC content of less than 250 grams per liter), limits on halogenated solvent content, and the absence of heavy metals and substances classified as toxic or very toxic. Although these criteria were in line with the ones underlying existing national eco-labels on paint products (Blue Angel [Germany] and Stichting Milieukeur [the Netherlands]), most of CEPE's criteria were not quantified, thus making it impossible to predict the selectivity of the trademark. As a matter of fact, several criteria had never been documented on a quantified technical and scientific basis. These concerns remained unresolved at the outset of the negotiations of the European criteria.

Product LCA and Criteria Negotiations

In spring 1991, France[7] proposed to base the preparatory phase (May 1991–June 1994) on a quantified LCA. An LCA was undertaken on a set of products that were representative of the different technologies under development or available in the marketplace (Ecobilan S.A. 1991a, 1991b, 1992a, 1992b, 1993a, 1993b, 1993c, 1993d); results are presented in Figure 8-2.

The LCA allowed the classification of the products into three families with increasing environmental impact (Figure 8-2). Its results emphasized the greenhouse effects related to TiO_2 production as well as atmospheric acidification and pollution related to VOCs released during paint application (Ecobilan S.A. 1992a). In short, results showed that using "microvacuum" technology (acrylic), high-dry-extract technology (glycerophthalic), or natural paints would generate environmental benefits.[8]

As of April 1993, the ad hoc European committee had proposed a list of eight areas of concern for future eco-labeling criteria (Ecobilan S.A. 1993a). Criteria relating to these concerns had to be documented throughout the course of the negotiation. Two major issues were the thresholds for the maxi-

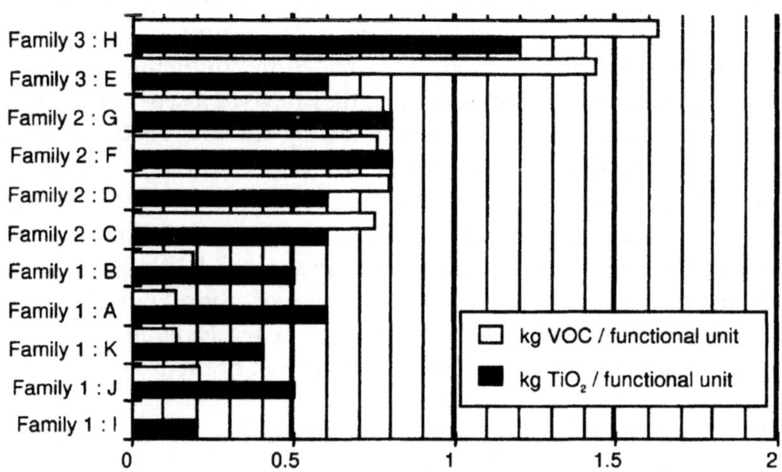

FIGURE 8-2. Volatile Organic Compounds (VOCs) and Titanium Dioxide (TiO_2) Content in 11 Paints Analyzed by S.A. Ecobilan with LCA, Displayed in Three Families of Products

Notes: Family 1 = "natural" paints (I), high-dry-extract paints (J), microvacuum paints with water (K), and conventional styrene–acrylic resin paints with water (A and B); Family 2 = conventional paints containing solvent (C, D, F, and G); Family 3 = pliolite paints containing solvent (E) and limed oil containing solvent (H).

Source: Ecobilan S.A. 1993d.

mum authorized quantities of TiO_2 and VOCs in eco-labeled products. Early on, it had been agreed that the TiO_2 criterion should take the form of a maximum "grams of TiO_2 per square meter covered" threshold.[9] A value was first proposed by France on the basis of its past experience. This value was further negotiated in several steps during the preparatory phase, first increasing under industry and TiO_2 producers' pressure, then decreasing under international pressure from the United Kingdom, the Netherlands, and Denmark (AFNOR 1993; Ecobilan S.A. 1993c). Reaching a final compromise on the TiO_2 criterion required cross-negotiation over the VOC criteria issue. In particular, Portugal and the paint industry, which were asking for more permissive values on both criteria, relented on the TiO_2 criterion to obtain a satisfactory VOC criterion.

Interestingly, the lack of LCA data on a chloride-based production process for TiO_2 was another obstacle to reaching a compromise. Outside expert advice was consulted to compare the environmental impact of the chloride process with that of the sulfates process. Despite the lack of information and without any further justification, experts stated that the environmental impact of chloride-processed TiO_2 "should not" strongly differ from that of sulfate-processed TiO_2. This consensus was not based on facts, given the difference in the by-products emitted by each process. Yet, the experts' answer was sufficient, at that time, to allow the ad hoc group to overlook the lack of information on the chloride process. The final TiO_2 criterion ended up being less stringent than the initial French proposal.

Determining the VOC criterion was by far one of the most difficult issues. A preliminary survey (European Commission 1993) concluded that this regulation should distinguish between different categories of paints and establish separate VOC requirements. Yet, adequate product categories had not been defined.

The E.U. negotiation proved to be difficult because of differences in national consumption patterns. Industry ceaselessly defended the idea that for the eco-label to be successful, consumers should find it available on the category of products that they usually purchase. Accordingly, it proposed to differentiate two or three categories of paints, allowing some products in each category to be eligible for the eco-label. Although this suggestion satisfied the southern European countries (for example, Portugal) that primarily used higher-polluting glycerophthalic products, northern European countries wanted the eco-label to be targeted at lower-polluting water-based products. This structural divergence was reflected in the successive proposal of VOC criteria made by the various countries during the preparatory phase (AFNOR 1993, 1994a, 1994b). This phase resulted in a relatively lax compromise that was close to the paint industry and Portugal's initial proposal.

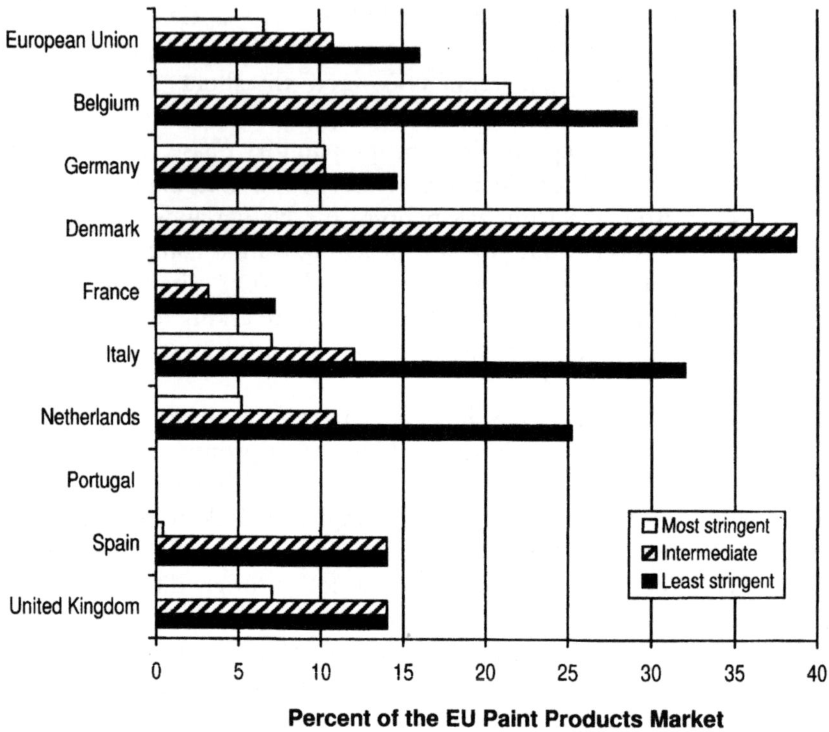

FIGURE 8-3. Selectivity and Market Share due to VOC Stringency

Later on, the forum of interest groups favored the northern countries' position and tightened the requirements (Consultation Forum 1994; European Commission 1994). The paint industry replied by disseminating information about the selectivity of the eco-label assessed under several VOC criterion scenarios (Figure 8-3).

Figure 8-3 shows the wide variation in market share by country that would be eligible for the eco-label. The most stringent assumption of the "class 1 requirements" (that is, for low-VOC products) fulfilled the commission's goal of selectivity, that is, 10–15% of the European market at its launching (Ecobilan 1994). Yet, because of differences among the national markets, fulfilling this goal proved to be incompatible with the legitimacy of the eco-label in some northern countries (for example, 40% of eco-labeled products would be perceived as unduly lax in Denmark) and product eligibility for the eco-label in southern countries where none of the paint sold would meet the criteria (for example, Portugal).

Hence, the reasons underlying the opposition between northern member states (opposing lax criteria) and southern member states (opposing selective criteria) was documented but not overcome. Austria and Sweden joined the negotiation during the final consultation of the Regulatory Committee. These additional member states changed the balance of power in a manner typical of European negotiations, in which

- northern countries (in this case, Austria, Denmark, Germany, the Netherlands, and Sweden) were in favor of strengthening environmental requirements, that is, a lower threshold for low-VOC products and suppression of (or a lower threshold for) high-VOC products;
- France was somewhere in the middle;
- southern countries (Portugal in this case) asked for lax environmental criteria; and
- industry joined forces with one of these last two groups (Portugal in this case).

Four revisions of the criteria (European Commission 1994, 1995a, 1995b, 1995c) were issued before reaching a final compromise (European Commission 1995d) that only slightly modified the criteria. The modifications involved

- the difference between the two classes: the change allowed more products to fall into the low-VOC class and was a concession to requests from Denmark, Germany, and Sweden to remove the high-VOC class. On the other hand, the requirement for the low-VOC class remained unchanged,[10] which clearly was a concession to requests from CEPE, France, and Portugal.
- the division of the high-VOC class into two subgroups: the VOC criterion was strengthened in one subgroup, whereas high-dry-extract paints could fall in the other subgroups and be eco-labeled.[11] This change satisfied the northern countries' request for a strengthening in the VOC criteria on the high-VOC class while allowing the possibility for the high-dry-extract paints to become eligible for the eco-label if they were to be marketed in the near future.

The final E.U. eco-labeling criteria presented in Table 8-3 were adopted in December 1995 and published in the *Official Journal of the European Commission* in January 1996 (European Commission 1996).

The final VOC criteria were more stringent than the self-proclaimed CEPE environmental standard, in part because of using LCA for their development, but they still mirrored industry's goals. The criteria distinguished between two classes of products: one that roughly covered acrylic products, and another that covered glycerophthalic products and allowed producers to eco-label forthcoming low-VOC innovations. Producers could thus preserve the current market differentiation while managing a smooth transition to new products.

TABLE 8-3. Commission's Decision for Indoor Paints and Varnishes, 1995

Criteria	*Applicable to*	*Requirement*
Classification		Class 1: paints with specular gloss ≤45 units, α = 60° Class 2: paints with specular gloss >45 units, α = 60°; varnishes
Labeling	All products	Labeling must clearly indicate that the product is intended for indoor use.
Content in white pigment (all)	Paints	≤40 g/m² of dry film, with 98% opacity
Production of white pigment (TiO_2)	Paints	Requirements limit solid waste production, discharges in water and air.
VOC content	All products	Class 1 paints: ≤30 g/l (minus water) Varnishes and class 2 paints: ≤200 g/l (minus water)
Aromatic hydrocarbons	All products	Class 1 paints: ≤0.5% (m/m) of the product Varnishes and class 2 paints: ≤1.5% (m/m) of the product
Water pollution coming from cleaning application tools	All products	The packaging should display recommendations for cleaning tools to limit water pollution.
Solid wastes, packaging, and product residues	All products	To limit solid waste (residues and can), the packaging shall display recommendations concerning product storage conditions after opening.
Pigments and other substances	All products	Paints should contain neither substances including cadmium, lead, chromium VI, mercury, or arsenic nor dibutyl-, dioctyl- or di-2-ethyl-hexyl phthalates.
Dangerous substances	All products	Ingredients (substances or preparations) used in paints and varnishes shall not be classified as carcinogenic, mutagenic, toxic for reproduction, toxic, or very toxic.
Hiding power	Paints	Paints shall have a hiding power (area to be covered) ≤7 m²/l.
Resistance to liquid	Varnishes	Varnishes shall have satisfactory resistance to water for 1 hour at ambient temperature.

Notes: Specular gloss is a measure of how shiny the paint appears; the measure is taken at a given angle (α). g = grams; m = meter; l = liter.

Source: European Commission 1996.

Conclusions

In this case study, the general conditions are favorable to the development of an expert label: the industry is homogeneous; the final demand for a green variant of the products is significant because the overlap of environmental, health, and usage characteristics makes it worthwhile for consumers to buy low-VOC products; and the pace at which these innovations diffused throughout the industry indicates that firms can easily finance them.

Theory tells us that under these conditions, stakeholders should be interested in developing an expert label and processing the available information to make it as sound as possible. Regulators also might find it worthwhile to invest in a product LCA if such data were not already available.

As a matter of fact, paint producers were clearly interested in developing a system of environmental labeling for their products, but other types of tensions—including some technical heterogeneity—limited the production and the exploitation of scientific and technical data. Member states had diverging interests because of differences in their consumption patterns and the selectivity of the eco-label in their home markets. Some industrial suppliers (such as TiO_2 suppliers that used chloride processes) did not find it in their best interest to produce LCA data. As a result, shortcuts to scientific assessment were taken at several steps in the process to overcome conflict and help achieve compromise.

The fact that the development of this eco-label was among the most successful of the E.U. eco-labels tends to confirm the theoretical results. Yet, a closer look at this case shows that tensions might come from actors other than the concerned industry, such as its suppliers. Other constituencies that may oppose or support the negotiations should be taken into account when deciding whether to invest in the production of new scientific data.

This case study also highlights the importance of the negotiation phase and the potential role of voluntary programs as an alternative to both command-and-control regulation and self-regulation. Indeed, the paint industry's involvement with the eco-label can be regarded as a double-edged sword. On the one hand, the paint industry fulfilled its goal by influencing the course of negotiation with the information it provided about paint products and their market shares. However, the final eco-labeling criteria are quantified and more stringent than the former CEPE environmental standard, and the shift from self-regulation to a public voluntary program resulted in more meaningful environmental requirements for green products.

Notes

1. Life-cycle analysis (LCA) is a method aimed at quantifying the various pollution emissions related to the whole life cycle of a product, that is, from cradle (inputs

and processing) to grave (dumping, recycling, or incineration). A general LCA matrix is indicative of the general kinds of pollutant emissions at each stage. LCA can then mix expert judgements and technical measurements focused on particular steps and emissions. This mixture corresponds to a streamlined LCA. A detailed LCA emphasizes technical measurements over expert judgement, trying to be exhaustive in the polluting emissions and life cycle steps that are quantified.

2. This section is based on Nadaï 2000.

3. Data for percentage of the E.U. paints and varnishes market are for 1992 (from the author's interview with Total Group S.A.).

4. An *experience good* is defined by microeconomists as a good whose quality can be assessed only after purchase, by experiencing the good. Typical examples are food products, meals at restaurants, and movies. Therefore, brand and reputation are key assets for the sellers of these goods.

5. Results for nonmethanic VOC emissions in 1994 (Belgium, Denmark, France, Germany, Luxembourg, and the Netherlands) were obtained from the European Environment Information and Observation Network (EIONET) website (http://www.eionet.eu.int/).

6. High-dry-extract paints include high-VOC (volatile organic compound) alkyd paint, low-VOC acrylic dispersion paint, and low-VOC/high-solid alkyd paint.

7. In the early 1990s, France was one of the member states systematically opposing E.U. eco-labeling criteria that were not based on quantified life-cycle analysis.

8. Microvacuum technology (Product K) allowed the reduction of concentrations of titanium dioxide and volatile organic compounds in water-based paints (Products A, B, and K), whereas high-dry-extract technology (Product J) reduced these concentrations for solvent-based paints (Products C, D, F, G, and J).

9. The criterion is based on an "input per unit of output" performance index. The quality of the coat is defined as 98% opacity.

10. It was kept at 30 instead of 10 grams per liter minus water, as requested by northern countries.

11. Microvacuum technology allowed the reduction of the concentrations of TiO_2 and volatile organic compounds in water-based paints (Figure 8-2, Family 1), whereas high-dry-extract technology reduced these concentrations for solvent-based paints (Figure 8-2, Family 2).

References

AFNOR (Association Française de Normalisation). 1992. Second Meeting of the French Mirror Group on Ecolabeling of Paints and Varnishes. December 9. Paris, France: AFNOR-Paris.

———. 1993. Third, Fourth, Fifth, Sixth, Seventh, and Eighth Meetings of the French Mirror Group on Ecolabeling of Paints and Varnishes. March 22, May 24, July 19, August 31, September 2, and December 9. Paris, France: AFNOR-Paris.

———. 1994a. Eighth Meeting of the European Ad Hoc Group on Ecolabeling of Paints and Varnishes. February 2. Paris, France: AFNOR-Paris.

———. 1994b. *Community Ecolabel for Paints and Varnishes: Remaining Discussion Points on Final Proposal.* February 22. Paris, France: AFNOR-Paris.

———. 1994c. *Community Ecolabel: Assessment and Attribution Procedures for Paints and Varnishes.* February 17. Paris, France: AFNOR-Paris.

Arora, S., and S. Gangopadhyay. 1995. Toward a Theoretical Model of Voluntary Overcompliance. *Journal of Economic Behavior and Organization* 28: 289–309.

CEE (Communauté Economique Européenne). 1992. Council Regulation no. 880/92 of March 23 concerning a community system of attribution of ecological label. *JO* L99(April 11).

———. 1994. *The System of Attribution of the EU Ecological Label: Information for Applicants.* mm6/94. Brussels, Belgium: European Commission.

Consultation Forum. 1994. *Report of the Consultation Forum for the European Ecolabel on the Criteria of Awarding the European Ecolabel to Indoor Paints and Varnishes.* June 20. Brussels, Belgium: European Commission.

Crampes, C., and L. Ibanez. 1996. The Economics of Green Labels. Working paper 96.36.439. November. Toulouse, France: GREMAQ.

Ecobilan S.A. 1991a. *L'écolabel Européen: Projet d'application aux peintures et vernis* (Volume 1: Approche des produits et du marché). October. Paris, France: Ecobilan S.A.

———. 1991b. *L'écolabel Européen: Projet d'application aux peintures et vernis* (Volume 2: La méthodologie pour l'analyse de cycle de vie). December. Paris, France: Ecobilan S.A.

———. 1992a. *L'écolabel Européen: Projet d'application aux peintures et vernis* (Volume 3: Les résultats de l'inventaire du cycle de vie). May. Paris, France: Ecobilan S.A.

———. 1992b. L'écolabel Européen: Projet d'application aux peintures et vernis. Working paper no. 3: Les principaux impacts environnementaux de quatre peintures. December. Paris, France: Ecobilan S.A.

———. 1993a. L'écolabel Européen: Projet d'application aux peintures et vernis. Working paper no. 4: Vers des critères. March. Paris, France: Ecobilan S.A.

———. 1993b. *L'écolabel Européen: Projet d'application aux peintures et vernis* (Volume 4: Phase d'interprétation—Les principales contributions aux problèmes écologiques dans le cycle de vie de quatre peintures). March. Paris, France: Ecobilan S.A.

———. 1993c. L'écolabel Européen: Projet d'application aux peintures et vernis. Working papers nos. 5, 6, and 7: Vers des critères. May, July, and August. Paris, France: Ecobilan S.A.

———. 1993d. *L'écolabel Européen pour les peintures et vernis* (Volume 5: Résultats de la phase d'extension: L'analyse de cycle de vie de onze peintures décoratives). October. Paris, France: Ecobilan S.A.

———. 1994. *L'écolabel Européen: Projet d'application aux peintures et vernis—Les critères d'écolabélisation basés sur l'inventaire de cycle de vie de onze peintures décoratives.* June. Paris, France: Ecobilan S.A.

European Commission. 1993. *Study of the Implementation of an EC Policy for Reducing VOC Emissions from Private and Architectural Use of Paints and Varnishes.* Brussels, Belgium: European Commission.

———. 1994. *Commission Decision of 1994 Establishing the Criteria for the Award of the Community Eco-Label to Indoor Paints and Varnishes,* Draft 3 (June 10), Draft

4 (July 12), Draft 5 (September 5), Final Draft—Revision 2 (December 7). Brussels, Belgium: European Commission.

———. 1995a. *Commission Decision of 1995 Establishing the Criteria for the Award of the Community Eco-Label to Indoor Paints and Varnishes.* Final Draft—Revision 3 (January 16). Brussels, Belgium: European Commission.

———. 1995b. *Comments Received by the Commission Concerning the Draft Criteria Document no. XI/455/94, Rev. 3, of 16 January 1995.* February. Brussels, Belgium: European Commission.

———. 1995c. *Commission Decision of 1995 Establishing the Criteria for the Award of the Community Eco-Label to Indoor Paints and Varnishes.* Minutes of meetings 1–12. February 14. Brussels, Belgium: European Commission.

———. 1995d. *Commission Decision of 1995 Establishing the Criteria for the Award of the Community Eco-Label to Indoor Paints and Varnishes.* Final (February 14). Brussels, Belgium: European Commission.

———. 1996. Commission Decision of the 15th of December Establishing the Criteria for the Attribution of the Community Ecolabeling Criteria for Indoor Paints and Varnishes. In *Official Journal of the European Communities.* Doc. 96/13/CE. January 6. Brussels, Belgium: European Commission.

Haldfield, G.K., and D. Thompson. 1998. An Information-Based Approach to Labeling Biotechnology Consumer Products. *Journal of Consumer Policy* 21: 551–578.

Karl, H., and C. Orwat. 2000. Economic Aspects of Environmental Labeling. In *The International Yearbook of Environmental and Resource Economics 1999/2000: A Survey of Current Issues,* edited by H. Folmer and T.H. Tietenberg. Aldershot, U.K.: Edward Elgar.

Kirchhoff, S. 1998. Overcompliance, Labeling, and Lobbying: The Case of Credence Goods. Working paper 98-25. November. College Park, MD: University of Maryland.

Linnemer, L., and A. Perrot. 1997. Une analyse économique des signes de qualité: labels collectifs et certification des produits. Working paper CREST 9732. July 1998. Paris, France: Institut National de la Statistique et des Etudes Economiques.

Mattoo, A., and H.V. Singh. 1994. Eco-labelling: Policy considerations. *Kyklos* 47(1): 53–65.

Nadaï, A. 1998. Competition on products environmental qualification (in French). *Revue d'Economie Industrielle* 83: 197–212.

———. 1999. The Conditions of Development of a Product Ecolabel, European Environment. *The Journal of Environmental Policy* 9(5): 202–212.

———. 2000. When Does Industry Co-operate to the Development of a Product Ecolabel? The EU Indoor Paints and Varnishes Ecolabel. Working paper series. Pittsburgh, PA: Carnegie Mellon University, Center for the Study and Improvement of Regulation.

Nadaï, A., and B. Morel. 2000a. *Product Ecolabeling: Looking Further into Policy Considerations.* Paper presented at a conference of the European Association of Environmental and Resource Economists, University of Crete, Rethymnon, Greece, June/July 2000.

———. 2000b. *Product Ecolabeling, Competition, and the Environment.* FEEM Working Paper 82-2000. Milan, Italy: Fondazione Eni Enrico Mattei.

Pesso, C. 1994. An Overview of the Life-Cycle Approach to Product/Process Environmental Analysis and Management. October. Working paper. Paris, France: Organisation for Economic Co-operation and Development.
Sedjo, R.A., and S.K. Swallow. 2000. Eco-labeling and the Price Premium. RFF discussion paper 00-04. Washington, DC: Resources for the Future.
Viscusi, W.K. 1994. Efficacy of Labeling of Foods and Pharmaceuticals. *Annual Review of Public Health* 15: 325–343.

9

Trade-Offs among Environmental, Human Health, and Quality-of-Life Impacts

JAMES N. FOLLIN AND PAUL S. FISCHBECK

Comparing complex problems that span environmental, human health and safety, and quality-of-life impacts requires making difficult trade-offs. Which is more important: saving wetlands or reducing the accident rate on a stretch of highway? Who should make these decisions: risk experts, vocal activists, or the lay public? Understanding how people value and trade off these attributes is a key component to communicating impact and risk information.

In this chapter, we describe the methodology, performance, and analysis of a ranking study that was conducted on a set of hypothetical road-building projects. Using a conjoint design, the experiment examined the relative importance of different ecological, environmental, human health and safety, and quality-of-life impacts. Results indicate that people were able to make consistent trade-offs among various impacts, reflecting their assessed concerns in their rankings of the projects. However, impact and project rankings were not predicted by demographic variables, and significant differences were found among groups of laypeople, environmentalists, and the technical community (professional engineers and scientists).

Background and Motivation

Many studies have attempted to ascertain the public's level of environmental concern and the relationship between environmental concerns and environmental behavior (for example, Hallin 1995; Dietz, Stern, and Guagnano 1998; Steg and Sievers 2000). Researchers have found limited correlation between the level of concern and actual actions to support the environment (Maloney and Ward 1973; Smythe and Brook 1980; Van Liere and Dunlap 1981; Buttel 1987; Ostman and Parker 1987; Dunlap 1991; Gigliotti 1992; Scott and Willits 1994; Tarrant and Cordell 1997). Dunlap (1989) identifies five reasons for the apparent disagreement: increasing government attention, the tendency for institutional (not individual) responsibility, differing values toward various environmental behaviors, lack of information about environmental behavior, and lack of strong leadership. Tarrant and Cordell (1997) suggest three explanations for the mismatch of attitudes and behaviors: the lack of congruence between the measures, reliance on "poor attitude measures" (Weigel 1983, 262), and failure to identify the external factors affecting behavior. The concern regarding external factors (for example, sociodemographics, knowledge, and politics) has been studied frequently (Weigel 1983; Stern 1992). In most cases, the sociodemographic and political variables have explained less than 10% of the variance in attitude or behavior (Wall 1995).

However, most of these studies have examined whether a behavior is completed, as opposed to comparing an environmental choice (such as saving wetlands) with a quality-of-life choice (like travel time) with a human health choice (for example, accident rate). In the study presented here, we assessed the factors affecting a specific choice of attributes, rather than a yes-or-no behavior—for instance, would people sacrifice 10 minutes of travel time to help save an endangered species? This kind of trade-off is similar to those expressed by Hardin (1968) in the "tragedy of the commons." Conflicts between individual good and community good also have been addressed by Dawes (1980) and Yamagishi (1994). Karp (1996, 112) states that "environmental protection presents a similar conflict between individual and collective preferences."

The risks or impacts examined in this study included attributes such as noise pollution, air pollution, historical preservation, and death rates from automobile accidents. Individuals were asked to rank different design alternatives for constructing a new road. Each alternative had a different combination of seven impact attributes (for example, travel time and wetland preservation). This setup is analogous to the process used to evaluate a federal road construction project in which different routes are compared. When a major federal project is studied, U.S. legislation does not determine which decisionmaking methods should be used or what impacts are more important. All that is required is that federal decisionmakers have "all" the necessary information for the decision (U.S. EPA 1978, Part 1500.1).

The study researched three questions:

1. Are people's rankings of alternatives (behavior) consistent with their rankings of impact attributes (concerns)?
2. Is there a relationship between stated concerns and demographics variables (for example, age, gender, income, and education)?
3. Do groups with similar interests vary significantly in their rankings of alternatives and impact attributes?

Study Design

A conjoint design provides the experimenter with a great level of statistical power to examine many different attributes with relatively few hypothetical alternatives (Roy 1990). The basis for a conjoint design is an orthogonal combination of attributes that is used to describe the different alternatives. Because the attributes are not correlated in the design, simple mathematical manipulations can determine the relative importance of the attributes to the respondents when ranking or rating the alternatives (Green and Krieger 1993). Care must be taken to ensure that the conjoint design creates only plausible combinations of attributes for each of the hypothetical alternatives.

Highway projects were selected for this study based on a review of 175 random environmental impact statements (EISs) (Follin 1998). They were the most prevalent major federal project (33 out 175) and involved various organizations (for example, the Federal Highway Administration, the U.S. Army Corps of Engineers, the National Park Service, the U.S. Coast Guard, and the Forest Service). Highway construction projects are understood by both laypeople and technical experts and have wide-ranging impacts, such as family displacement, noise pollution, accident rates, and loss of wetlands habitat.

In this study, eight alternatives for new highway construction to reduce congestion in a downtown area were created and presented to the participants, each described by a set of seven impact attributes that spanned health and safety, environmental, and quality-of-life risks selected from the following.

- Human health and safety risks:
  - accident rate
- Environmental risks:
  - level of air pollution
  - effect on species habitat
  - effect on wetlands
- Quality-of-life risks:
  - family and business displacement
  - effect on historic buildings

- noise levels
- travel time
- visual or aesthetic quality

These attributes were based on the impacts found in reviewing the EISs. Across the designs, impact attributes were set at one of two levels (for example, travel time through the city was set at 35 or 45 minutes). These levels were set so as to be realistic and allow for trade-offs between designs. Pilot studies with verbal protocols were used to determine appropriate values. Participants were asked to rank the eight alternatives based on their personal preference (they answered the question "Which highway design would you most like to see built?").

The nine attributes were divided into two groups of seven (the most permitted by a conjoint design with eight alternatives) creating two separate surveys, A and B. The surveys had five attributes in common (travel time, accident rate, effect on wetlands, displacement of family or business, and effect on endangered species habitat). Survey A included noise levels and effect on historical buildings, whereas Survey B included visual quality and air quality. Each attribute had two different levels of activity. Participants received either Survey A or Survey B.

Detailed Study Description

The survey consisted of three pages and eight alternative cards. The first page provided background on the highway project and explained the seven attributes that were used to describe each design alternative. Each attribute was discussed in a short paragraph that gave both a numerical and a relative description of how the attribute changed among the different alternatives. For example, one description read, "Some alternatives greatly reduce the number of sharp turns and dangerous intersections. The current accident rate is small (three fatal accidents on the road in 20 years), but calculations indicate that there would be a 5% reduction in accidents with these safer alternatives." This information allowed the participant to determine how important the accident-rate attribute was to him or her. The alternatives were given on individual cards made from heavy stock to encourage shuffling and sorting of alternatives. Figure 9-1 is an example alternative card from Survey A. The alternative setups for Surveys A and B are presented in Tables 9-1 and 9-2, respectively.

Participants were asked to order the cards, using whatever method or decisionmaking skills they wished, from best (most preferred) to worst (least preferred). This ranking was transcribed onto the survey sheet.

Alternative B
Travel Time Reduced to 35 minutes
Endangered Species No impact on striped long-necked snail
Accident Rate No change in accident rate
Displacements No displacements
Noise Level Long-term rise of 5–10 dBa
Wetlands Potential loss of 100 acres
Historic Buildings Several buildings would have to be destroyed

FIGURE 9-1. Alternative B Card from Survey A

Subsequent to a series of basic demographic questions (age, gender, income level, and education), participants were asked to state how many people (of 100 average citizens) would agree with their most preferred highway alternative. A high value would indicate that the participant felt that there was one obvious choice for the highway design; a low value would indicate that the participant felt that he or she was different from the majority of citizens.

Next, participants were asked to rank the importance of the attributes that they used in their alternative rankings. Each of the seven attributes was given a rank from 1 (most important) to 7 (least important). Again, participants

TABLE 9-1. Design of Alternatives for Survey A

	Alternative							
Attribute	*A*	*B*	*C*	*D*	*E*	*F*	*G*	*H*
Travel time: –, 45 minutes; +, 35 minutes	–	+	–	+	+	–	+	–
Endangered species: –, 50% loss of habitat; +, no loss of habitat	+	+	–	–	–	–	+	+
Accident rate: –, no reduction; +, 5% reduction	–	–	–	–	+	+	+	+
Displacements: –, 15–20 families and 3 businesses; + none	+	+	–	–	+	+	–	–
Noise level: –, rise of 5–10 decibels; +, no increase	+	–	–	+	+	–	–	+
Wetlands: –, potential loss of 100 acres; +, no loss	+	–	+	–	+	–	+	–
Historic buildings: –, several buildings destroyed; +, no impact	+	–	–	+	–	+	+	–

TABLE 9-2. Design of Alternatives for Survey B

	Alternative							
Attribute	*I*	*J*	*K*	*L*	*M*	*N*	*O*	*P*
Travel time: –, 45 minutes; +, 35 minutes	–	+	–	+	+	–	+	–
Endangered species: –, 50% loss of habitat; +, no loss	+	+	–	–	–	–	+	+
Accident rate: –, no reduction; +, 5% reduction	–	–	–	–	+	+	+	+
Displacements: –, 15–20 families and 3 businesses; +, none	+	+	–	–	+	+	–	–
Visual disturbance: –, affects local parks; +, no change	+	–	–	+	+	–	–	+
Wetlands: –, potential loss of 100 acres; +, no loss	+	–	+	–	+	–	+	–
Air quality: –, no impact; +, improved downtown air quality	–	+	+	–	+	–	–	+

were asked how many people (of 100 average citizens) would agree with their most important attribute.

Finally, participants were asked six questions dealing with their attitudes toward various subjects related to the environment and development:

- Are *historic sites* adequately protected from commercial development?
- Is there a correct balance between *business* or industrial development and concern for the environment?
- Is enough attention being paid to *global warming* concerns?
- Do you believe the current *speed limits* are appropriate?
- Do environmental policies and *regulation* affect economic development?
- Do today's policymakers adequately consider *long-term* consequences?

These questions are similar to, but independently derived from, the questions of the New Environmental Paradigm (Dunlap and Van Liere 1978), the Environmental Concern Scale (Weigel and Weigel 1978), and the Awareness of Consequences Scale (Stern, Dietz, and Kalot 1993). Stern (1992) and Stern and Oskamp (1987) suggest that these scales share significant common variance; thus, a general environmental scale may be appropriate. The questions chosen are more similar to the ranking task, not only to check on concern level but also to extend the environmental concerns to items such as traffic and historic buildings.

The respondents recorded their answers using a nine-point scale where for the first question, 1 signified "Too much protection" and 9 signified "Too little protection"; 5 was labeled "Just right."

Survey Results

Surveys A and B were conducted during the summer of 1998. The surveys were randomly administered and took approximately 20 minutes to complete. A total of 75 people completed Survey A, and 73 people completed Survey B. Once they started, all participants completed the survey. Approximately 75% of the people polled said that the ranking exercise was not more than moderately difficult.

Participants

To ensure that technical, environmental, and lay subjects were well represented in the data set, participants were recruited from various sources in southwestern Pennsylvania, including the Bettis Laboratory in West Mifflin, the Pittsburgh Audubon Society, the Carnegie Mellon University Earth Club, and the Allegheny County Jury Pool. Participants were divided into the three categories based on membership in an environmental group, employment in a technical (engineering or scientific) profession, or neither. No payment was given for participation from the jury pool. Candy bars were offered to other participants after they completed the survey.

Demographics

The participants for the two surveys were not significantly different in any major demographic category. The average participant age was 45 years old; 57% were male, and 43% were female. Education levels ranged from some high school to graduate degrees, and the most common level was a college degree. The most common income was between $50,000 and $100,000. Approximately 20% claimed to be members of environmental groups, and approximately 39% identified themselves as members of the technical community. (For additional details about the demographics, see Follin 1998.)

Ranking Results

Two different rankings were directly assessed from each participant: an alternative ranking (summarized in Table 9-3 for Survey A and Table 9-4 for Survey B) and an attribute ranking (summarized in Table 9-5 for Survey A and Table 9-6 for Survey B). Because of the conjoint design, it is possible to use an individual's alternative rankings to infer attribute ranking. Likewise, by making several simplifying assumptions, it is possible to use an individual's attribute ranking to infer alternative ranking. Comparisons of an individual's two alternative rankings (one directly assessed and one inferred) and two attribute rankings (one directly assessed and one inferred) can be used to measure consistency. The following sections outline the steps necessary to determine the inferred rankings.

Inferred Attribute Rankings

Because of the conjoint design, for each attribute, four of the eight alternatives have the "better" attribute value and four have the "worse" attribute value. For example, using Survey A (see Table 9-1), four alternatives have the faster travel times (B, D, E, and G), whereas the remaining four have the slower travel times (A, C, F, and H). By adding the ranks given to the four alternatives with the faster driving times, it is possible to determine the importance to the subject of better driving times. For example, if faster driving time is most important to a subject, he or she would rank the four alternatives with the faster times in the first four positions; the sum of the ranks would be 10 (that is, 1 + 2 + 3 + 4). However, if faster travel times are not important, a subject might rank the alternatives with faster times at the bottom; the sum of the ranks would be 26 (that is, 5 + 6 + 7 + 8). These sums of ranks are a measure of attribute importance and can be ranked to infer attribute rankings (see Table 9-7 for Survey A and Table 9-8 for Survey B).

To determine an overall attribute ranking for the two surveys, two methods are possible: individual rankings of attributes can be averaged and then ranked, or individual attribute importance scores can be averaged and then ranked. For Survey A, both methods yielded the same attribute ranking. For Survey B, minor differences resulted from the higher level rankings.

The direct and inferred attribute rankings were compared by using two methods: agreement of the most important attributes and calculation of the Spearman rank-order correlation. For 74% of the participants, the most important attribute was the same in both the direct and inferred attribute rankings. For 58% of the participants, the two most important attributes were the same between the two rankings. These values indicate a high consistency between the two attribute rankings. (If the two attribute rankings had been completely random, agreement would have been 14% and 3%, respectively.)

The results of the Spearman rank-order test were similar. For Survey A, the average correlation was 0.67, and for Survey B, it was 0.65.[1] Correlations greater than 0.714 are significantly different from zero at the 0.05 level. This value was exceeded by 75 of 146 people (51% of the sample). Only a few of the participants appeared to be unable to process the information in a consistent fashion and had negative rank-order correlations (7 of 146). One of these people expressed concern with the complexity of the task and the large number of attributes and alternatives.

Note that in some cases, people may have used only three of the attributes to decide on their alternative rankings (eight alternatives can be uniquely ranked using three binary attributes). Thus, this inferred technique does not necessarily provide a complete attribute ranking. Therefore, a more accurate appraisal of agreement may be provided by using the direct attribute ranking infer alternative ranking. This process is described in the next section.

Inferred Alternative Rankings

Inferred alternative ranking is slightly more complex than direct attribute ranking and requires the assumption of a linear multiattribute decision model for each subject; that is, each subject's value for an alternative is a weighted sum of its attributes. The ranks of the attributes were used to determine the weights (or multipliers) of the attribute values. Three different weighting methods were used.

- Linear weighting method: the attribute ranked 1 was assigned a weight of 7, the attribute ranked 2 was assigned a weight of 6, the attribute ranked 3 was assigned a weight of 5, and so forth.

TABLE 9-3. Direct Alternative Rankings from Survey A

	Ranking		
Alternative	*Times first*	*Overall mean*	*Mean rank*
A	43	2.36	1
B	2	4.13	4
C	0	6.57	8
D	2	5.40	6
E	17	3.22	2
F	0	4.79	5
G	10	3.25	3
H	1	6.24	7

TABLE 9-4. Direct Alternative Rankings from Survey B

	Ranking		
Alternative	*Times first*	*Overall mean*	*Mean rank*
I	28	2.75	2
J	5	3.75	3
K	0	5.78	7
L	0	6.84	8
M	32	2.10	1
N	1	5.52	6
O	5	4.00	4
P	2	5.26	5

TABLE 9-5. Direct Attribute Rankings from Survey A

	Ranking		
Attribute	*Times first*	*Overall mean*[a]	*Mean rank*
Accident rate	6	4.19	5
Displacements	18	3.08	1
Endangered species	12	3.90	4
Historic buildings	5	3.76	3
Noise levels	2	5.01	7
Travel time	16	4.49	6
Wetlands	16	3.49	2

[a] Because of ties in the rankings, totals may not be equal in each case.

TABLE 9-6. Direct Attribute Rankings from Survey B

	Ranking		
Attribute	*Times first*	*Overall mean*[a]	*Mean rank*
Accident rate	13	3.97	3
Air quality	4	4.12	4
Displacements	26	2.62	1
Endangered species	10	4.36	5
Travel time	7	4.82	7
Visual quality	1	4.73	6
Wetlands	13	3.44	2

[a] Because of ties in the rankings, totals may not be equal in each case.

TABLE 9-7. Inferred Attribute Rankings from Survey A

	Ranking		
Attribute	*Times first*	*Overall mean*[a]	*Mean rank*
Accident rate	4	4.39	7
Displacements	27	2.41	1
Endangered species	12	3.30	4
Historic buildings	10	3.22	3
Noise levels	4	4.12	6
Travel time	13	3.79	5
Wetlands	17	3.00	2

[a]Because of ties in the rankings, totals may not be equal in each case. If there was a tie, the tying scores were given the higher rating. Therefore, the mean scores are higher for the calculated rankings than for the self-reported rankings.

TABLE 9-8. Inferred Attribute Rankings from Survey B

	Ranking			
Attribute	*Times first*	*Overall mean*	*Mean rank*	*Mean weights*
Accident rate	8	4.21	7	5.5
Air quality	4	3.99	4	5.5
Displacements	28	2.49	1	1
Endangered species	12	3.30	3	3
Travel time	7	4.11	6	4
Visual quality	4	4.10	5	7
Wetlands	23	2.67	2	2

- Inverse weighting method: the attribute ranked 1 was assigned a weight of 1, the attribute ranked 2 was assigned a weight of ½, the attribute ranked 3 was assigned a weight of ⅓, the attribute ranked 4 was assigned a weight of ¼, and so forth.
- Three-attribute weighting method: the attribute ranked 1 was assigned a weight of 1, the attribute ranked 2 was assigned a weight of ½, the attribute ranked 3 was assigned a weight of ¼, all other attributes were assigned a weight of zero and were not included in the model.

An alternative ranking can be inferred for each weighting method by ranking the weighted-sum scores. These weighted-sum scores are calculated for each alternative by multiplying the weights by 0 for a low attribute value (for example, the slow commute time) and by 1 for the better attribute value (for example, the fast commute time) and then adding all attributes. The three different weighting methods will lead to different weighted sum scores and possibly different rankings of alternatives.

Spearman rank-order correlations between the direct and the three inferred alternative ranks were calculated for each participant. Correlations above 0.643 are significantly different than zero at the 0.05 level. Both the three-attribute model (mean correlation of 0.796) and the inverse model (mean correlation of 0.783) performed well, especially compared with the linear model, with its more uniform weights (mean correlation of 0.304). Only 2 of 146 participants had negative correlations with the three-attribute model.

A direct comparison of the inverse method versus the three-attribute method revealed that the three-attribute model provided the better score in more than half the cases. In Survey A, the three-attribute method was better in 43 cases, results of the two methods were identical in 12 cases, and the inverse method was better in 17 cases. In Survey B, the three-attribute method was better in 35 cases, results of the two methods were identical in 12 cases, and the inverse method was better in 26 cases. A simple binomial test indicated that the three-attribute test was better at the $p = 0.001$ level of significance.

Individuals appeared to make their alternative ranking decisions based on the values of two to three attributes. They established a preference for these attributes and ranked the alternatives logically.

Environment and Development Attitudes

Table 9-9 summarizes responses to the six questions about attitudes toward development and the environment. Note that every extreme position (that is, scores of 1 or 9) was held by at least one participant (for example, at least one participant felt that there was too much emphasis on the environment, and

TABLE 9-9. Responses to the Environment and Development Attitude Questions

Question	*Meaning of 1*	*Meaning of 9*	*Mean±SD*	*Minimum*	*Maximum*
"History"	Too much protection of historic sites	Too little protection of historic sites	5.98±1.52*	1	9
"Business"	Too much emphasis on business	Too much emphasis on environment	4.15±1.81*	1	9
"Warming"	Too little attention to global warming	Too much attention to global warming	4.76±2.27	1	9
"Speed"	Speed limits are too fast	Speed limits are too slow	5.42±1.70**	1	9
"Regulation"	Policies and regulation help economic development	Policies and regulation hurt economic development	5.90±1.63*	1.5	9
"Long-term"	Too little long-term emphasis by policymakers	Too much long-term emphasis by policymakers	3.29±1.74*	1	8.9

Note: SD = standard deviation. Mean is significantly different from 5 (one-sample *t*-test): $^{*}p < 0.001$; $^{**}p = 0.004$.

at least one other felt that there was too much emphasis on business). For this group, the average score for each question was significantly different from neutral (that is, a score of 5) except for the global warming question, for which there was a strong bimodal response (that is, half believing that too little attention was being paid and half believing that too much attention was being paid to the issue). On the surface, participants appeared to be in significant agreement that overall, current policies

- put too little focus on the protection of historical sites,
- place business interests above those of the environment,
- underemphasize the long-term perspective,
- implement regulations that hurt economic development, and
- set speed limits too low.

However, when group differences are examined, results indicate that the answers to the questions cannot be represented by a normal distribution and are quite often bipolar. Significant differences were found among groups (laypeople, environmentalists, and the technical community), as shown in Table 9-10. In almost all cases, attitudes noted by each group differed from those noted by the other groups. The laypeople often scored between the

TABLE 9-10. Group Differences in the Environment and Development Attitude Questions (Environmentalists, Laypeople, and Technical Community)

Question	*Environmentalists vs. laypeople*	*Environmentalists vs. technical community*	*Laypeople vs. technical community*
"History"	*	**	*
"Business"	**	**	**
"Warming"	**	**	**
"Speed limits"	NS	NS	*
"Regulation"	NS	**	**
"Long-term"	**	**	NS

Notes: $*p < 0.05$; $**p < 0.01$; NS = not significant.

environmentalists and technical community. The positions held by the groups were consistent with prior stereotypes (for example, the environmentalists felt that there was too little emphasis on the environment). The only category in which the three groups were in moderate agreement was on speed limits. The most significant differences were between the technical people and the environmentalists. A principal component factor analysis of five questions (because of the relative agreement, speed limit was omitted) found that the questions were adequately described by only one factor. The factor breaks down into a pro-environment (low score) and pro-development (high score) dimension that explains approximately 53% of the variance in the five questions.

Predictions of Agreement with Others

The survey asked the participants to identify how many people out of 100 (equivalent to asking for the percentage) would agree with their most important attribute and best alternative choices. There were two topics of interest regarding these predictions:

- Are the predictions of agreement with the general public accurate?
- Are the predictions correlated with the actual agreement?

Table 9-11 summarizes the results of these predictions and provides mean agreement percentages for both surveys. Overall, people believed that approximately 40% of the public would agree with both their best alternative choice and their most important attribute choice. (Note that a random-choice process would result in 14% agreeing with their selections.) However, the attribute ranked "most important" most often was selected only 25% of the time for Survey A (Table 9-5) and 36% of the time for Survey B (Table 9-6). The alternative ranked "best" most often was selected approximately 50%

TABLE 9-11. Predicted and Actual Agreement of Respondents

	Attribute agreement (%)			*Alternative agreement (%)*	
Survey	*Predicted*	*Actual (direct)*	*Actual (inferred)*	*Predicted*	*Actual*
A	45	19[a]	20[a]	43	40
B	42	22[a]	24[a]	43	36[b]

[a] Different from predicted value at the 0.001 significance level.
[b] Different from predicted value at the 0.01 significance level.

of the time on both surveys (Tables 9-3 and 9-4). These percentages serve as upper bounds on the level of agreement; in fact, for individuals who had not picked the most popular attribute or alternative, the level of agreement would be much lower. Statistical tests comparing the predicted agreement with actual agreements show that these differences are significant (Table 9-11). People believe that others will be more like them than they actually are, especially when it comes to predicting how others view the importance of impact attributes. People are better at predicting which alternative will be popular than predicting which impact will be viewed as most important.

A second analysis was done to determine whether people who expected higher agreement with others actually had higher agreement (and conversely, whether people who felt they were different than most were in fact different). In this case, no relationship was identified for Survey A and only weak relationships for Survey B, implying that the accuracy of individuals in predicting the opinions and preferences of others is quite poor. This result has design implications for the process by which agreements are reached across stakeholders.

Interpretation of Results

Returning to the three questions posed earlier, in this section we explore the implications of this study in relationship to broader policy questions. Applications and limitations are discussed.

Are Alternative Rankings Consistent with Attribute Rankings?

The participants in this study appeared to have a high level of consistency between their ranking of alternatives and stated preferences of impact attributes. Whether measured as the agreement between direct and inferred attribute rankings or direct and inferred alternative rankings, at least for this simplified task, it appears to be possible to rank alternatives by asking individuals their preferences for underlying impact attributes and then applying a relatively simple weighted-linear model. If this application holds true across more complex alternatives, then it would be possible to predict how individuals would rank alternatives without having to ask them to do so directly.

The benefits of such a procedure could be significant. Protocols for assessing impact attributes could be developed and given to a relevant population before knowing the complete set of possible alternatives that will be considered. Of course, before assessing an individual's preferences for the attributes, the range of attribute values that the alternatives span would have to be determined. In the simple case presented here, only binary levels of impacts were modeled. Real-world policy problems would require a much more thorough assessment. However, after the relative preferences are elicited, a decision model could be used to rank alternatives as they are formulated. Policymakers could determine a rough ordering of the alternatives. A final limited set of alternatives then could be presented and discussed in detail. Stakeholder concerns could be predicted and mitigated, reducing the likelihood of contentious and unproductive debates. Alternatives that are "nonstarters" could be eliminated before they grab unwarranted attention. If disparate attribute preferences are detected early in the process, an innovative set of alternatives could be constructed that would potentially satisfy multiple parties.

Can Demographic Variables Be Used to Predict Stated Concerns?

Because individuals were generally consistent in how they applied their preferences for the impact attributes in ranking alternatives, if these preferences were predictable based on demographic variables, a greatly simplified decision model could be developed. To test these relationships, general linear models were run using the calculated attribute weight as the dependent variable and the following independent variables:

- gender (two categories: male = 1, female = 2)
- age (four categories: 1, 2, 3, 4—older is higher number)
- education (five categories: 1, 2, 3, 4, 5—more education is higher number)
- income (four categories: 1, 2, 3, 4—more income is higher number)
- environmentalist (no = 0, yes = 1)
- technical background (no = 0, yes = 1)
- pro-environment/pro-development factor

Separate models were run for each of the attributes for each survey.

Although several variables had coefficients in the expected direction, none of demographic information predicted the attribute importance scores. This result is consistent with many of the environmental concerns papers (for example, Wall 1995). People's attitudes and preferences are complex, with many other factors coming into play, such as life experiences. Despite this

TABLE 9-12. Group Differences in Direct Attribute Rankings (Environmentalists, Laypeople, and Technical Community)

	Group		
Attribute	*Environmentalists vs. technical community*	*Environmentalists vs. laypeople*	*Laypeople vs. technical community*
Accident rate	NS	*	NS
Displacements	NS	NS	*
Endangered species	NS	**	NS
Travel time	**	*	NS
Wetlands	**	**	NS

Notes: *$p < 0.05$; **$p < 0.01$; NS = not significant.

lack of success, results shown in Table 9-12 indicate significant difference between the attribute choices of laypeople and the environmentalists. Based on our analysis, this difference cannot be attributed to any one variable but rather to an entire "package" of variables.

Are There Significant Differences in How Groups Rank Alternatives and Attributes?

Even though most subjects made reasonable rankings on the basis of simple attribute-preference models, subjects did not necessarily agree as to which attributes were important and which alternatives were best. Tables 9-3 and 9-4 show that in neither survey did one alternative capture more than 57% of the first-place ranks. In both surveys, six of the eight alternatives were ranked best by at least one person. The results are more diffuse when looking at the direct attribute rankings (see Tables 9-5 and 9-6). Here, no attribute received more than 35% of the first place rankings, and every attribute was ranked most important by at least one person. In this section, differences in rankings are analyzed from the group perspective.

This analysis is complicated by the different impact attributes (and therefore different alternatives) used in the two surveys. An alternative analysis must be done by separately for each survey, limiting the sample size and the statistical power of the tests. When comparing the alternative rankings in Survey A, there were no significant differences (at the $p < 0.05$ level) between laypeople and the technical community or between laypeople and the environmentalists. However, three differences were noted between the technical community and the environmentalists in Survey A. There were no differences in Survey B between the alternative rankings among the three groups.

However, the level of concern (as expressed by attribute choice) shows a larger set of differences. If the laypeople (nontechnical and nonenvironmen-

TABLE 9-13. Mean Direct Attribute Rankings for the Five Common Attributes

	Respondent's group		
Attribute	*Environmentalists (n = 29)*	*Technical community (n = 49)*	*Laypeople (n = 70)*
Accident rate	3.57	3.18	2.87
Displacements	2.57	2.59	2.12
Endangered species	2.68	3.22	3.49
Travel time	4.18	3.14	3.49
Wetlands	1.82	2.86	3.03

Note: n = number of respondents.

talist) were the baseline, then the other two groups differed in certain areas. In Survey A, the environmentalists differed (at the $p < 0.05$ level) in one attribute as self-reported (wetlands) and in a different attribute as calculated (endangered species). The technical community differed in two attributes as self-reported (displacements and historic buildings) but none as calculated. In Survey B, the environmentalists differed in three attributes as self-reported (accident rate, travel time, and wetlands) and in one category as calculated (visual quality). The technical community differed in one attribute as self-reported (visual quality) and in no attribute as calculated. However, more similarities than differences were apparent among the groups.

Because the two surveys share five common attributes (accident rate, displacements, endangered species, travel time, and wetlands), these attributes can be ranked using all the data. Table 9-13 shows the mean of the direct attribute ranking (where 1 is most important, 5 is least important). Table 9-12 shows the results of *t*-tests comparing the three groups. The major finding is that the environmentalists are different from the general public in four attributes (all but displacements), whereas the technical community is different from the general public in only one attribute (displacements). Note that for all but one attribute (travel time), laypeople's average rank fell between the other groups. The environmentalists and technical community had more extreme positions. However, laypeople ranked travel time as more important than any other group.

Conclusions

Several conclusions can be drawn from this simple study. The rankings of the alternatives as inferred from the attributes closely followed the direct rankings of the alternatives. Thus, the participants' stated concerns were clearly reflected in their behavior. The strength of relationship is robust and provides for much stronger predictions than simple population averages. This conclusion is different from those of several previous studies (Buttel 1987; Dunlap 1991; Gigliotti 1992; Scott and Willits 1994; Tarrant and Cordell 1997). The difference may be attributable to the limited the size of this exercise and the use of binary attribute levels. However, because of the conjoint design, this exercise removed an incongruity between the concerns (attributes) and the behaviors (chosen alternatives) that has been discussed by Tarrant and Cordell (1997) and illustrated by Ajzen and Fishbein (1980); Shetzer, Stackman, and Moore (1991); Scott and Willits (1994); and Weigel and Weigel (1978). Tarrant and Cordell (1997, 620) state that although "researchers are more often concerned with predicting behavior across a range of situations, understanding the attitude–behavior relationship at the general (rather than the specific) level may have the greatest utility" (Schwartz and Tessler 1972). However, for specific projects, such as road construction, the specific level may be appropriate and important. The predictability of simple decision models built from ranked concerns begins to support their application for certain domains of problems.

The fact that people's concerns and preferences for this problem were not predicted from basic demographic variables highlights the importance of not assuming stereotypes. People's concerns and actions are based on much more than gender and education (for example, opinions of older, well-educated, wealthy men are not homogeneous). Assuming that demographic variables are predictive could lead to gross overconfidence.

Overconfidence in being able to predict how others will act was also demonstrated with the self-assessed question regarding agreement with others. People sincerely felt that more than 42% of the population would agree with what they believed and their choice of alternative. The fact that it was less than 20% for some questions indicates that in a meeting context, people are going to be surprised by the "hostility" and "lack of cooperation" toward their position. The fact that this discrepancy between predicted and actual agreement was greater for concerns (reduction of accident rates versus loss of wetlands) than for the selection of the best alternative indicates that this misperception is at a fundamental level. However, the demonstrated better calibration in predicting the ranks of alternatives could be an artifact of the alternative descriptions used in this study.

Environmentalists in this study expressed different beliefs about the six questions to elicit information about attitudes toward development and the environment. For land-use questions, these differences can lead to testy disagreements about the correct emphasis that should be placed on business opportunities or the environment. Historically, these disagreements have led to litigation. Advocacy groups are responsible for approximately 37% of the litigation involved with EISs (CEQ 1991–1995). When the five common attributes from the two surveys were examined, the environmentalists were significantly different from the laypeople in four of the five attributes.

For this particular study, the attitudes of technical community (scientists and engineers) toward development and the environment also were different from those of the laypeople (that is, five of the attitude six questions were different). These differences could be important in land-use decisions, because the technical group collects and analyzes underlying data and risk information. By taking a perspective different from that of the general public, the technical group could easily influence which information is collected as well as how it is measured and presented. This realization emphasizes the need for public involvement in the earliest phases of a project when crucial issues are being scoped out. In this study, differences between the technical community and laypeople were much less when the five common attributes were compared (that is, only one displacement of five was significantly different). However, the technical group held more extreme positions for all the attributes except travel time.

The differences between the three groups and the lack of appreciation that these differences exist shows the need to involve laypeople, not only environmental advocacy groups and the technical community. Steg and Sievers (2000, 265) state, "Understanding individual differences in risk perception and risk judgments might facilitate the development of effective environmental risk-management strategies. Public measures might be more effective when they address the main determinants of the behavior and when the beliefs and perceptions of the target group are taken into account."

Can the findings from this study be applied in real life? The key conclusion is that policymakers have to conduct fairly detailed interviews or surveys to receive proper information. They cannot rely on the average of a large sample. Nor can they rely on the vocal activist to represent the general public. Finally, policymakers must be careful to ensure that the studies performed by the technical community provide the information required by the general public. Therefore, it is imperative that laypeople be involved in the decisionmaking process from start to finish.

Note

1. Because of ties with the inferred attribute ranking, in many cases, the rank order could not be 1, even if the rankings were completed using a consistent method. For example, if a participant used only three attributes in his or her alternative ranking, then the other four attributes would all be tied with a neutral score. In that case, the Spearman rank-order statistic would be 0.86, as opposed to 1.0. This "artificial" upper bound on the correlations is a limitation of the method.

References

Ajzen, I., and Fishbein, M. 1980. *Understanding Attitudes and Predicting Behavior.* Englewood Cliffs, NJ: Prentice-Hall.

Buttel, F.H. 1987. Age and environmental concern: A multivariate analysis. *Youth and Society* 10: 237–256.

CEQ (Council on Environmental Quality) 1991–1995. Annual reports. Washington DC: CEQ.

Dawes, R.M. 1980. Social dilemmas. *Annual Review of Psychology* 31: 169–193.

Dietz, T., Stern, P.C., and Guagnano, G.A. 1998. Social structural and social psychological bases of environmental concern. *Environment and Behavior* 30: 450–471.

Dunlap, R.E. 1989. Public opinion and environmental policy. In *Environmental Politics and Policy,* edited by J.P. Lester. Durham, NC: Duke University Press, 87–134.

———. 1991. Public opinion in the 1980s: Clear consensus, ambiguous commitment. *Environment* 33: 10–15, 32–37.

Dunlap, R.E., and Van Liere, K.D. 1978. The "new environmental paradigm": A proposed instrument and preliminary results. *Journal of Environmental Education* 9: 10–15.

Follin, J.N. 1998. Environmental risks, decision making, and public perception: A case study involving environmental impact statements. Doctoral dissertation. Pittsburgh, PA: Carnegie Mellon University.

Gigliotti, L.M. 1992. Environmental attitudes. *Journal of Environmental Education* 24(1): 15–26.

Green, P.E., and Krieger, A.M. 1993. Conjoint analysis with product positioning applications. In *Marketing Handbooks in Operations Research and Management,* edited by J. Eliashberg and G.L. Lilien. New York: North-Holland, Chapter 10.

Hallin, O. 1995. Environmental concern and environmental behavior in Foley, a small town in Minnesota. *Environment and Behavior* 27(4): 558–578.

Hardin, G. 1968. The tragedy of the commons. *Science* 162: 1243–1248.

Karp, D.G. 1996. Values and their effect on pro-environmental behavior. *Environment and Behavior* 28(1): 111–133.

Maloney, M.P., and Ward, M.P. 1973. Let's hear from the people. *American Psychologist* 28: 583–586.

Ostman, R.E., and Parker, J.L. 1987. Impact of education, age, newspapers, and television on environmental knowledge, concerns, and behaviors. *Journal of Environmental Education* 19: 3–9.

Roy, R. 1990. *A Primer on the Taguchi Method.* New York: Van Nostrand.

Schwartz, S.H., and Tessler, R.C. 1972. A test of a model for reducing measured attitude-behavior discrepancies. *Journal of Personality and Social Psychology* 24: 225–236.

Scott, D., and Willits, F.K. 1994. Environmental attitudes and behavior: A Pennsylvania survey. *Environment and Behavior* 28(2): 239–260.

Shetzer, L., Stackman, R.W., and Moore, L.F. 1991. Business-environment and the new environmental paradigm. *Journal of Environmental Education* 22(4): 14–221.

Smythe, P.C., and Brook, R.C. 1980. Environmental concerns and actions: A social-psychological investigation. *Canadian Journal of Behavioral Science* 12: 175–186.

Steg, L., and Sievers, I. 2000. Cultural theory and individual perceptions of environmental risks. *Environment and Behavior 32*(2): 250–269.

Stern, P.C. 1992. Psychological dimensions of global environmental change. *Annual Review of Psychology* 43: 269–302.

Stern, P.C., and Oskamp, S. 1987. Managing scarce environmental resources. In *Handbook of Environmental Psychology* (Volume 2), edited by D. Stokols and I. Altman. New York: John Wiley & Sons, 1043–1088.

Stern, P.C., Dietz, T., and Kalot, L. 1993. Value orientations, gender and environmental concern. *Environment and Behavior* 25(3): 322–348.

Tarrant, M.A., and Cordell, H.K. 1997. The effect of respondent characteristics on general environmental attitude-behavior correspondence. *Environment and Behavior* 29(5): 618–637.

U.S. EPA (Environmental Protection Agency). 1978. Council on Environmental Quality. *Code of Federal Regulations,* Parts 1500–1517, Title 40.

Van Liere, K.D., and Dunlap, R.E. 1981. Environmental concern: Does it make a difference how it's measured? *Environment and Behavior* 13(6): 651–676.

Wall, G. 1995. General versus specific environmental concern: A Western Canadian case. *Environment and Behavior* 27(3): 294–316.

Weigel, R.H. 1983. Environmental attitudes and the prediction of behavior. In *Environmental Psychology: Directions and Perspectives,* edited by N.R. Feimer and E.S. Geller. New York: Praeger.

Weigel, R.H., and Weigel, J. 1978. Environmental concern: The development of a measure. *Environment and Behavior* 10: 3–5.

Yamagishi, T. 1994. Social dilemmas. In *Sociological Perspectives on Social Psychology,* edited by K.S. Cook, G.A. Fine, and J. House. Boston, MA: Allyn and Bacon, 311–334.

10

The Use of Public Risk Ranking in Regulatory Development

MICHAEL L. DEKAY, H. KEITH FLORIG, PAUL S. FISCHBECK, M. GRANGER MORGAN, KARA M. MORGAN, BARUCH FISCHHOFF, AND KAREN E. JENNI

Government agencies charged with regulating or otherwise managing health, safety, and environmental risks must somehow set priorities for addressing the many risks in their domain of responsibility. Although legislation and legislature-approved budgetary authority determine the broad outlines of an agency's mission, considerable latitude remains in the attention and resources that an agency devotes to particular problem areas. Over the past 10 or 15 years, risk management agencies have begun to experiment with more open prioritization processes in which input is sought from external experts, the lay public, and other stakeholders (Stern and Fineberg 1996).

The U.S. Environmental Protection Agency (EPA) has been especially active. In 1986, its staff ran the groundbreaking *Unfinished Business* project (U.S. EPA 1987), which compared the agency's actual allocation of regulatory attention with a ranked list of the risks that senior staff considered most important. This effort was followed by two agency-wide risk-ranking projects conducted by EPA's Science Advisory Board (U.S. EPA 1990, 2000). In addition, the EPA's Regional and State Planning Bureau has supported approximately 50 local and regional "comparative risk" projects (U.S. EPA 1993), the results of which have been summarized elsewhere (Minard and Jones with Paterson 1993; Davies 1996; Delhagen and Dea 1996; Feldman 1996; Feldman, Perhac, and Hanahan 1996; Minard 1996; WCCR 1996; Feldman, Hanahan, and Perhac 1997; Jones 1997; Jones and Klein 1999; Konisky 1999; GMI n.d.).

Elsewhere in the U.S. government, the Occupational Safety and Health Administration (OSHA) has undertaken a broad priority-setting exercise that involves considerable input from stakeholders in labor and industry as well as from the scientific community (OSHA 1996). The Federal Aviation Administration (FAA) has begun an effort to prioritize its safety initiatives (McKenna 1998), and a committee of the Institute of Medicine has recommended greater and more systematic public participation in the establishment of research priorities at the National Institutes of Health (Institute of Medicine 1998). Outside the United States, several countries have pursued rankings of their environmental priorities (for example, Environment Canada 1993; New Zealand Ministry for the Environment 1996), and others have expressed interest in such ranking.

When done well, comparative risk efforts can simultaneously make decisionmaking about risk management more public and more systematic by providing an appropriate structure for integrating the public's values with relevant scientific information. As noted in the introduction to this volume and elsewhere (for example, Davies 1996), there is much to be gained from keeping the "big picture" in mind while focusing on the specifics of the various risks being evaluated. Of course, incorporating public opinion in a systematic manner is only one of several possible motivations that agencies might have for sponsoring risk-ranking exercises (Delhagen and Dea 1996; Feldman, Perhac, and Hanahan 1996; Jones 1997). They also might hope to promote risk communication, foster dialog among stakeholders, or—more cynically—delay politically difficult choices. Nonetheless, these risk-ranking efforts represent a major shift in the way that risk management agencies integrate science and values to inform their policy decisions.

Despite the number and scope of such activities, risk ranking is still in an early stage of methodological development. In 1993, following a recommendation by the Carnegie Commission on Science, Technology, and Government (1993) for the wider use of risk ranking as an input to decisionmaking about risk management, the Office of Science and Technology Policy in the Executive Office of the President asked several research groups to propose methods for the federal government to rank risks within and across agencies. In response, members of our research group proposed a set of procedures that involve eliciting explicit preferences from lay groups (M.G. Morgan and others 1996). In this chapter, we describe the conceptual and methodological challenges faced in developing the method and report results from four studies designed to assess its validity and usefulness.

Overview of the Carnegie Mellon Approach to Risk Ranking

Our proposed method for ranking risks has five interdependent steps (Figure 10-1). First, the risks to be ranked are grouped into a manageable number of

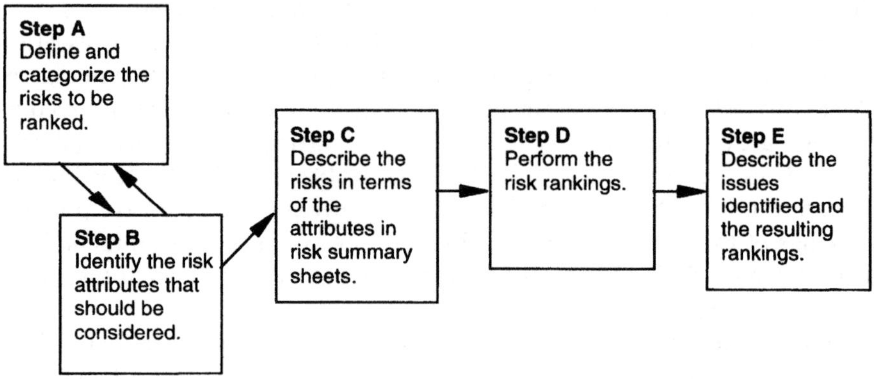

FIGURE 10-1. Steps in the Proposed Risk-Ranking Method
Source: Adapted from Florig and others forthcoming.

categories (Step A). Although risk management agencies typically are responsible for thousands of specific hazards, it is infeasible for laypeople (or even staff) to compare and rank such a large number. We refer to the resulting risk categories simply as "risks" to avoid cumbersome language.

Second, a set of attributes is selected as the basis for characterizing the risks (Step B). These attributes include both quantitative measures such as expected mortality and more qualitative measures such as controllability. Because the way that risks are categorized depends in part on which attributes are used to describe them, these first two steps are likely to involve an iterative process.

Third, each risk is characterized in terms of each attribute. This information is combined with narrative descriptions to create *risk summary sheets* that provide concise, systematic information in a format that facilitates risk comparisons (Step C). The technical literature and risk experts are used to ensure the accuracy and completeness of each risk summary sheet.

Fourth, the risk summary sheets are used by groups of laypeople to rank the risks (Step D). In this process, participants produce both individual and group rankings. Before any group meetings, individual rankings are elicited to help participants articulate their own values. The groups are intended to provide an airing of different views to help participants evaluate and refine their own opinions. Individual rankings also are collected after the group ranking to track the effect of the group discussion on individual judgements and to assess the extent of individual dissention from the group ranking. In full versions of the method, participants complete both a *holistic ranking procedure,* in which they rank the risks directly, and a *multiattribute procedure,* in which they judge the relative importance of the different risk attributes and these judgements are used to generate an implied ranking of the risks

(Keeney and Raiffa 1976; von Winterfeldt and Edwards 1986). Assessing participants' opinions in two ways provides an internal consistency check and may help allay concerns regarding the dependence of the results on the particular method used to obtain the rankings. In contrast to Steps A–C, in which risk experts play a key role, the actual rankings in Step D are performed by groups of laypeople. The rationale for this division of labor is that the characterization of risk requires expert technical knowledge, whereas the ranking involves social value judgements that are more appropriately left to members of the public or their representatives (Hammond, Harvey, and Hastie 1992). Both good science and considered values are necessary for informed risk ranking; neither is sufficient individually.

Finally, when the rankings have been completed, a "thick description" of the results is prepared (Step E) so that decisionmakers can interpret them appropriately (for example, with knowledge of which risks were contentious).

In developing this method, we have chosen to follow the Carnegie Commission, EPA, and others in ranking risks rather than risk management options. Although one would ultimately prefer to rank risk management options (Graham and Hammitt 1996), doing so would greatly complicate the ranking task, in part because the number of possible risk management options far exceeds the number of risks. Also, in cases where risks are more stable over time than management options, a ranked list of risks may have a longer shelf life. For example, sulfur air pollution has been a concern for many decades, during which time the technologies to control or remove it from the air have undergone frequent changes. Nonetheless, such a ranking is only one factor that decisionmakers need to consider when setting risk management priorities. Risks with middle and low ranks still deserve action if they can be effectively reduced at small cost. Conversely, if little can be done to reduce a highly ranked risk, managers should not spend resources on it that could provide more protection if invested elsewhere. However, the high rank could signal the need to invest in research that could eventually make cost-effective management possible.

In contrast to most of the state and local comparative risk exercises conducted to date, we have decided to focus on lay participants drawn from the general public rather than on "stakeholders" representing particular groups. (For an example of how the results of a ranking exercise might differ for laypeople, environmentalists, and members of the technical community, see Chapter 9.) Our hope was to develop a method that would be broadly applicable to citizens willing to commit to learning about risk issues, despite their lack of prior involvement. Actual applications would depend on the setting. Sometimes, stakeholder rankings might provide a consensus that translates into political action. Other times, the goal might be to stimulate broader public discussion involving individuals who are not bound by their groups' public positions. We view our approach as being generally consistent with the National Research Council's analytic–deliberative process (Stern and Fineberg 1996; see also CSA 1997; Royal Commission on Environmental Pollution 1998) and with approaches that use stakeholders for the initial elicitation of concerns and evaluative criteria but use citizen groups for the final evaluation process (Renn and others 1993; Renn 1999).

Development of the Centerville Middle School Test Bed

Although this project builds on theory and empirical results from behavioral social science, any proposed methodology for ranking risks requires empirical evaluation. To that end, we developed an experimental test bed concerning risks at the hypothetical Centerville Middle School (CMS). A school environment was selected because it involves a wide range of physical, chemical, biological, and social risks and because most adults have had extensive personal experience in the school setting, as former students and perhaps as parents.

In the information provided to participants, CMS is described as a public school that serves 430 seventh-, eighth-, and ninth-grade boys and girls and is located in a suburban residential community in the Midwestern United States. The two-story brick building was built in 1971 and is in a good state of repair. The first floor includes administrative offices, 14 classrooms, a cafeteria, an auditorium, a library, and utility space; the second floor includes 22 classrooms, several special purpose rooms, a gymnasium, and a swimming pool. Cooking and heating are provided with natural gas. The school grounds are fenced and include a parking lot, athletic fields, a track, basketball hoops, and some playground equipment. A two-lane suburban street, a four-lane divided expressway, railroad tracks, and a high-voltage power line are located nearby. The neighborhood has little crime, and the school grounds are monitored by local police as part of their regular patrol route.

The sections that follow describe the design challenges addressed in applying our proposed method to the CMS test bed. Florig and others (forthcoming) provide additional detail on CMS, including links to a sketch and floor plan of the building and a map of the surrounding community. M.G. Morgan and others (2000) and Florig and others (forthcoming) provide additional detail on the categorization of risks, the selection of attributes, and the development of the risk summary sheets.

Categorization of Risks

The problem of categorizing risks for ranking has received little formal attention in EPA's comparative risk exercises. They usually start with a list of environmental problems published in its guidebook (U.S. EPA 1993), in which the basis for risk categorization is not described. In EPA's list, drinking water supplies and groundwater are singled out as resources that can be degraded, but air is not. Some point sources of pollution are categorized according to the entity responsible for creating the risk, but others are not. Municipal wastewater and solid waste are distinguished from industrial wastewater and solid waste, but a similar public–private distinction is not made for sources of air pollution. Accidental chemical releases are distinguished from routine releases, but this distinction is not made for other hazards. Several categories are defined by the physical or chemical agent directly responsible for harm (for example, radon, "other" radiation, particulate matter, sulfur dioxide, ozone, pesticides, lead, or carbon dioxide), but hundreds of other agents are not explicitly listed. Finally, Resource Conservation and Recovery Act (RCRA) hazardous waste sites are distinguished from Superfund sites, presumably because they are administered differently, even though other characteristics of the risks from the sites are very similar. A survey of participants in comparative risk projects revealed similar concerns, including claims that "problem areas were too broad, lacked common measurement criteria, and failed to distinguish past and present activities" (Feldman 1996, 12).

Clearly, the results of risk-ranking efforts can be very sensitive to the way that risks are initially grouped (for example, small risks may escape attention if viewed separately, but not if aggregated; conversely, large risks may be divided into oblivion). Using mixed criteria for defining categories (for example, affected resources in some cases and human health endpoints in others) can result in double counting. Such logical problems are perhaps less troubling if comparative risk exercises put "a higher premium on using the list as a risk-communication tool rather than on its intellectual consistency" (Minard 1996, 46) or use the project as a vehicle to promote stakeholder interaction (Delhagen and Dea 1996; Feldman, Perhac, and Hanahan 1996; Jones 1997), although confusing definitions still may erode support. Ultimately, it is difficult to imagine sound regulatory policy based on incoherent or inappropriate categories.

Some insights into the problem of categorization can be gleaned from the research literatures in psychology (Barsalou 1983; Medin and Smith 1984; Medin and Ortony 1989; Keil 1989; Komatsu 1992), natural science (Sokal 1974), and risk analysis (Cvetkovich and Earle 1985; Webler and others 1995). Komatsu (1992) makes a key distinction between "similarity-based" and "explanation-based" approaches to categorization. *Similarity-based approaches* place a particular instance into a category based on the instance's similarity to an idealized category member (a "prototype"), to other specific category members ("exemplars"), or to both. *Explanation-based approaches* link category members together according to some underlying scenario, knowledge structure, or purpose. Barsalou (1983) provides the example of a category consisting of children, pets, photo albums, family heirlooms, and cash—the set of things one should grab when leaving the house in the event of a fire. Cvetkovich and Earle (1985, 7–8) note that the explanation-based approach to categorization (or the "constructivist" approach, to use their term) "considers classification systems as aids to thinking and communicating and assumes that there is no single generally best classification system. The quality of a classification system depends upon how well the functions of analysis and communication are performed. Thus, whether one system is better than another depends upon the specific aims of its user."

Given the wide variety of dimensions along which risks can be considered similar or dissimilar, similarity-based categorization of risks is inherently ambiguous. In an explanation-based approach, on the other hand, the goals of risk managers may be translated into categories relatively directly. We adopted the explanation-based approach while attempting to meet the additional goals that risk categories be logically consistent, compatible with administrative systems, equitable, and compatible with human cognition (Fischhoff 1995; M.G. Morgan and others 2000). Table 10-1 provides addition detail on these desirable, but sometimes conflicting, goals.

Our specific approach conceptualizes the explanations underlying categorization strategies in terms of causally linked risk processes (M.G. Morgan 1981). These processes begin with human or natural activities (for example, the release of a pollutant or an accident-initiating event) that can give rise to environmental loadings. The loadings, in turn, lead to exposure and effects processes, which are then perceived and valued by people. Fischhoff and others (1978a), Cvetkovich and Earle (1985), and Webler and others (1995) have adopted similar strategies for risk categorization, although they do not consider the role of categorization in risk ranking per se. Figure 10-2 illustrates these causally linked processes and, using air pollution risks as an example, shows criteria that could guide categorization at each stage. Each element, in turn, could be expanded to include particular secondary categorization factors. For example, the kind of facility creating an environmental loading (for example, a power plant or steel mill) might be supplemented by the facility's organizational or legal status (for example, profit, nonprofit, cooperative, or municipal).

Categories are likely to be most useful when focused on the point in the chain of risk processes at which management intervention is most commonly or efficiently applied. For risk management organizations that concentrate on the early stages of the chain, risks should be sorted according to human activities, initiating events, and environmental loadings. However, this categorization may not directly address the risk endpoints that concern many people. These endpoints might require secondary divisions, such as separat-

TABLE 10-1. Desirable (but Sometimes Conflicting) Characteristics of an Ideal Risk-Categorization System for Ranking Risks

Categories for risk ranking should be
Logically consistent
Exhaustive, so that no relevant risks are overlooked
Mutually exclusive, so that risks are not counted twice
Homogeneous, so that all risk categories can be evaluated on the same set of attributes
Administratively compatible
Compatible with existing organizational structures and legislative mandates
Relevant to management, so that risk priorities can be mapped into risk-management actions
Many, so that regulatory attention can be finely targeted
Compatible with existing databases, to make the best use of available information in the preparation of risk summary sheets
Equitable
Fairly drawn, so that the interests of various stakeholders, including the general public, are balanced
Compatible with cognitive constraints and biases
Chosen with an awareness of inevitable framing biases
Simple, so that risk categories are easy to communicate
Few, so that the ranking task is tractable
Free of the "lamp-post" effect, in which more-understood risks are categorized more finely than less-understood risks

Source: Adapted from M.G. Morgan and others 2000.

ing risks that have chronic and acute effects, that have significant catastrophic potential, or that have global versus local effects (for example, chlorofluorocarbons and greenhouse gases versus other trace pollutants). Because they involve very different regulatory authorities, indoor air pollutants and radionuclei might be considered separately from other air pollutants.

When categorizing risks for the CMS test bed, we considered several organizational principles for grouping the risks:

- agent (for example, radon, infectious disease, or school bus accident)
- activity giving rise to the hazard (for example, transportation, recreation, or education)
- location (for example, classroom, athletic field, or commuting route)
- pathway (for example, ingestion, inhalation, or physical trauma)
- endpoint (for example, injury, disease, or death)
- group at risk (for example, students, teachers, or maintenance workers)
- entity responsible for creating the risk (for example, students, school management, industry, or nature)

Risk process

Possible associated categorization

by pollutant source	by pollutant species	by pathway	by health or ecological end point	by normative attributes

Illustration for air pollution

power plants, steel mills	SO_x, NO_x, PM	outdoor air, indoor air	lung disease, agricultural yield	age, sex, race, income; ecological status

FIGURE 10-2. Examples of Alternative Categorization Schemes that Could Be Used at Different Points along the Causal Chain of Risk Processes

Note: Examples given are for air pollution risks, but the basic categorization strategies can be applied to any risk domain.

Source: Adapted from Morgan and others 2000.

- entity most responsible for managing the risk (for example, students, parents, State Department of Education, or County Department of Health).

Primarily because of data limitations, we restricted the problem to risks to students and risks arising from action, but not from inaction (for example, not having antismoking programs or quality counseling services). Following the above logic, our categorization scheme began with the agents (say, automobile accidents) that would initiate the processes leading to the undesirable consequences (such as deaths and injuries). Many of the risks considered have multiple determinants. For example, the risks from hazardous materials transport on nearby highways or railroads may be caused by inadequate maintenance of the infrastructure or vehicle (a social or economic agent), drowsy or inattentive operators (a biological or psychological agent), or a specific chemical agent. For chemicals, we focused on the proximal agent (hazardous chemicals) but aggregated across chemical types (to keep the number of categories manageable) and differentiated hazardous materials transport from on-site storage of swimming pool chemicals (which were included in the more general category "accidents").

Additional distinctions separated risks that seemed substantially different. For example, we divided infectious diseases into "common" and "less common" categories (for example, colds versus pneumonia and meningitis) and separated commuting in school buses from independent commuting (for example, by car, by bike, or on foot) because of the different accident rates and responsible parties.

In the interest of simplification, we ignored interactions between different risks (for example, that school bus riders may be at greater risk of catching an infectious disease). Finally, our focus on risks at school obviously downplayed the more distal determinants of some risks (for example, the driving forces behind suicide or intentional injury). Table 10-2 shows the resulting list of 22 risk categories.

Selection of Risk Attributes

Although most (but not all) comparative risk projects have attempted to describe risks in terms of a common set of attributes to facilitate comparison (for example, cancer risks, noncancer risks, or risks to special populations [Jones 1997]), the attributes selected have varied substantially across projects. Even when attributes are obvious (for example, deaths and injuries), the metrics for capturing them still require policy choices (Crouch and Wilson 1982; Fischhoff, Watson, and Hope 1984). For example, should

TABLE 10-2. Risk Categories for the Centerville Middle School Test Bed

Accidental injuries (excluding sports)
Airplane crashes
Allergens in indoor air
Asbestos
Bites and stings
Building collapse
Common infectious diseases
Commuting to school on foot, by bike, or by car
Drowning
Electric and magnetic fields from electric power
Electric shock
Fire and explosion
Food poisoning
Hazardous materials transport
Intentional injury
Self-inflicted injury or harm
Lead poisoning
Less common infectious diseases
Lightning
Radon gas
School bus accidents
Team sports

Source: Adapted from Florig and others forthcoming.

expected mortality be reported as the number of deaths expected per year in a geographic area or in an occupation, or as the chance in a million that an average person in that area or occupation will be killed?

The set of attributes that people care about has been the subject of extensive research (Fischhoff and others 1978b; Slovic, Fischhoff, and Lichtenstein 1980, 1985, 1986; Slovic 1986). As a first cut, this research suggests grouping attributes under two headings, or factors, on the basis of their intercorrelations. One of these factors reflects knowledge of the risk and captures attributes such as "known to science," "known to those exposed," "observability," "newness," and "immediacy of effects." The other factor, which has been called "dread," reflects attributes such as "controllability," "catastrophic potential," "severity," "risk to future generations," "equity," and "voluntariness." A third factor that reflects societal and personal exposure and is correlated with mortality and morbidity sometimes emerges in these studies.

Because the attributes associated with each factor are highly correlated, any of them could represent the factor. We selected two attributes from each factor to allow participants to attend to the one most relevant to their personal concerns and to get a richer feeling for the factor as a whole. Our choice was guided by the strength of the normative argument for relying on the attribute, its comprehensibility to participants, and the quality of the data for assessing it (Fischhoff and others 1978b; Jenni 1997). Specifically, we chose two attributes to represent the knowledge factor ("quality of scientific understanding" and "time between exposure and health effects") and two attributes to represent the dread factor ("greatest number of deaths in a single episode" and "ability of student or parent to control exposure"). To these, we added three other attributes related to deaths. Two of these captured the expected risk of death to the average CMS student using different metrics ("number of deaths per year" and "chance in a million of death per year for the average student"), whereas the third addressed the equity issue by noting that not all students face the same level of risk ("chance in a million of death per year for the student at greatest risk"). We also added three attributes for illnesses and injuries: "number of illnesses and injuries per year," "number of disability days per year," and "number of hospital days per year."

On the basis of several individual and group risk-ranking exercises, we made two major revisions to this list. First, because participants desired more information on the nature of injuries and illnesses, we dropped the three original attributes in favor of a two-by-two division distinguishing severity ("less serious" versus "more serious") and duration ("short term" versus "long term"). The four new attributes reported the number of cases of illnesses and injuries in each of these categories. Second, because some participants were confused by our presentation of uncertainty in the estimates for deaths, illnesses, and injuries, we created a third attribute to represent the knowledge factor. Labeled "combined uncertainty in death, illness, and injury," it was a weighted average of the uncertainties for these attributes. It was expressed quantitatively, as a number similar to a coefficient of variation, and qualitatively, as "low," "medium," or "high." These revisions yielded a final set of 12 attributes: 4 focused on mortality, 4 focused on morbidity, and 4 focused on other factors.

Development of Risk Summary Sheets

Although all comparative risk projects have provided information to participants regarding the risks to be ranked, these materials have seldom (if ever) been designed in a way that facilitates comparisons on all of the attributes under consideration. To make informed decisions, individuals must have adequate information, presented in a consistent, comprehensible format.

To this end, we prepared a summary sheet for each CMS risk, designed to be concise, to allow subjects to learn about many risks in a short time; nontechnical, to foster comprehension; and consistent in terms of information and formatting, to facilitate comparisons. We put this information on pages small enough for participants to perform the ranking by sorting and re-sorting the risk summary sheets themselves. The first page of the 6.5 × 10-inch design is shown in Figure 10-3. At the top of the page is a simple identifying label, followed by a short qualitative description of the risk. The lower half of the page is a summary table with values for each of the 12 attributes, along with high and low estimates for death and illness or injury.

The second page (not shown) provides a general discussion of the risk, including what is known and not known about it, relative to the attributes. At the top of the third page (not shown) is a qualitative and quantitative discussion of the risk in the specific context of CMS, including comparisons with relevant risk norms (for example, the concentration of radon at which EPA recommends mitigation). At the bottom of that page is a description of what actions the school has taken to deal with the risk. This information is intended to provide a realistic context; make it clear that the exercise concerns residual risks; and focus the ranking on how serious the risks are, regardless of the feasibility or cost of additional risk management.

In preparing these sheets, we drew on the available literature (for example, U.S. Congress, OTA 1995), supplemented by interviews with technical and school experts. Estimating mortality and morbidity often required both modeling and judgement. References, assumptions, and calculations were recorded in separate technical documents that served as internal references. Two technically trained risk analysts and two nonexperts reviewed each risk summary sheet. The nonexpert reviews included tape-recorded read-aloud protocols to allow us to identify miscommunications. We also asked nonexpert reviewers to offer suggestions for improving the presentations.

Accidental Injuries (excluding sports)

Summary:

Each year, 1 in 4 middle-school children seek medical treatment for accidental injury. About 23% of these injuries are sports-related. The rest result mainly from motor vehicle accidents, being struck or cut by some object, and falls. About 10-15% of all accidental non-sports injuries to middle school students occur in school or in the school yard. Although most injuries are minor, serious accidents occur often enough to make accidental injuries a leading cause of death, disability, hospitalization, and missed school days among children in the U.S. In addition to the risk of physical injury from accidents, CMS students are at smaller risk of injury from accidental exposure to chlorine gas used to treat the swimming pool.

Accidental Injury risk for Centerville Middle School*

	Low estim.	**Best estimate**	High estim.
Student deaths			
Number of deaths per year	.003	**.006**	.009
Chance in a million of death per year for the average student	7	**14**	21
Chance in a million of death per year for the student at highest risk	7	**14**	21
Greatest number of deaths in a single episode		**10 - 20**	
Student illness or injury			
More serious long-term cases per year	.002	**.004**	.009
Less serious long-term cases per year	.05	**.10**	.25
More serious short-term cases per year	.12	**.25**	.60
Less serious short-term cases per year	5	**10**	25
Other Factors			
Time between exposure and health effects		**immediate**	
Quality of scientific understanding		**high**	
Combined uncertainty in death, illness, injury		**0.3 (low)**	
Ability of student/parent to control exposure		**low**	

*See "Notes on the Numbers" for definitions and explanations of assumptions.

FIGURE 10-3. Example of the First Page of a Risk Summary Sheet

Notes: Additional information about the risk is included on the two inside pages. One risk summary sheet was prepared for each of the 22 risks at Centerville Middle School (CMS).

To facilitate consideration of all risk attributes, we also provided participants with an 11 × 17-inch chart with all 22 risks ranked separately, according to each attribute. For example, the list for "number of deaths per year" had the risk with the highest expected mortality rate (commuting to school on foot, by bike, or by car) at the top and the risk with the lowest expected mortality rate (lead poisoning) at the bottom. In addition to facilitating comparisons along each attribute, the chart emphasized the fact that risks that rank high on one attribute might rank low on another.

Empirical Studies Using the Centerville Middle School Test Bed

In addition to the materials development work described above, we have completed four studies involving the CMS test bed.

The first study involved 112 Carnegie Mellon undergraduates and assessed the contribution of the text and table portions of the risk summary sheets to individuals' rankings of 12 of the 22 risks (K.M. Morgan 1999). Participants received text-only materials (a risk label plus a one-paragraph description), table-only materials (the attribute table from the risk summary sheet, with a generic risk label), text-plus-table materials (the risk label, the one-paragraph description, and the attribute table, but none of the additional narrative from the summary sheets), or full summary sheets. Participants also received information describing the CMS context, including the sketch, map, and floor plans. After reading all of the materials in the packet, participants ranked the risks according to their level of concern.

The second study involved 24 laypeople from the Washington (District of Columbia) and Pittsburgh (Pennsylvania) areas and used the full set of 22 risks (K.M. Morgan, Fischbeck, and DeKay 1999). Both the holistic ranking procedure (in which participants ranked the risks directly) and several different multiattribute procedures (in which participants indicated the relative importance of the different attributes and this information was used to construct implied rankings) were used. Specifically, each participant's attribute weights were used to construct concern scores for each risk by using the following formula:

$$\text{Concern}_j = \sum_{i=1}^{n} w_i \times v_i(x_{ij})$$

where j is a risk, i is an attribute, n is the number of attributes, w_i is the weight for attribute i, v_i is the value function for attribute i, and x_{ij} is risk j's level on attribute *i*. In a complete multiattribute assessment, the value functions are assessed separately for each attribute, and the attribute weights are assessed by using one of several techniques (for example, "swing weights"; see Keeney and Raiffa 1976; von Winterfeldt and Edwards 1986). We elicited this information but also investigated several simpler, less time-consuming approximations. The concern scores were computed for each risk and then ranked for comparison with the holistic rankings.

All of the rankings collected in this study were from individual participants. Participants generated holistic rankings before providing the inputs necessary to construct the multiattribute rankings. The multiattribute rankings based on the complete assessment were shown to the participants, who then resolved the discrepancies between their holistic and multiattribute rankings to produce their final rankings of the 22 risks. Two informative analyses were performed on these judgements. The first used multiple regression to assess the extent to which participants used the results of the multiattribute exercise when generating their final rankings. The second assessed the extent to which a complete multiattribute utility assessment could be replaced by simpler estimation procedures without loss of fidelity. The level of consistency between the holistic and multiattribute rankings within individuals was used as a measure of convergent validity for this purpose; the higher the correlation between holistic and multiattribute rankings, the better.

The third study involved 218 risk management professionals enrolled in a short course entitled "Analyzing Risk: Science, Assessment, and Management" at the Harvard School of Public Health (K.M. Morgan and others forthcoming). The study assessed (a) the consistency between holistic and multiattribute rankings at the individual and group levels, (b) participants' satisfaction with the ranking procedures and the results, and (c) the agreement among individuals and among groups. Participants were divided into 43 groups, each of which considered 9 or 11 of the 22 risks. Participants ranked the risks individually, then in groups, and again as individuals. The final individual rankings were included to assess changes of opinion and participants' dissention from their groups' rankings.

The fourth study involved 86 laypeople from the Pittsburgh area, split among 11 groups, and used the full set of 22 risks (Fischbeck and others 2000). We used several procedures in these sessions, looking for the one that captured the most information without unduly burdening the participants. In the standard procedure, participants completed their individual holistic rankings, provided information necessary to construct individual multiattribute rankings, and then reconciled the differences between these two rankings to generate individual revised rankings. They then repeated these procedures as groups and again as individuals (to assess changes of opinion and participants' dissention from their groups' rankings).

Several key results from these four studies provide insight into the validity and usefulness of our proposed method for ranking risks. First, with the undergraduate participants in the first study, agreement among participants' holistic rankings (as assessed by the mean Spearman correlation [r_s] among rankings) was higher when the risk summary sheets included the attribute tables. Agreement was highest for the table-only condition (mean $r_s = 0.48$), second highest for the full-summary-sheet condition (mean $r_s = 0.44$), third highest for the text-plus-table condition (mean $r_s = 0.35$), and least for the text-only condition (mean $r_s = 0.15$). (Participants in the other studies—all of whom received full summary sheets—typically had higher correlations among initial holistic rankings, ranging from 0.6 to 0.7, depending on the risk set.) When average rankings were created for each of these four conditions, the table-only ranking was very similar to a ranking by expected mortality ($r_s = 0.94$). Average rankings for the full-summary-sheet and text-plus-table conditions were very highly correlated with each other ($r_s = 0.99$) but somewhat less correlated with expected mortality ($r_s = 0.78$ and 0.76, respectively). The average ranking from the text-only condition was essentially unrelated to expected mortality ($r_s = -0.11$) or to the rankings from the other conditions ($r_s = -0.10$ to 0.22). Thus, participants who are provided with both the text and table appear to incorporate both sources of information into their rankings of the risks and are more likely to agree regarding the appropriate ranking.

Second, high levels of agreement between holistic and multiattribute rankings indicate that the multiattribute ranking procedure can be greatly simplified without loss of fidelity. Using the correlation with individuals' holistic rankings as a measure of convergent validity, simple multiattribute models that rely on estimation procedures for single-attribute utility functions and attribute weights perform as well as (or better than) more difficult and time-consuming elicitation procedures. Specifically, in one procedure that appears to work quite well, the weight for each attribute is based on the reciprocal of the attribute rank (Barron and Barrett 1996) and the value function for each attribute is based on the ranking of the risks' levels on that attribute. This result should encourage the use of multiattribute techniques in real-world comparative risk projects, because the procedure reduces the time and effort required from both participants and facilitators.

Third, we have used this simple procedure in our more recent risk-ranking exercises and have observed mean correlations of about 0.6 between individuals' initial holistic and multiattribute rankings. These correlations typically increase as the result of group discussion, but this result may depend on the set of risks under consideration. Consistency between the holistic and multiattribute procedures is similarly high when group rankings are considered. These results suggest that participants and groups have internally consistent views of the relative riskiness of hazards or that they construct internally consistent views as they work with the risk summary sheets.

Fourth, multiple-regression results indicate that when given the chance, both individuals and groups revise their initial holistic rankings on the basis of their constructed multiattribute rankings. These revisions occur even among individuals and groups who have already been exposed to the attribute information when generating their holistic rankings. Thus, the multiattribute procedures may help participants to focus on the attributes in a way that is missing from the holistic procedure.

Fifth, individuals are quite satisfied with their groups' decisionmaking processes and would strongly support using these procedures and the resulting rankings in risk management decisionmaking. Although participants often disagree with some aspects of their groups' rankings, their final individual holistic rankings usually are closer to their groups' holistic rankings than to their own initial individual holistic rankings. Thus, participants support their groups' rankings implicitly as well as explicitly.

Sixth, the level of agreement among individuals (as assessed by the mean Spearman correlation among individuals' holistic rankings) increases from about 0.6 to more than 0.8 as a result of the group discussion. Across groups considering the same set of risks, agreement is also quite high, with a mean correlation of about 0.8, regardless of whether the groups are composed of risk management professionals or laypeople. Such high levels of agreement indicate that the rankings that result from this process are quite replicable.

Seventh, regression analyses indicate that (a) individuals and groups pay substantial attention to expected mortality and expected illnesses and injuries and (b) they also incorporate other more qualitative risk attributes into their rankings. However, different individuals and groups do not always agree on which of these attributes should be included (or even which way they should be weighted). More work is needed to determine whether these attributes are viewed similarly in tasks that involve risk perception and risk ranking.

Conclusions

Since EPA's initial effort some 15 years ago (U.S. EPA 1987), numerous comparative risk projects have been conducted at all levels of government. These projects represent a real step forward for two reasons: because they strive to evaluate a variety of risks in a systematic way that allows regulators to see the forest as well as the trees, and because they recognize the interdependence of science and values in creating sensible policy. As we have noted, however, the methods used to conduct and document these risk-ranking efforts leave substantial room for improvement. The research reported in this chapter suggests several specific ways by which the state of the art can be advanced.

First, much more attention needs to be paid to risk categorization, given the sensitivity of results to these early choices. Poor choices can increase the cognitive burden on participants and distort the results of the ranking exercise. It is often most useful to focus initial categorization efforts on those points in the chain of risk processes where risk management is most feasible. Thus, comparative risk projects may have greater influence on risk management decisions if the risk categories are based on activities, initiating events, and environmental loadings rather than on subsequent environmental and health effects.

Second, the information describing the risks should be provided in a concise, easy-to-understand format that facilitates direct comparisons of risks in terms of their important attributes. These attributes need to be determined at an early stage of a comparative risk project, so that suitable data can be gathered. Characterizing risks in a clear and consistent manner might have communication and management value, even if the materials are never used in formal risk-ranking exercises.

Third, comparative risk projects should strive to include multiattribute procedures. Doing so takes greater advantage of the effort spent on specifying the attributes of the risks, by increasing the chance that participants will attend to these details. The value of multiattribute procedures can be seen in the fact that our participants and groups revised their holistic rankings on the basis of constructed multiattribute rankings and expressed satisfaction with the process and the resulting rankings. It appears that these benefits can be obtained even with simplified multiattribute procedures that require less time and effort on the part of participants and facilitators.

Fourth, individuals' and groups' rankings and opinions should be assessed at several points during the exercise, because doing so provides insight into the quality of the risk-ranking process and results. These assessments measure participants' and groups' internal consistency, participants' implicit and explicit satisfaction, the sensitivity of rankings to procedural changes, and the stability of decision processes and resulting rankings across groups.

In our view, these improvements would greatly enhance the validity and usefulness of comparative risk projects and increase the likelihood that meaningful and legitimate policy decisions would follow from the results of such efforts. Indeed, the rankings obtained in our studies of risks to middle school students have several desirable features that would afford them credibility as inputs to risk management decisionmaking: rankings are highly correlated within and across individuals, levels of consistency and agreement increase as the exercise proceeds, and independent groups that consider similar sets of risks come to very similar conclusions about how those risks should be ranked.

Although our studies have focused on ranking health and safety risks in a rather limited domain, this promising approach may be expanded and adapted to address issues in a much wider range of settings. Following the lead of numerous comparative risk projects, an obvious extension is to incorporate ecological, social, and other quality-of-life risks and their attributes into a more integrated risk framework. The fundamental issues are similar to those already discussed for health and safety risks, but several vexing practical concerns also must be addressed. For example, the science regarding such risks is in many ways less mature, more complex, and more uncertain than that for health and safety risks and therefore may be more difficult to capture accurately in concise, comprehensible risk summary sheets. This problem is compounded by the fact that we know relatively little about the attributes that are appropriate for describing such risks.

Assuming for the moment that these problems can be solved, at least on the ecological side (DeKay and Willis 2000; Willis and others 2000), it is tempting to imagine numerous other domains in which risks to health, safety, and the environment could be ranked against each other. For example, how would the FAA go about evaluating all of the risks associated with air transportation? Of course, these sorts of exercises do not have to be limited to government agencies. How would the managers of an individual industrial facility (or a firm, or an entire industry) evaluate the risks associated with their operations? Although it certainly would be a huge challenge to compile and condense the data necessary to describe the relevant risks, we believe that a model of the type depicted in Figure 10-1 would be appropriate for such an undertaking.

Finally, it would seem desirable to move beyond "simply" ranking risks to ranking options for risk management. However, ranking management options presents numerous difficulties, some of which were mentioned earlier in this chapter. The problem that the number of risk management options far exceeds the number of risks might be solvable through creative categorization of the options, particularly if some options can be used to address several different risks simultaneously. But that is far from given. A popular alternative is to compute cost-effectiveness estimates for a few management options for an individual risk (for example, comparing different therapies for treating or preventing a particular medical condition), but the cumulative application of this approach may never amount to the systematic, programmatic comparison of risk management activities that is the holy grail of researchers and policymakers alike. At this stage, simplification of the grand prioritization problem is absolutely necessary. And of the possible ways in which the problem might be simplified, ranking risks as opposed to risk management options seems a very reasonable choice, particularly if risk-ranking exercises are conducted in a way that maximizes their usefulness to risk management decisionmakers.

Acknowledgements

This work was supported by an EPA Science to Achieve Results (STAR) fellowship and by grants from National Science Foundation (SRB-9512023 and SES-9975200), EPA (R8279200-1-0), Electric Power Research Institute (W02955-12), Alcoa Foundation, and Chemical Manufacturers Association.

We thank Jun Long, Claire Palmgren, Patti Steranchak, and Henry Willis for advice and assistance.

The views expressed are those of the authors.

References

Barron, F.H., and B.E. Barrett. 1996. Decision Quality Using Ranked Attribute Weights. *Management Science* 42: 1515–1523.

Barsalou, L.W. 1983. Ad Hoc Categories. *Memory and Cognition* 11: 211–227.

Carnegie Commission on Science, Technology, and Government. 1993. *Risk and the Environment: Improving Regulatory Decision Making.* New York: Carnegie Corporation.

Crouch, E.A.C., and R. Wilson. 1982. *Risk/Benefit Analysis.* Cambridge, MA: Ballinger.

CSA (Canadian Standards Association). 1997. *Risk Management.* Report CSA-850. Ottawa, Ontario, Canada: CSA.

Cvetkovich, G., and T.C. Earle. 1985. Classifying Hazardous Events. *Journal of Environmental Psychology* 5: 5–35.

Davies, J.C. 1996. *Comparing Environmental Risks: Tools for Setting Government Priorities.* Washington, DC: Resources for the Future.

DeKay, M.L., and H.H. Willis. 2000. Public Perceptions of Environmental Risks. Presented at the Annual Meeting of the Society for Judgment and Decision Making, New Orleans, LA, November 18–20.

Delhagen, E., and J. Dea. 1996. *Comparative Risk at the Local Level: Lessons from the Road.* Boulder, CO: Western Center for Environmental Decision-making.

Environment Canada. 1993. *Environmental Issue Ranking: A Proposed Priority Setting Methodology for Environment Canada.* Ottawa, Ontario, Canada: Conservation and Protection Service, Ecosystem Sciences and Evaluation, Economics and Conservation Branch.

Feldman, D.L. 1996. *Environmental Priority-Setting through Comparative Risk Assessment.* Knoxville, TN: University of Tennessee, Energy, Environment, and Resources Center.

Feldman, D.L., R. Perhac, and R.A. Hanahan. 1996. *Environmental Priority-Setting in the U.S. States and Communities: A Comparative Analysis.* Knoxville, TN: University of Tennessee, Energy, Environment, and Resources Center.

Feldman, D.L., R.A. Hanahan, and R. Perhac. 1997. *Subnational Comparative Risk Projects: An Analysis of their Risk Management Phase.* Knoxville, TN: University of Tennessee, Energy, Environment, and Resources Center.

Fischbeck, P.S., M.L. DeKay, B. Fischhoff, M.G. Morgan, H.K. Florig, C.R. Palmgren, and H.H. Willis. 2000. Evaluating a Risk-Ranking Methodology. Presented at the Annual Meeting of the Society for Risk Analysis, Arlington, VA, December 3–6.

Fischhoff, B. 1995. Ranking Risks. *Risk: Health, Safety and Environment* 6: 189–200.

Fischhoff, B., C. Hohenemser, R.E. Kasperson, and R.W. Kates. 1978a. Can Hazard Management Be Improved? *Environment* 20: 16–20, 32–37.

Fischhoff, B., P. Slovic, S. Lichtenstein, S. Read, and B. Combs. 1978b. How Safe Is Safe Enough? A Psychometric Study of Attitudes towards Technological Risks Benefits. *Policy Sciences* 9: 127–152.

Fischhoff, B., S. Watson, and C. Hope. 1984. Defining Risk. *Policy Sciences* 17: 123–139.

Florig, H.K., M.G. Morgan, K.M. Morgan, K.E. Jenni, B. Fischhoff, P.S. Fischbeck, and M.L. DeKay. Forthcoming. A Deliberative Method for Ranking Risks (I): Overview and Test-Bed Development. *Risk Analysis.*

GMI (Green Mountain Institute for Environmental Democracy). n.d. *Comparative Risk Documents.* http://www.gmied.org/PUBS/papers/crdocs/crdocs.html (accessed December 10, 2000).

Graham, J.D., and J.K. Hammitt. 1996. Refining the CRA Framework. In *Comparing Environmental Risks: Tools for Setting Government Priorities,* edited by J.C. Davies. Washington, DC: Resources for the Future, 93–109.

Hammond, K.R., L.O. Harvey, and R. Hastie. 1992. Making Better Use of Scientific Information: Separating Truth from Justice. *Psychological Science* 3: 80–87.

Institute of Medicine. 1998. *Scientific Opportunities and Public Needs: Improving Priority Setting and Public Input at the National Institutes of Health.* Washington DC: National Academy Press, Committee on the NIH Research Priority-Setting Process.

Jenni, K.E. 1997. *Attributes for Risk Evaluation.* Doctoral Dissertation, Department of Engineering and Public Policy. Pittsburgh, PA: Carnegie Mellon University.

Jones, K. 1997. *A Retrospective on Ten Years of Comparative Risk.* Montpelier, VT: Green Mountain Institute for Environmental Democracy.

Jones, K., and H. Klein. 1999. Lessons from 12 Years of Comparative Risk Projects. *Annual Review of Public Health* 20: 159–172.

Keeney, R.L., and H. Raiffa. 1976. *Decisions with Multiple Objectives.* New York: Cambridge University Press.

Keil, F.C. 1989. *Concepts, Kinds and Cognitive Development.* Cambridge, MA: MIT Press.

Komatsu, L.K. 1992. Recent Views of Conceptual Structure. *Psychological Bulletin* 112: 500–526.

Konisky, D.M. 1999. Comparative Risk Projects: A Methodology for Cross-Project Analysis of Human Health Risk Rankings. RFF Discussion Paper 99-46. Washington, DC: Resources for the Future.

McKenna, J.T. 1998. Industry, FAA Struggle to Steer Agenda. *Aviation Week and Space Technology* April 27: 60–61.

Medin, D.L., and A. Ortony. 1989. Psychological Essentialism. In *Similarity and Analogical Reasoning,* edited by S. Vosniadou and A. Ortony. New York: Cambridge University Press, 179–195.

Medin, D.L., and E.E. Smith. 1984. Concept and Concept Formation. *Annual Review of Psychology* 35: 113–138.

Minard, R.A., Jr. 1996. CRA and the States: History, Politics, and Results. In *Comparing Environmental Risks: Tools for Setting Government Priorities*, edited by J.C. Davies. Washington, DC: Resources for the Future, 23–61.

Minard, R.A., Jr., and K. Jones, with C. Paterson. 1993. *State Comparative Risk Projects: A Force for Change.* South Royalton, VT: Northeast Center for Comparative Risk.

Morgan, K.M. 1999. *The Development and Evaluation of a Method for Risk Ranking.* Doctoral Dissertation, Department of Engineering and Public Policy. Pittsburgh, PA: Carnegie Mellon University.

Morgan, K.M., P.S. Fischbeck, and M.L. DeKay. 1999. Assessing a Multi-attribute Model for Ranking Risks. Presented at the Annual Meeting of the Institute for Operations Research and the Management Sciences, Philadelphia, PA, November 7–10.

Morgan, K.M., M.L. DeKay, P.S. Fischbeck, M.G. Morgan, B. Fischhoff, and H.K. Florig. Forthcoming. A Deliberative Method for Ranking Risks (II): Evaluation of Validity and Agreement among Risk Managers. *Risk Analysis.*

Morgan, M.G. 1981. Choosing and Managing Technology-Induced Risk. *IEEE Spectrum* 18(12): 53–60.

Morgan, M.G., B. Fischhoff, L. Lave, and P. Fischbeck. 1996. A Proposal for Ranking Risk within Federal Agencies. In *Comparing Environmental Risks: Tools for Setting Government Priorities*, edited by J.C. Davies. Washington, DC: Resources for the Future, 111–148.

Morgan, M.G., H.K. Florig, M.L. DeKay, and P.S. Fischbeck. 2000. Categorizing Risks for Risk Ranking. *Risk Analysis* 20: 49–58.

New Zealand Ministry for the Environment. 1996. *Towards Strategic Environmental Priority Setting: Comparative Risk Scoping Study.* Wellington, New Zealand: Ministry for the Environment.

OSHA (Occupational Safety and Health Administration). 1996. *The OSHA Priority Planning Process.* http://www.osha.gov/oshinfo/priorities/index.html (accessed December 10, 2000).

Renn, O. 1999. A Model for an Analytic-Deliberative Process in Risk Management. *Environmental Science and Technology* 33: 3049–3055.

Renn, O., T. Webler, R. Horst, P. Dienel, and B. Johnson. 1993. Public Participation in Decision Making: A Three-Step Procedure. *Policy Sciences* 26: 189–214.

Royal Commission on Environmental Pollution. 1998. *Setting Environmental Standards.* London, U.K.: Her Majesty's Stationery Office.

Slovic, P. 1986. Perception of Risk. *Science* 236: 280–285.

Slovic, P., B. Fischhoff, and S. Lichtenstein. 1980. Facts and Fears: Understanding Perceived Risk. In *Societal Risks Assessment: How Safe Is Safe Enough?* edited by R.C. Schwing and W.A. Albers. New York: Plenum Press, 181–216.

———. 1985. Characterizing Perceived Risk. In *Perilous Progress: Managing the Hazards of Technology*, edited by R.W. Kates, C. Hohenemger, and J.X. Kasperson. Boulder, CO: Westview Press, 91–125.

———. 1986. The Psychometric Study of Risk Perception. In *Risk Evaluation and Management*, edited by V.T. Covello, J. Menkes, and J. Mumpower. New York: Plenum Press, 3–24.

Sokal, R. 1974. Classification: Purposes, Principles, Progress, Prospects. *Science* 185: 1111–1123.

Stern, P.C., and H.V. Fineberg. 1996. *Understanding Risk: Informing Decisions in a Democratic Society.* Washington, DC: National Academy Press, National Research Council, Committee on Risk Characterization, Commission on Behavioral and Social Sciences and Education.

U.S. Congress, OTA (Office of Technology Assessment). 1995. *Risks to Students in School.* OTA-ENV-633. Washington DC: U.S. Government Printing Office.

U.S. EPA (Environmental Protection Agency). 1987. *Unfinished Business: A Comparative Assessment of Environmental Problems.* Washington, DC: U.S. EPA, Office of Policy Analysis.

———. 1990. *Reducing Risk: Setting Priorities and Strategies for Environmental Protection.* Washington, DC: U.S. EPA, Science Advisory Board.

———. 1993. *Guidebook to Comparing Risks and Setting Environmental Priorities.* Washington, DC: U.S. EPA, Office of Policy, Planning, and Evaluation.

———. 2000. *Toward Integrated Environmental Decision-Making.* Washington, DC: U.S. EPA, Science Advisory Board, Integrated Risk Project Steering Committee.

von Winterfeldt, D., and W. Edwards. 1986. *Decision Analysis and Behavioral Research.* New York: Cambridge University Press.

WCCR (Western Center for Comparative Risk). 1996. *A Review of Ecological Assessments from Five Comparative Risk Projects.* Boulder, CO: WCCR.

Webler, T., H. Rakel, O. Renn, and B. Johnson. 1995. Eliciting and Classifying Concerns: A Methodological Critique. *Risk Analysis* 15: 421–436.

Willis, H.H., M.L. DeKay, P.S. Fischbeck, B. Fischhoff, H.K. Florig, M.G. Morgan, and C.R. Palmgren. 2000. Extension of the Carnegie Mellon Risk-Ranking Test Bed to Include Environmental and Ecological Factors. Presented at the Annual Meeting of the Society for Risk Analysis, Arlington, VA, December 3–6.

PART III

Uncertainty and Technology

Uncertainty permeates regulatory decisionmaking. Prescriptive codes that outline specifically how to design, build, and operate a system safely have an air of certainty, but how safe are the systems actually? Aging, unplanned uses, and poor maintenance can erode well-intentioned safety levels. With new and evolving technologies, the problem of controlling risks becomes even more difficult. Improved sensors and measurement techniques can identify previously unknown risks. New materials and designs can open opportunities for better performance and reduced costs, but at what risk? Any balancing of benefits and risks must acknowledge the inherent uncertainties that surround both. The numerous sources of these uncertainties range from a lack of scientific knowledge to an inability to predict human behavior. The muddying effects of uncertainty complicate already difficult value-laden trade-offs among choices. This section explores how technologies and uncertainty interact and create new challenges for policymakers.

In Chapter 11, Notarianni and Fischbeck investigate the steps necessary to adopt a performance-based fire code. Although such a code is touted as a way to stimulate creative designs and engineering systems that increase safety and reduce costs, how does an engineer or architect prove that a design is safe enough? Determination of "acceptable" risk levels is highly dependent on how people behave in the presence of fire, something that is unlikely to be well understood or predicted. Taking a very different perspective, Ponce and others demonstrate in Chapter 12 that with new technologies, it is not always people's behavior that determines the risk; it can be their genetic codes.

Should otherwise qualified workers be blocked from certain jobs based only on genetic testing? This question sets up a discussion of the far-reaching societal implications of genetic screening.

In Chapter 13, Corbett and Fischbeck first identify international shipping as a major source of global pollution and then demonstrate how different technologies can be used to reduce it. Determining which technology is "best" depends on many factors, including cost and benefit trade-off ratios, stakeholder perspective, geographic scope, and time frame. Uncertainty about the significance of the problem, the cost and performance of emissions control devices, and the effectiveness and compliance of different policies complicates the analysis. In Chapter 14, Macauley and Brennan investigate new technologies that are used not to reduce actual emissions but to improve the assessment and monitoring of environmental compliance. Locations that are currently prohibitively expensive to monitor could be checked using remote sensors, but how is the relevant information processed, and with what accuracy? This technology advance could lead to whole new classes of regulations.

These chapters raise numerous questions for research and discussion. While reading, you might consider questions such as these:

- When are performance-based regulations easy to verify? What types of problems are difficult to verify? When are computer simulations good enough?
- How should performance standards be written when uncertainty about human behavior is significant? Should they assume typical behavior of an average person?
- If it is appropriate to use genetic testing to screen job applicants based on potential occupational health risks, why not screen on their ability to do the job?
- How certain do genetic tests have to be to be used in job screening? Should a person be denied a job if he or she has a 2% increase in risk over another applicant?
- Because emissions control costs for transportation systems are passed on to consumers and comprise only a few percent of the final product price, why should much stronger controls not be mandated?
- Should emissions control technologies or emissions standards be mandated? Which would cost less, which would be easier to monitor, and which would be more effective?
- How could remote sensing lead to a more cost-efficient policy for pollution control (dollars per ton removed)? Why might industry be in favor of increased surveillance?

11

Performance with Uncertainty

A Process for Implementing Performance-Based Fire Regulations

KATHY NOTARIANNI AND PAUL S. FISCHBECK

Fire poses a significant societal risk for the United States, causing close to 5,000 deaths a year (Hall 1996) and costing $8 billion annually in property losses (National Fire Data Center 1992). If all fire-related costs are included (for example, paying for the fire protection services, additional construction costs, and fire insurance), then the annual costs for the United States exceed $120 billion (Hall 1998; Meade 1991).

In an effort to reduce the cost of fire safety, stimulate innovation, and increase design flexibility, the U.S. fire and building code system—following the lead of several other countries (for example, Building Code of Australia)—is currently undergoing a regulatory transformation from a system of prescriptive codes to one that uses performance-based codes. The current prescriptive codes specify fire protection requirements based on the construction of the building and its intended use (for example, an elementary school must have a certain type of fire sprinklers in specific locations). In contrast, performance-based standards specify the results, not the means, of regulatory compliance (for example, at least five minutes of safe egress time must be provided). Potentially, these changes will have many benefits. With performance-based codes, the debate will focus around the actual objective or risk. Societal goals, design objectives, and performance requirements will be explicit and open to public debate. Society's expectations of the level of safety provided in buildings will become part of the discussion, unlike under prescriptive-based codes, in which discussions more likely center on design-specific issues with no direct link to regulatory objectives.

Performance codes and standards work best when actual performance can be evaluated to determine compliance—for example, the output performance of a pump or motor can be verified easily with laboratory testing. However, in fire safety, direct measurement of the performance of a building or building systems is not usually possible. To do so, full-scale prototypes of the building would have to be built and then burned under various scenarios (Brannigan and Smidts 1998). Even if this option were not cost prohibitive, it would be impossible to determine and test all possible fire scenarios. Performance evaluators are limited to using computer models of buildings and fire behavior. This situation is further complicated by many other factors, such as poorly substantiated loss statistics, incomplete and inadequate fire models, difficulty in predicting human behavior in fires, the presence of many stakeholders with competing objectives, the multidisciplinary nature of the fire problem, the fragmented nature of the construction industry, the lack of technically trained building code officials, and our legal system.

Uncertainty plays a major role in all aspects of a performance-based design. There is uncertainty in the fire model's physics, in the input parameter to the model, in people's values when it comes to fire safety, and in human behavioral responses to fire. Identification, characterization, and a rigorous methodology are needed to handle these uncertainties. The method must satisfy the needs of designers, code officials, code developers, policymakers, and researchers. Accounting for unknowns and variations is necessary to provide confidence in the final design.

However, in addition to choosing the right model, a systematic process involving multiple stakeholders is needed to evaluate designs. In this chapter, we discuss the importance of uncertainty, identify the barriers that currently hinder the adoption of new standards, and propose a process by which performance-based standards can be applied to fire protection systems.

Identifying the Types of Uncertainty

Uncertainty can play an important role in the development and implementation of fire safety regulations. Beyond being just another step in the process of getting a building approved, properly determining and documenting a level of confidence in the design will have numerous benefits. The treatment of uncertainty is key to ensuring and maintaining an appropriate level of public safety while allowing the flexibility necessary to reduce costs. This is true for all fire safety engineering calculations, whether conducted to meet a performance-based code, aid in the establishment of a prescriptive requirement, or compare performance options with their prescriptive counterparts.

The quantitative treatment of uncertainty will facilitate cooperation among stakeholders by increasing the overall understanding of risks and costs. Distributions of outcomes are a much richer description of what is possible than the typical point value answers. Although stakeholders and policymakers still must determine how much risk to accept, with thorough uncertainty analyses, these decisions will be informed and free of the uneasiness that typically surrounds acceptance of deterministic performance calculations.

In this section, we describe five general categories of uncertainties that should be considered in a complete performance-based design.

Scientific Uncertainties

Scientific uncertainties are caused by both lack of knowledge (in the underlying physics, chemistry, fluid mechanics, or heat transfer of the fire process) and the necessary approximations required for operational practicality of a fire model or calculation. There are roughly five subcategories of scientific uncertainties: theory and model uncertainties, data and input uncertainties, calculation limitations, level of detail, and representativeness of the design fire scenarios.

Uncertainties and Variability in Human Behavior

Human behavioral uncertainties concern both how people act in a fire and how their actions should be considered during steps in the design process (in defining project goals, selecting performance criteria, and developing and evaluating candidate designs). For example, human behavior in response to a fire alarm must be modeled in terms of time to respond to the alarm and the kind of response: Does the person immediately begin to evacuate the building? Does he or she take the stairs or the elevator? What factors enter into that choice? Does the person try to fight the fire? Does he or she stop to gather personal possessions or to call a neighbor? During egress, does a person use the best exit or the most familiar one? How long does a person take to start to exit? What range of occupant characteristics (such as age and handicaps) should be considered?

For example, two very different scenarios could ensue from a cooking fire:

- A grease fire from cooking sets off a smoke detector, which alerts the occupant, who reacts and properly extinguishes the fire while it is still small.
- The occupant forgets about a pot simmering on a burner, takes a sleeping aid, and goes to bed; the overheated pot ignites, and the fire spreads to one or more adjacent items.

Uncertainties and Variability in Risk Perceptions and Values

There is both variability and uncertainty in the way people perceive and value risk. Capturing differences that people have in their perceptions and values related to risk is a necessary step in the design process. Research has shown that even though people typically view consequences from voluntary risks less severely than equal consequences resulting from an unknown or involuntary risk, there is variability (Starr 1969). For example, whereas some people would agree that an increase in risk to fire fighters (people who accept risk as part of their job) is justifiable if a corresponding decrease in risk to the public could be achieved, others would not.

Few studies have been conducted that clearly demonstrate how society values fire safety risks at the level needed to support performance-based trade-offs. (Some work on incorporating risk concepts and identifying levels of acceptable risk is discussed in Meacham 2000.) It is important to identify where value judgments enter into a performance-based calculation and to make any assumptions explicit regarding the impacts of different values on the final design.

Uncertainties in the Life-Cycle Use and Safety of Buildings

Many factors change over the lifetime of a building. The uncertainties surrounding future use, occupancy, and other factors contribute to the difficulty in conducting a structured performance-based design. Even daily fluctuations in these design parameters can affect the safety of a building. For example, a building or area of a building that is normally occupied 24 hours a day may become unoccupied (or occupied by very different people) for extended periods of time as a result of extraneous factors (business closing, maintenance, or renovation). The characteristics of the occupants (for example, elderly or handicapped) can lead to very different design considerations.

Other changes that may affect the life-cycle safety of the building are fire service characteristics, such as distance of the building to the nearest firehouse and expected response time of the fire department.

Equity and Incorporation of Societal Values

There is uncertainty as to how risks and benefits will be distributed across the population and the many possible subgroups. Most projects entail many stakeholders (building owner, design engineer, architect, and code official) and the public (users of the building and its neighbors); how should society value the wants and needs of each party when making design decisions for fire protection? Should risks to children and risks to adults be treated differently? If, by changing the focus or importance of different subgroups, the acceptability of a design changes, then this sensitivity must be made explicit. All stakeholders should appreciate the values that are being incorporated in the performance-based design process.

Obstacles in Performance-Based Fire Regulation

Aside from the considerable obstacle of confronting and integrating uncertainty into the design and evaluation process for a fire protection system, several other significant barriers make the successful adoption of performance-based standards difficult at the current time. In this section, we highlight these obstacles.

Performance Criteria Are Not Established

Currently, no national guidelines are available for setting performance criteria. Discussions occur on four levels:

- What criteria should be included (temperature, CO concentration, or smoke density)?
- For each criterion, what threshold values should be used?
- For each threshold value, how should time be incorporated (peak exposure at 5 minutes or total dose after 10 minutes)?
- What probability threshold should be used (95% chance of not exceeding threshold for CO concentration in five minutes)? Should different criteria be used with different subpopulations (such as the sick, the elderly, or the handicapped)?

Unspecified Selection Process for Design Fires

Design fires are defined fire challenges (for example, a grease fire on the stove or a smoldering cigarette fire on the sofa). Because it is impossible to evaluate physically the performance of building systems in response to all design fire scenarios that might occur, how can a reasonable set be selected? Instead of generating a large sample of design fires, should "worst case" or "reasonable worst case" fires be used? How are worst-case scenarios determined? How likely is a particular scenario? Should designs be able to handle the one-in-a-million fire?

Assumptions about Human Behavior

To evaluate a design, assumptions must be made about how people will behave during the fire. For example, two of the stated assumptions built into one internationally used egress model are that 100% of the occupants are readily mobile and that occupants begin leaving the building immediately on hearing an alarm (Portier, Peacock, and Reneke 1996). Another assumption implies that an equal number of people egress through each available exit. More typically, actual human behavior is to use the door one normally uses to enter and exit the building.

Limited and Poorly Understood Fire Models

Limitations of fire models are not well documented or widely understood. For example, computer fire models do not model fire directly and predict fire effects based on user-selected input data. Because many existing fire model and calculation methods were originally developed as research tools, the model conditions—defined as "fundamental requirements for the model's validity" (Brannigan and Smidts 1998)—often are unknown or unstated. Estimates provided by a model are technically credible only when model conditions have not been violated, but it may be difficult to determine whether this has happened.

Fire Models Do Not Explicitly Incorporate Uncertainty

Even when the model is used within its intended limitations, fire model outputs are point values that do not reflect inherent input uncertainties (for example, uncertainties in fire growth rates and initial conditions). Without knowledge of the uncertainty surrounding a prediction, it is impossible to be certain of a design's acceptability.

Modelers Do Not Have the Necessary Expertise

Problems can also occur when fire protection design engineers are required to work in domains outside their expertise. "Conservative" assumptions made by well-intentioned engineers may not be as conservative as intended. For example, a design engineer intending to be conservative may assume that tenability would be violated when any one of a set of individual criteria (such as temperature or carbon monoxide concentration) exceeded a minimum value. However, toxicity experts might argue that temperature and gas interactions cause tenability to be violated even when each of the individual species is in an "acceptable" range.

Similarly, a design engineer may assume that the time to react to an alarm for a resident is "conservatively" set equal to the travel time needed to go from one remote corner of the unit to the other most remote corner of the same unit. However, this assumption may not be conservative, because even a fully ambulatory occupant may stop to gather belongings, rescue a pet, call a neighbor, and so forth. Both of these conservative assumptions have been included in recent engineering studies.

Reliability of Systems Is Not Modeled

The last barrier identified is the uncertainty surrounding both the reliability of a given fire protection device or system characteristic and the lack of standardized methods to incorporate reliability into performance-based engineering calculations and decisions based on these calculations. For example, we may be uncertain about the reliability of a given fire suppression system. Sometimes a fire suppression system is proposed as an alternative to passive fire protection, such as compartmentalization. However, these two alternatives have different reliabilities. There is no universally agreed-upon method that describes how to account for these differences.

Proposed Method for Performance-Based Evaluation of Fire Protection Designs with Uncertainty

In this section, we propose a process for completing a performance-based evaluation of a building design incorporating uncertainty. At present, no method exists for the quantitative treatment of uncertainty in fire safety engineering calculations. The process that we propose (Table 11-1) is both rigorous and comprehendible. It integrates uncertainty analysis into the performance-based design process and is easily generalized to a broad range of fire safety engineering calculations. It incorporates ways of handling value judgments and demonstrates how results can be displayed graphically to stimulate discussion among the stakeholders.

However, the process is not without costs. It relies on complex computer models, requires many decisions to be made, and is built on a probabilistic understanding of the uncertainties. If done correctly and thoroughly, the process can be far more difficult to implement than verifying the typical prescriptive code used today. But, by laying out each step explicitly, the challenges involved with performance-based standards can be appreciated, and a road map of necessary research programs can be outlined.

Steps 1–3: Define the Scope, Goals, and Objectives

The first step in the performance-based design process is to define the scope of the project, that is, to identify the boundaries of the analysis. This definition includes occupant and building characteristics as well as current and

TABLE 11-1. Steps in the Performance-Based Design Process with Uncertainty

Step	*Procedure*
1	Define the project scope (uncertainties related to life-cycle use and safety of buildings).
2	Define goals (uncertainties related to equity and incorporation of societal values).
3	Define stakeholder and design objectives (risk perception and uncertainty about values).
4	Develop a probabilistic performance statement (criteria, threshold, probability, and time).
5	Develop a distribution of scenarios for design fires.
	a. Select the calculation procedure(s).
	b. Identify uncertain input parameters.
	c. Define the distribution model input parameters (including correlations).
	d. Select the sampling method and determine the number of scenarios needed.
6	Develop candidate designs.
7	Evaluate candidate designs.
	a. For each design and outcome criterion, run all fire scenarios and create cumulative distribution functions.
	b. Determine the sensitivity to elements of the probabilistic performance statement.
	c. Evaluate the base case (optional).
	d. Evaluate the importance of uncertainty.
8	Determine whether the design meets all requirements of the probabilistic performance statement.
9	Select a final design.
10	Prepare the design documentation.

future uses of the building. Because of uncertainty regarding how the building may eventually be used, it may be reasonable to consider a broad range of optional activities in the analysis. The number and type of the different uses considered are related to the time horizon applied in the building's life-cycle analysis. By increasing the scope of the analysis, only more robust (and probably more expensive) safety systems will be acceptable.

Given a scope, the second step in the design process is to identify and document fire safety goals of various stakeholders (building owner, design engineer, architect, code official, and the public or end user). These goals include acceptable levels of protection for people and property as well as the importance of factors such as continuity of operations, historical preservation, and environmental protection. The identification of competing interests and goals by the different stakeholders is important, and negotiating a satisfactory balance may not be easy. With performance-based design reviews, these steps are explicit. If the acceptability of a design is sensitive to the values, attitudes, or risk perceptions assumed, then this sensitivity should be acknowledged in the design documentation.

The third step in the design process is the development of objectives, that is, design goals that have been further refined into values quantifiable in engineering terms. Objectives might include mitigating the consequences of a fire expressed in terms of dollar values, loss of life, or maximum allowable conditions, such as the extent of fire spread, temperature, or spread of combustion products. Uncertainties arise in risk perceptions and values; there is both uncertainty and variability in the way people perceive and value risk.

Capturing those differences that people have in their perceptions and values related to risk is a necessary step in the design process. For example, it may be a goal of the stakeholders to protect historical features of a building or to protect against business interruption or loss of operating capability. Stakeholders with different values may see these needs differently. It is important to identify where value judgments enter into a performance-based calculation and to make explicit any assumptions regarding values and the impact of different values on the final design.

Step 4: Develop a Probabilistic Performance Statement

The fourth step in the design process is to develop a probabilistic statement or statements of performance, the criteria by which to judge the acceptability of the design. These criteria are a further refinement of the design objectives and contain numerical values with which the expected performance of the candidate designs can be compared. Each probabilistic design statement contains a minimum of four elements: performance criteria, threshold values for those criteria, time limits, and a probability. For example, one objective may be to maintain tenable gas concentrations in the corridor. A corresponding probabilistic design statement for life safety might specify, "The design must allow for a 0.9 probability of having four minutes or more before a temperature of 65 °C is reached in the corridor." Thus, all four elements are included: probability, time, performance criteria, and threshold value. A location is also specified.

Many issues must be addressed when establishing probabilistic statements of performance. For example, which criteria should be evaluated (layer height, temperature, levels of asphyxiate gases, heat flux, obscuration, flashover, or some combination)? Which asphyxiate gases should be considered (carbon monoxide, hydrogen cyanide, carbon dioxide, or reduced oxygen), and which interactions should be modeled (for example, lower concentrations of carbon monoxide are more lethal when carbon dioxide concentration are high, because people breathe more rapidly)? Several fire toxicologists believe that the observed effect of the exposure of animals and humans to the products generated by burning materials can be explained by the impact of the combined effect of a few key gases released during combustion. The combined effect of these gases is expressed using a parameter called fractional effective dose (FED) (Purser 1995, 1999). FED calculates the time to a toxic dose of combustion gases, assuming that the total observed effect equals the sum of the effects of each of the component parts. Under this assumption, if a person receives 50% of a lethal dose of one gas and 50% of a lethal dose of another, death will occur. However, the equations that are used to calculate these values vary greatly between researchers. Which one is correct?

How should time be integrated into the analysis? Exposure to a high concentration for a short period of time may be survivable, but exposure to a much lower concentration for a long period of time can be fatal. There is disagreement in the literature as to which values cause negative health effects. For example, should the limit be 25,000 parts per million (ppm) for one minute, 2,000 ppm for five minutes, or a simple threshold of 1,500 ppm? All three limits have been proposed for incapacitation criteria. The negative consequences must be defined, that is, whether the threshold values represent incapacitation or lethality. The probability limit establishes the level of acceptable risk to the stakeholders. Setting the time to untenability value requires understanding behavioral patterns of people in fire as well as making value judgments as to which subpopulations the value is trying to protect.

More complicated performance statements that include multiple locations, time values, untenability criteria, and probability levels are often desirable, for example:

- "The design must allow for a 0.9 probability of having four minutes or more before untenability based on a temperature of 65 °C is reached AND a 0.9 probability of having six minutes or more before 100 °C is reached."
- "The design must have greater than or equal to a 0.95 probability of 65 °C in the apartment hallway AND a less than or equal to a 0.1 or more probability of 100 °C in the building corridor."
- "The design must provide for a 0.9 probability of providing four minutes before 65 °C is reached and a 0.9 probability of having eight minutes or more before untenable gas conditions are reached."

Location statements can vary considerably based on the goals and concerns of the code's authors. Untenability can be evaluated as a minimum anywhere in any building or only along the egress path. Different analyses can give very different results in terms of design acceptability. Adding sprinklers only along the egress path could have large impact for one analysis but not for the other.

Step 5: Develop a Distribution of Design Fire Scenarios

Generating a set of realistic input scenarios that represent statistically both the kinds of fires and the frequency at which they occur in a given occupancy type is critical to a performance-based analysis. The "input scenario generator" should integrate information about the uncertainty, variability, and correlation structure of relevant factors. It can be done in one of two ways: by generating a large set of "equally likely" scenarios with a proportionally large number of "typical" cases and a few "worst-case" scenarios, or by generating a smaller number of representative scenarios and weighting their likelihoods explicitly so that the representative typical scenario is given much more weight than the representative worst-case scenario. The first method requires more model runs but perhaps fewer and less-subjective assessments. Such procedures are discussed in the following five substeps.

Step 5a: Select the Calculation Procedures. Because the complexity of a design fire scenario depends on the fire model used, the first step in this process is to select the calculation procedures to be used in the performance-based design analysis. A range of calculation tools and models is currently available. The *Fire Protection Handbook* provides a good overview of the various kinds of fire models (Beyler 1991).

Fire models (categorized as shown in Figure 11-1) can be used to predict a hazard, predict a risk, reconstruct a fire, interpolate between or extrapolate beyond test results, or evaluate a parametric variation. The model or type of model selected depends on several factors, including the application of interest (discussed in Nelson 1991). Each application may have purpose at some stage of the performance-based design process. For example, a deterministic fire model may be needed to estimate the buildup of heat and products of combustion throughout the building; special purpose models may be used to calculate the time to activate a sprinkler or detector, the time needed to egress a building safely, or the time to structural failure.

The quantitative treatment of uncertainty discussed here can be applied to any kind of fire model. Uncertainty and variability of the various models' inputs can be captured; however, scientific uncertainty inherent in the tools themselves is not addressed by our method. For example, deterministic enclosure fire models share many limitations. One such limitation shared by fire calculation procedures is the absence of a fire growth model. These models all require the user to specify the fire in terms of the rate of energy and mass released by the burning item as a function of time. Such data are obtained by measurements taken in large- and small-scale calorimeters, or from room burns. Potential sources of uncertainty include measurement errors related to the instrumentation, the degree to which radiation feedback

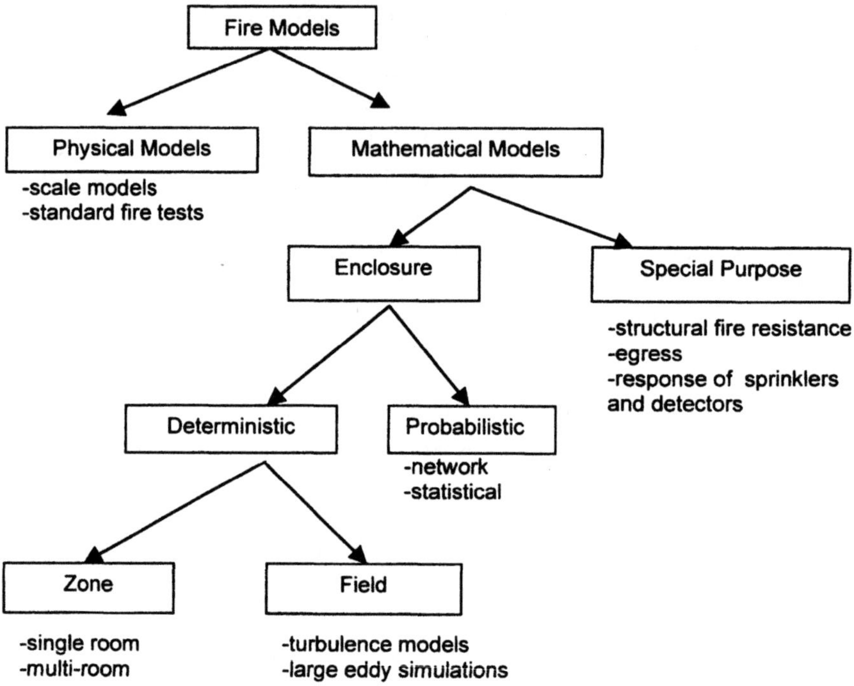

FIGURE 11-1. Categories of Fire Protection Models

affects burning rates and combustion chemistry, uncertainties caused by scaling factors used to extrapolate small-scale data to a full-size item, and uncertainties related to the representativeness of the item burned to the item ignited during an actual fire (for example, a couch or chair burned in a calorimeter will not necessarily represent the chair or couch in the building being designed).

Other shared scientific uncertainties, particularly among zone fire models, result from the following features of the models:

- All zone fire models divide each room into a small number of control volumes, each of which is assumed internally uniform in temperature and composition.
- Empirical correlations are used for flow and entrainment coefficients.
- Knowledge is lacking regarding post-flashover chemistry.
- Models to predict concentrations in each room use user-specified hydrocarbon ratios and species yields.
- Entrainment coefficients are empirically determined values; although small errors in these values will have a small affect on the fire plume or the flow in the plume of gases exiting the door of that room, in multiroom

compartment models, errors are multiplicative as the flow proceeds through many compartments, possibly resulting in a significant error in the farthest rooms.
- No generally applied design or analysis procedures exist for the interaction of automatic sprinklers and the plume.

Uncertainties arise when physical processes are modeled on the basis of empirically derived correlations or simplifying assumptions are made. These factors lead to uncertainties in the results. For most fire models and calculation procedures, very different answers can result depending on the calculation limitations and level of detail of the modeling. Computational limitations lead to simplifications and approximations that induce uncertainty. Also, models often are used outside of their intended applications because of poor documentation or misunderstanding. Several researchers (Lantz 2000; Siu, Droguett, and Mosleh 1999) suggest methods to aid in the identification of a model or set of models on which predictions may be based and to aid in quantifying the uncertainties inherent in using these models.

Step 5b: Identify Uncertain and Crucial Input Parameters. After a calculation procedure is chosen and candidate designs have been selected, the input parameters to treat as uncertain must be determined. Depending on the complexity of the design and the models selected, there could easily be hundreds of potential variables. Ideally, only parameters or combinations of parameters with uncertainty great enough to change decisions regarding the final design are treated as uncertain. Unfortunately, it is not always possible to know a priori which of the input parameters possess crucial uncertainty. Therefore, this winnowing process can be time-consuming the first time a design for a specific type and size building is evaluated; initially, many inputs are treated as uncertain. However, once a full-blown study has been completed, the set of uncertain inputs can be thinned considerably, and future performance-based evaluations will not have to repeat the process. Step 7d, discussed below, expands on this discussion.

Step 5c: Define Distributions of and Correlations among Other Input Parameters. Design fire scenarios take into account possible fire events: heat release rates, characteristics of the material burning and of the building, and other relevant information such as weather conditions. The uncertainty and variability surrounding each variable must be captured in the mathematical description of that variable. Any and all available knowledge regarding the value of that parameter should be incorporated in the input scenario generator: empirically measured values, known variations, and statistically compiled data. For example, for a given occupancy type, the National Fire Protection Association (NFPA) publishes statistical data on the percentage of fires that start in each room of potential fire origin. This information should be incorporated into the random scenario generator so that the generator mimics these statistics. Distributions can be constructed for variables such as temperature, wind, and relative humidity from regional data published by the National Weather Service. Methods for quantifying measurement uncertainty (ANSI 1997) are used to capture uncertainty and variability in empirically measured parameters such as rates of production of products of combustion. In many cases, where hard data do not exist and are not possible to create, expert elicitation is needed to quantify the uncertainty.

It is very important to model correlation among variables so that the input scenario generator will not generate unrealistic scenarios. For example, if the design incorporated a weather module, a month of the year would be randomly selected. For that given month, a value is sampled from an outdoor temperature distribution based on National Weather Service data for that region. The selected outdoor temperature is correlated to other weather variables (external pressure, wind, and relative humidity). Logically, the likelihood of windows and doors being open, indoor temperature and pressure, and initial fuel temperature will be correlated to these values. The likelihood of the windows being open should be much higher in August than in December.

Step 5d: Select the Sampling Method and Determine the Number of Scenarios. A single fire scenario is created by sampling one value from each of the uncertain input parameters. To do this, a sampling method must be selected. Because the goal is to determine the safety of designs by their associated distributions of untenability times, the sampling method should efficiently capture the full range of times with the smallest sample needed. For the types of problems generally encountered, the methods that are "less random" but that ensure a spread of inputs (Latin Hypercube sampling or the use of a quasi-random generator) are preferred to straight Monte Carlo methods (Law and Kelton 1991). This way, the tails of the output distributions get filled in quickly, which is important because many policy criteria will probably involve high percentiles (such as 95th or 99th) of untenability distributions.

There is no simple answer as to how many scenarios are needed to capture the distribution of untenability times. The actual number will depend on the model's complexity, the type, distribution, and correlation of the inputs, and the desired accuracy of the output distribution. In many cases, the limitation on the sample size will be determined by the run time of the fire models themselves. In a recent experiment, single runs of design fire scenarios of a relatively simple apartment design took several days in CFAST, a fire model from the National Institute of Standards and Technology (Notarianni 2000).

As a general rule, the more runs that can be gathered, the better or more refined the results, but there are certainly diminishing returns for continued sampling. Many techniques that can evaluate the stability of the output distribution can help determine when adding additional samples will have little impact on what is known about the outputs of interest (Law and Kelton 1991).

Step 6: Develop Candidate Designs

To evaluate a design to determine whether it meets a performance-based standard, the candidate design must be quantified in a form that matches the inputs to the fire models. A candidate design includes basics (such as building dimensions and locations of doors and windows) as well as proposed fire protection systems (such as sprinklers and smoke detectors), construction features and building materials, and operational procedures that could affect the fire performance of the building.

Step 7: Evaluate Candidate Designs

Each candidate design must be evaluated using each design fire scenario to determine whether it will meet the elements of the probabilistic statements of performance. How this is done requires multiple steps. In the following subsections, we outline the key elements that must be considered.

Step 7a: Calculate a Distribution of Values for Each Outcome Criterion. A single value will be determined for each outcome criterion calculated for each design fire scenario run. Multiple scenarios will provide not only a range of values for criteria of interest but also their entire cumulative probability distributions. In addition, most of these criteria vary over time. Each scenario will predict a different curve of the key outcome criteria versus time. For example, if upper-layer temperature is the criterion of interest, then four design fire scenarios would produce four different curves of the upper-layer temperature versus time. Figure 11-2 shows a representative graph of the value of outcome criterion A plotted against time from ignition. For any given design, each design fire scenario would determine a different curve. For each time step, a distribution of the output criteria can be determined. In addition, each fire scenario's curve has different characteristics (for example, magnitude of the peak value, time to the peak value, and time to exceed critical threshold). Distributions of these values also can be determined.

A useful way of displaying these distributions of output criteria is by their cumulative distribution function (CDF). It is done by graphing the value of the criterion against its rank order. For example, for *n* design fire scenarios, *n*

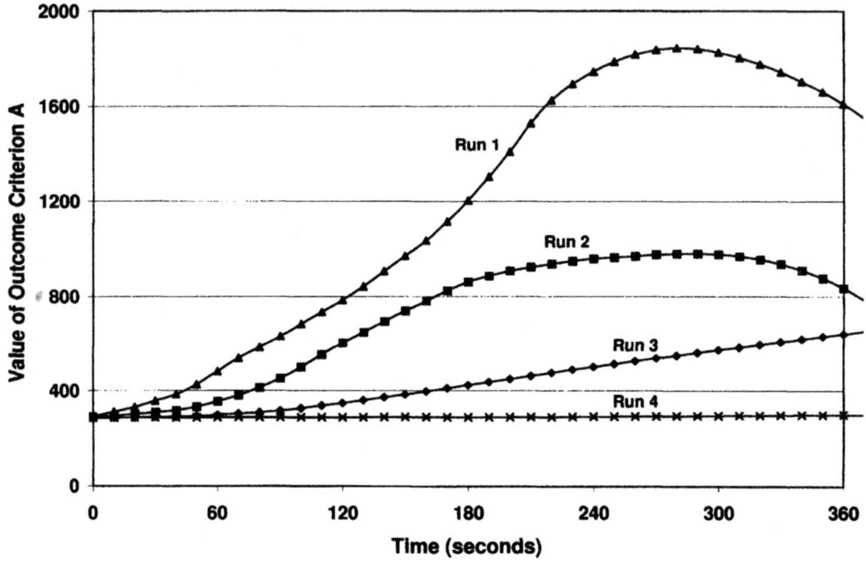

FIGURE 11-2. Variation in Prediction of Time Series Values of Outcome Criterion A

values of a given criterion are generated. These values are then sorted in descending order. The largest value is graphed versus $1/n$, the second largest against $2/n$, and so on, and the smallest value against n/n (which is equal to 1).

One example of a CDF is shown in Figure 11-3. The time to reach a threshold value of one or more of the tenability criteria (that is, a value determined to cause injury or death) can be determined from the time series predictions. The threshold value may be a particular temperature or carbon monoxide level, or a parameter used to represent some synergistic effect of a combination of the tenability variables. One value of time to untenability is obtained for each scenario run. The sorted set of values determined by the design fire scenarios provides a CDF of the outcome criteria.

The CDF shown in Figure 11-3 is interpreted as follows. The chance that time to a critical value of criterion A is 30 seconds or more is almost certain (a probability of 1.0); the chance that the time to this value is 120 seconds or more is 75%, that it is 180 seconds or more is 50%, and that it is 390 seconds or more is 10%. So, if the performance criterion stated that the chance of having at least two minutes for criterion A had to be 90%, then this design would not meet the standard.

It must be remembered that multiple untenability criteria could be used to evaluate a design and that each criterion will have its own CDF. How these multiple CDFs are evaluated must be determined in the probabilistic performance statement.

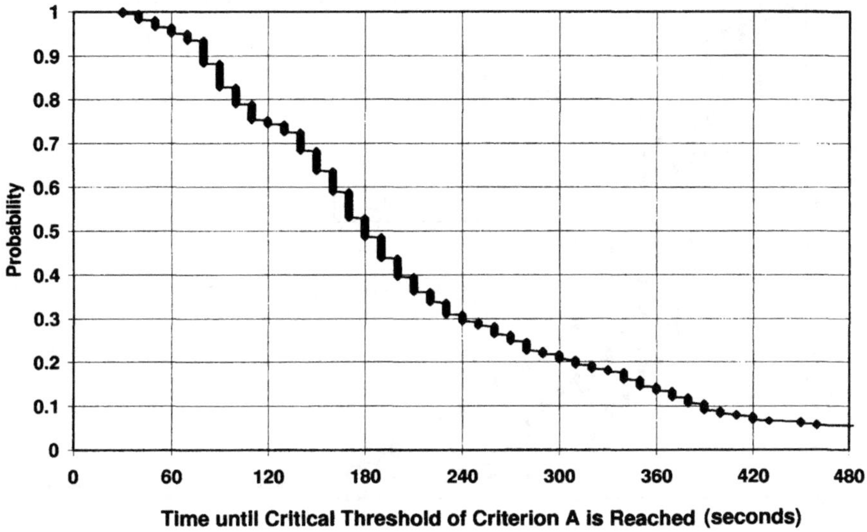

FIGURE 11-3. CDF of Time to Critical Value for a Criterion of Interest

Step 7b: Determine the Sensitivity of Outcome Criteria to Elements of the Probabilistic Performance Statement. The sensitivity of key outcome criteria to each of the elements of the probabilistic performance statement on which a design is judged must be known before policy and good design practice can be established. Elements such as criteria, threshold values, probabilities, and times to threshold values are neither mandated nor agreed upon by fire safety and health professionals and the public. Therefore, major conclusions as to the acceptability of a design should be verified to determine whether they are sensitive to minor changes in the probabilistic statement of performance. Is the design evaluation robust? Answering this question might require checking for times to untenable temperature and levels of carbon monoxide, carbon dioxide, and oxygen reduction. It might include checking for synergistic effects of the presence of these substances. It also might be appropriate to evaluate for heat flux and visibility.

The same design may be judged on two performance criteria or by two critical values of the same performance criterion. Figure 11-4 shows an example of time to untenability based on different values of upper-layer temperature. This kind of presentation could be used to determine the effect on time to untenability by selecting a group of tenability criteria or by including different sets of components in the specification of tenability criteria.

This type of evaluation is a good way to focus discussions among stakeholders as to what the tenability criteria need to be, what effects the selection of different threshold values of tenability criteria will have, what probability

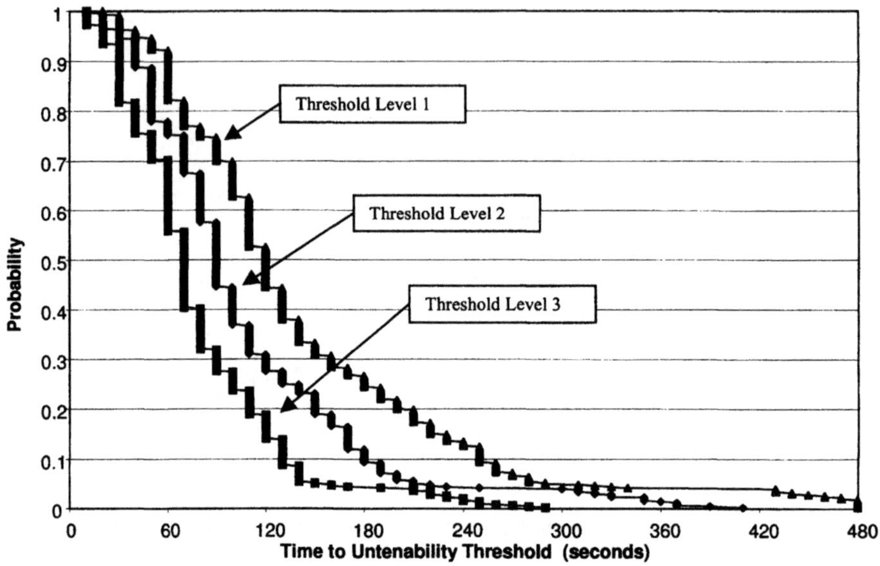

FIGURE 11-4. CDFs Comparing Three Different Untenability Threshold Values for a Criterion of Interest

level is acceptable to the stakeholders, and how to select the final design. At the end of this step, final performance criteria must be selected for use in judging the acceptability of designs and choosing a final design.

Step 7c: Compare Proposed Designs to the Reference Design. Depending on the needs and the scope of the project, it may be helpful to compare a candidate design to a reference (or base case) design. The reference design can be one that meets the prescriptive code, the design that includes the fire protection options currently in the building, or the design with no active fire suppression systems. The purpose of having a base case is to benchmark the effects of fire on the building and the building conditions against each of the designs.

In Figure 11-5, the results of multiple scenario runs are used to show the probability of safe egress graphed against the time to untenable conditions for two different designs. Designs 1 and 2 may represent two different performance designs. Reiss (1998) discusses the need for this comparative approach. The graph shows two design curves that exhibit crossover. Whereas design 1 provides a slightly higher probability of tenability out to 50 seconds, design 2 provides much higher probabilities of tenability at longer times.

Another way that the acceptability of a design is judged is by comparing the level of safety provided by the design to the level of safety provided by the

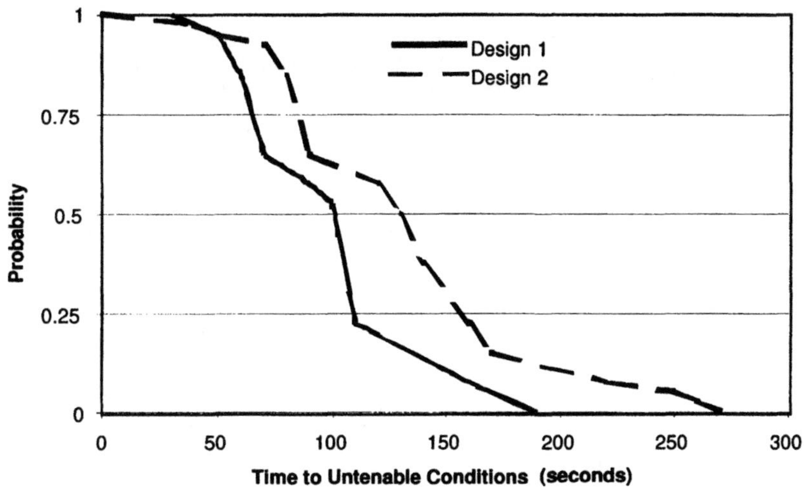

FIGURE 11-5. Comparison of CDFs of Time to Untenable Conditions for Two Different Designs

corresponding prescriptive design. Although never considered, significant uncertainty also is associated with the prescriptive design. Prescriptive codes mandate certain building materials and schemes for fire detection and fire suppression. However, uncertainty and variability remain in the weather, ventilation conditions, human behavioral aspects, and where and how the fire will start. Using the same basic design fire scenarios for these variables, a CDF for the prescriptive code can be generated and compared with the CDF for the performance code.

By using identical sets of fire scenarios for two building designs, the designs can be compared directly for each scenario. This kind of analysis can highlight the specific types of fires for which one design provides more protection (for example, kitchen fires) and can provide insight into how to improve a design. For example, Figure 11-6 is a CDF of the difference in time to untenability provided by design 1 minus the time to untenability provided by design 2 for each fire scenario. It shows that the chance that design 1 will provide a longer time to untenable conditions than design 2 is 25%. Conversely, the chance that design 2 will provide a longer time to untenability than design 1 is 75%, and the chance that the difference will be 100 or more seconds better is 25%.

Step 7d: Evaluate the Importance of Uncertainty. To be efficient, the analysis should focus on the input variables with the greatest uncertainty importance. If an input variable has trivial impact on the output criteria of interest,

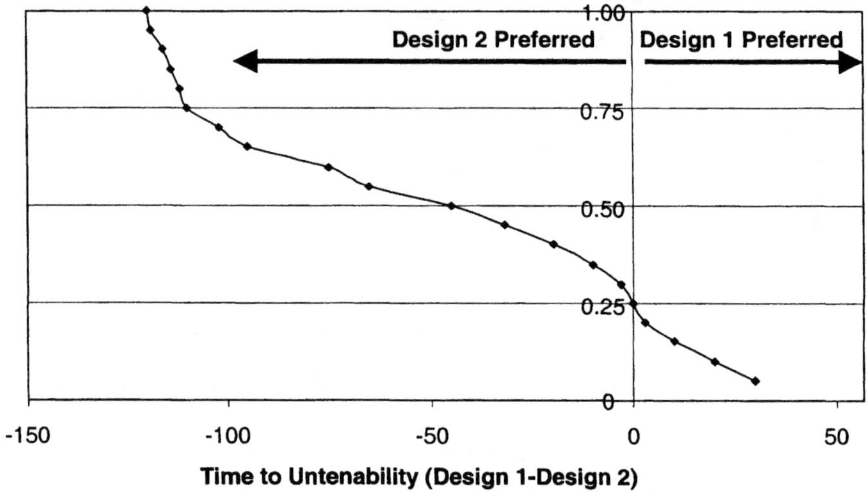

FIGURE 11-6. CDF of Time to Untenability (Design 1–Design 2)

it can be set to a nominal value without affecting the design evaluation process. Some input variables can be screened subjectively (that is, they obviously will have little impact), but unfortunately, because of the complexity and nonlinearity of the models involved in a typical evaluation, it may not be possible to exclude many variables quickly. Aside from subjective techniques, formal methods for doing this screening include correlation of input distributions to output criteria distributions and differential sensitivity (Morgan and Henrion 1992).

By calculating the correlation between the input and output distributions, input distributions that have correlations close to zero can be set to their nominal values. A simple statistical test can be used to determine the value closest to zero that renders a variable uninteresting (Hald 1952). Running more scenarios improves the sensitivity of this analysis. In addition, it is important to consider how different combinations of inputs may interact to cause significant impacts on the output criteria. To solve these problems, differential analyses that measure the effect of a percent change in the standard deviation of an input variable (or variables) on the output criteria's standard deviation can be used. Unfortunately, the computational effort needed for differential analyses is much greater. As a result, correlations are typically the methods of choice. In a model of an apartment building, for example, Notarianni (2000) set approximately 200 of 400 input variables to their nominal values based on expert opinion, and approximately 175 of the remaining 200 inputs were found to be unimportant based on correlation analyses. The number of input distributions that need to be considered can be further simplified by combining several into one variable (for example, instead of having distributions for length, width, and height of a room, a single distribution for room volume may be sufficient). Room volume may be correlated with key outcome criteria such as peak temperature or time to peak temperature.

Uncertainty importance analysis can be used to simplify future evaluations by determining the input uncertainties that are most crucial. It can simplify the process for a class of buildings and demonstrate where additional research would be effective in reducing uncertainty and ensuring a safer, more predictable building. It must be remembered, however, that *correlation* is not *causation*. Thus, any apparent strong correlation that is counterintuitive should be investigated with sound engineering judgment.

Step 8: Judge the Design's Acceptability on the Basis of All Elements of the Probabilistic Performance Statement

The hard part in evaluating a design with a performance-based standard is in the specification of the probabilistic performance statement. Besides selecting the criteria, individual behavior must be modeled. The time needed to egress a building safely often is represented in the literature as the time to detect the fire, plus the time to react, plus the time to travel to a safe place. Thus, for a design to be safe, the following expression must be true:

$$\text{time}_{\text{egress}} = \text{time}_{\text{detect}} + \text{time}_{\text{react}} + \text{time}_{\text{travel}} < \text{time}_{\text{untenability}}$$

It is very difficult to predict human behavior in terms of reaction time and travel time in a fire event. There is both variability due to age and health of the individual and uncertainty as to individual goals and concerns (whether the person tries to fight the fire, locate valuables, rescue pets, or notify other people about the fire). Because they are difficult to predict, egress times are best considered as societal and policy decisions that can vary considerably from design to design. For example, more time should be mandated for a healthcare facility, where patients are likely to be nonambulatory or asleep at the time of the fire, than in an office building, where occupants are generally awake and ambulatory.

Steps 9 and 10: Select a Final Design and Prepare the Documentation

Candidate designs that satisfy the probabilistic performance statement(s) may be considered for selection as the final design. When more than one candidate design meets all requirements of the probabilistic performance statement, other factors (such as cost and preference) are considered. When considering multiple designs or designs with very different features, a multi-criteria decision analysis model may be developed to aid in selecting the final design.

Proper documentation of a performance design is critical and should be written so that all parties involved understand what is necessary for implementation, maintenance, and continuity of the fire protection design. The Society of Fire Protection Engineers guide to performance-based design (SFPE 1999) suggests that the documentation have four parts: the fire protection engineering design brief, the performance design report, detailed specifications and drawings, and the building operations and maintenance manual. It is important that the performance-based design report convey the expected hazards, risks, and expected performance over the entire building life. It should include the project scope, goals, and objectives; probabilistic design statements; a discussion of the design fires and design fire scenarios; and any critical design assumptions.

Conclusions

Performance-based standards and regulations are becoming more popular across various applications. In many areas, the transition from prescriptive to performance-based standards is smooth and natural. However, for domains where rare events are the concern, it is very difficult to evaluate designs directly using performance-based standards. Computer models that capture performance during rare catastrophic events are needed. Unfortunately, the available computer models often do not provide the sophistication needed to complete a proper evaluation. Fire protection is a prime, but not unique, example. The requirement that all oil tankers be double hull or "equivalently safe" is another example. What is the definition of *equivalently safe*? Safety has many dimensions (probability of spill, average spill size, 95th percentile of spill size, environmental impact, loss of life, and so forth); which one or ones are correct?

In these domains, because of the large uncertainties inherent in the problem, the steps necessary to complete the analysis require many tough decisions to be made, and in the case of fire protection, these decisions will have to be made at a very local level using poorly documented models by people not practiced in conducting uncertainty analyses. Understanding the complexities and risks involved in applying performance-based standards in these domains is an important step in improving the overall process and giving decisionmakers the intended flexibility of new standards.

References

ANSI (American National Standards Institute). 1997. *U.S. Guide to the Expression of Uncertainty in Measurement.* National Conference of Standards Laboratories, Boulder, CO.

Beyler, C. 1991. Introduction to Fire Modeling, in *Fire Protection Handbook*, edited by A. Cote, pp. 10-82–10-85. National Fire Protection Association, Quincy, MA.

Brannigan, V., and C. Smidts. 1998. Performance Based Fire Safety Regulation under Intentional Uncertainty, in *Human Behavior in Fire: First International Symposium*, edited by J. Shields, pp. 411–420. University of Ulster, Fire Safety Engineering Research and Technology Centre, Belfast, Northern Ireland.

Hald, A. 1952. *Statistical Theory with Engineering Applications.* John Wiley & Sons, New York.

Hall, J.R. 1996. *The U.S. Fire Problem Overview Report Through 1994: Leading Causes and Other Patterns and Trends.* National Fire Protection Association, Quincy, MA.

———. 1998. *The Total Cost of Fire in the United States.* National Fire Protection Association, Quincy, MA.

Lantz, R. 2000. Model Validity Defined and Applied to the Problem of Making Legitimate Predictions from Fire Protection Engineering Models, in *Third International Conference on Performance-Based Codes and Fire Safety Design Methods, Lund, Sweden.* Society of Fire Protection Engineers, Bethesda, MD.

Law, A.M., and W.D. Kelton. 1991. *Simulation Modeling and Analysis.* McGraw-Hill, Inc., New York.

Meacham, B. 2000. A Process for Identifying, Characterizing, and Incorporating Risk Concepts into Performance-Based Building and Fire Regulation Development. Ph.D. Thesis, Clark University, Worcester, MA.

Meade, W. 1991. *A First Pass at Computing the Cost of Fire Safety in a Modern Society.* The Herndon Group, Chapel Hill, NC.

Morgan, M.G., and M. Henrion. 1990. *Uncertainty: A Guide to Dealing with Uncertainty in Quantitative Risk and Policy Analysis.* Cambridge University Press, New York.

National Fire Data Center. 1992. *Fire in the United States: 1983–1990.* U.S. Fire Administration, Emmitsburg, MD.

Nelson, H. 1991. Application of Fire Growth Models to Fire Protection Problems, in *Fire Protection Handbook*, edited by A. Cote, pp. 10-109–10-112. National Fire Protection Association, Quincy, MA.

Notarianni, K.A. 2000. The Role of Uncertainty in Improving Fire-Protection Regulations. Ph.D. Dissertation, Department of Engineering and Public Policy, Carnegie Mellon University, Pittsburgh, PA.

Portier, R., R.D. Peacock, and P.A. Reneke. 1996. *FASTLite: Engineering Tools for Estimation of Fire Growth and Smoke Transport.* National Institute of Standards and Technology, Gaithersburg, MD.

Purser, D.A. 1995. Toxicity Assessment of Combustion Products, in *SFPE Handbook of Fire Protection Engineering*, edited by P. DiNenno, pp. 1-200–1-245. National Fire Protection Association, Quincy, MA.

———. 1999. Assessment of Time to Loss of Tenability due to Smoke, Irritants, Asphyxiants, and Heat in Full-Scale Building Fires—Effects of Suppression and Detection on Survivability, in *Fire Suppression and Detection Research Application Symposium*, pp. 100–139. The Fire Protection Research Foundation, Orlando, FL.

Reiss, M. 1998. Global Performance-Based Design: Is It the Solution? in *Second International Conference on Performance-Based Codes and Fire Safety Design Methods, Maui, HI*, pp. 191–195. Society of Fire Protection Engineers, Bethesda, MD.

SFPE (Society of Fire Protection Engineers). 1999. *The SFPE Engineering Guide to Performance-Based Fire Protection Analysis and Design of Buildings.* SFPE and National Fire Protection Association, Quincy, MA.

Siu, N., E. Droguett, and A. Mosleh. 1999. Model Uncertainty in Fire Risk Assessment, in *SFPE Symposium on Risk, Reliability, and Uncertainty in Fire Protection Engineering.* Society of Fire Protection Engineers, Baltimore, MD.

Starr, C. 1969. Social Benefit vs. Technological Risk, *Science* 165(September 19): 1232–1238.

12

Genetic Testing and the Workplace

A Case Analysis of Chronic Beryllium Disease

RAFAEL A. PONCE, SCOTT M. BARTELL, ELAINE M. FAUSTMAN, AND TIMOTHY K. TAKARO

The use of genetic information in occupational settings is not new and in fact has declined over the past two decades (Gochfeld 1998). The earliest use relates to the practice of excluding women from certain types of jobs (Stellman 1999). In 1991, the U.S. Supreme Court ruled it an illegal discriminatory practice in *United Autoworkers vs. Johnson Controls*, a case related to lead exposure at a battery manufacturing plant (499 U.S. 187, 207). Attempts have been made to exclude workers with genetic conditions that might increase their risk of disease with specific exposures. The most common is the sickle-cell trait, a genetic abnormality in hemoglobin most frequently seen in workers of African descent (U.S. Congress, OTA 1983). Low-oxygen atmospheres such as found at high altitudes can theoretically induce a painful sickle-cell crisis caused by excessive sludging in the blood. The U.S. Air Force used this trait to exclude enrollees from its academy but has abandoned the practice. Clearly, genetics play an important role in risk assessment, for both workers and the general public. Determining when a particular genetic variant (*polymorphism*) can and should be used to protect individuals is a complex issue.

Because a link between genetic information on individual risk and occupational exposures has enormous potential for improving occupational health, information regarding individual genetic composition has been proposed in targeting interventions to reduce occupational illness (Barrett and others 1997). When applied in the workplace, it is hoped that this information can be used in the design of workplace environments, medical surveillance programs, and industrial hygiene practices that can either reduce or prevent occupational disease. However, specific use of genetic polymorphism information in occupational health programs is uncommon because of uncertainty in epidemiological associations between exposure and illness; an incomplete mechanistic understanding of the role of specific proteins (and their function) in given diseases; and the serious social, legal, and ethical implications of potential widespread use of genetic information as the basis for public or worker policy (Holtzman 1997).

As with many modern-day technologies, we are faced with a new type of moral conundrum: Is it ethical to limit access to specific occupations to individuals who are susceptible to otherwise preventable diseases based on their genetic makeup? Such a question naturally leads to other basic questions. Under what conditions should such occupational restrictions be imposed? Who has access to this information? Who makes the employment decision, the employer or the worker? Who is liable if the worker becomes ill? How certain must the disease risks be before occupational health policies based on genetic information are acceptable?

In this chapter, we briefly review the biological basis of genetic testing and the analytic criteria used in evaluating the utility of genetic biomarkers for use in the development of medical and health policy. We examine these analytic criteria in the context of occupational beryllium exposure as a case example. In particular, we examine those factors that limit the development of genetic biomarkers with high predictive power. We then discuss means for assessing the potential economic and social benefits of using genetic information for occupational health policy. We close with a discussion of potential options for using the genetic marker for beryllium disease and the implications of using this information in an occupational setting.

The Biology of Genetic Testing

Advances in molecular biology have resulted in a complete mapping of the human genome. Although this feat has numerous implications for individual health and public policy, understanding the potential use of this information requires an understanding of the biological basis of the disease in question and the quantitative criteria used to assess the strength of association between the genetic factor and the risk of disease. To begin this discussion, we briefly review the biological basis of genetic testing.

Compete mapping of the human genome began with the initial identification of the four deoxyribonucleic acid (DNA) building blocks—adenine (A), guanine (G), thymine (T), and cytosine (C)—and inquiry into how these nucleotides serve as hereditary material in living organisms. DNA is a polymer of these four nucleotides. DNA nucleotide sequencing, which identifies the pattern of nucleotide polymerization, has revealed functionally segregated coding and noncoding regions of DNA. DNA coding regions, also called *genes,* ultimately act as templates for protein synthesis. (In contrast, noncoding DNA regions have no clearly identified purpose but may be important in gene regulation and maintenance of the coding regions.)

The information contained in the DNA sequence of genes is used by all living cells to produce proteins. Just as DNA is a polymer of DNA nucleotides, proteins are polymers of amino acids. Although the details of this process are beyond the scope of this chapter, it is important to realize that the genetic code is specified by the relationship between the sequence of DNA nucleotides in coding regions (that is, genes) and the sequence of amino acids that form proteins. This relationship is critical because proteins are the primary structural and functional components of all cells. Proteins are involved in the most fundamental processes of all cells, such as maintaining cellular integrity, metabolism, inter- and intracellular communication, and cell division.

Whereas the DNA nucleotide sequence determines the individual's genotype, the effect of an altered amino acid sequence on the function of the associated protein determines the phenotype. Because there is a direct relationship between the DNA nucleotide sequence and the amino acid sequence of a given protein, variation in the genetic sequence can result in functional variation in the resultant protein. In some cases, a change in a single critical DNA nucleotide can dramatically alter the function of the resultant protein. The association between critical variations in DNA sequence and alterations in protein function underlies individual disease risk. The strength of this relationship and the role the protein plays in the disease process are the fundamental basis for genetic susceptibility to disease.

Medical Efficacy of Susceptibility Biomarkers

Genetic information can stimulate the development of new medicines and other medical treatments, but its most efficacious use is to prevent illness, particularly when the disease is related to environmental interactions. Genetically susceptible individuals may wish to avoid exposure to certain toxicants or high-risk behaviors when their disease likelihood is particularly high, the consequences of disease are severe, and the social and economic costs of avoidance are low. (Alternatively, individuals at high risk may choose to remain in high-risk occupations for financial gain.) In occupational health programs, required or voluntary screening for genetic susceptibility and heightened medical surveillance for susceptible workers have been proposed as risk management tools. Other forms of behavior modification and risk management based on genetic information are possible, such as the use of engineering controls or personal protective equipment, administrative controls such as risk avoidance and other behavior modification (for example, smoking cessation for asbestos-exposed workers), and medical treatment based on knowledge of the particular biochemical function of a given polymorphism.

Clinical research into the role of single-gene alterations in hereditary breast cancer, retinoblastoma, and hereditary nonpolyposis colon cancer has revealed high predictive value from the determination of individual genetic status (*genotyping*). In these cases, DNA mutations are inherited, and the prevalence of the mutation among population members is generally low. Thus, whereas the individual risk of developing the disease may be very high among people who have the mutation, the attributable risk to the general population is usually relatively low (Caporaso 1996). In cases where the sensitivity and specificity of disease outcomes to a genetic polymorphism are known, the predictive value of the genetic test generally can be estimated (Weinstein and Fineburg 1980; Bartell and others 2000); however, in diseases with long latency periods, this information may be unreliable and difficult to collect.

Genetic susceptibility information as a basis for directing occupational health programs stands in contrast to medical decisionmaking that uses hereditary mutations in single genes. Although single-gene mutations may be necessary and sufficient for disease, making their determination useful in clinical medicine, genetic variants conferring heightened susceptibility are neither sufficient nor necessary for causing disease, and some are expressed only in the presence of an environmental influence, such as an occupational exposure. In contrast to single-gene variants associated with familial cancers that may have a high relative risk, the relative risks associated with susceptibility genes may be low. However, when the prevalence of the polymorphism in given population is high and many members are exposed to the environmental agent, the population risk of disease can exceed that associated with single-gene mutations despite low individual (relative) risks (Caporaso 1996).

In addition to variations in single genes, a growing body of literature indicates that, for certain environmental exposures, individuals with combined susceptibility gene variations may be at greater risk than that predicted by addition of the relative risks associated with variants in each gene alone (for example, Badawi and others 1996). Although simultaneous genotyping information on multiple genes may be used to explore these relationships, this approach is currently uncommon. Because the relative risks associated with genetic variants associated with disease susceptibility are likely to be modest, uncertain, and strongly influenced by exposure and response misclassification, carefully controlled epidemiological studies with relatively large study populations will be required for their determination. Taken as a whole, the expected differences between susceptibility biomarkers used in occupational health and those used in clinical medicine suggest that the criteria used to develop susceptibility biomarker–based screening programs in industry may be different from those used to develop clinical diagnostic tools.

Analytic Criteria Used to Evaluate Genetic Biomarkers in Health and Medicine

Identifying genetic factors that underlie disease occurrence, progression, and variability among humans is a principal goal of modern-day molecular research in epidemiology directed toward understanding interactions between individual genetic status and environmental disease. The target of this research is to identify associations between inter-individual differences in genetic composition (specifically, in gene nucleotide sequence) and disease risk. By identifying associations between critical genetic nucleotide sequence variants (known as polymorphisms or genetic biomarkers) and disease, public health researchers are learning about the biochemical processes involved in disease development. These investigations already have revealed several specific genetic variants that appear to confer susceptibility to occupational cancers induced by benzene, vinyl chloride, radiation, asbestos, pesticides, and other toxic substances (Smith and others 1994; Hayes 1995; Ross 1996; Semenza and Weasel 1997; Xu and others 1998; Costa and others 1999; Moran, Siegel, and Ross 1999; Smith 1999; Furlong and others 2000).

Several panels on genetic testing have proposed broad criteria for establishing acceptable genetic tests applied to occupational health (U.S. Congress, OTA 1983, 1990). Integral to the definition of acceptability has been a requirement that the use of genetic tests conform to identified social, ethical, and legal norms and have positive medical utility. Although these panels have defined a framework in which the implications of developing and using particular genetic tests should be considered, specific criteria useful for defining "acceptability" are lacking. Ideally, such methods should allow health professionals and decisionmakers to balance the medical utility of the genetic biomarkers against socioeconomic, legal, and ethical considerations and should incorporate uncertainty in the available information.

In considering available criteria for examining the medical utility of a particular genetic test, information is typically available only in statistical terms because of variability in effects. We rely most heavily on the concepts of positive predictive value (PPV) and negative predictive value (NPV). PPV describes the estimated proportion of individuals who test positive for a given variant (polymorphism) and develop the disease. It is conveniently expressed as the equation PPV = Pr[D+|T+], which reads as "the positive predictive value is equal to the probability of having disease given a positive (genetic) test." NPV describes the proportion of individuals who test negative for a given genetic test and remain free of disease (NPV = Pr[D–|T–]). The proportion of individuals with the marker who will remain disease-free (the *false-positive* fraction) may be estimated by 1 – PPV or 1 – Pr[D+|T+]. Similarly, the proportion of individuals without the genetic marker who develop disease (the *false-negative* fraction) may be estimated by 1 – NPV or 1 – Pr[D–|T–]. In addition to PPV and NPV, we make use of sensitivity (Pr[T+|D+]), specificity (Pr[T–|D–]), baseline disease prevalence rate (Pr[D+]), prevalence of the genetic biomarker in the population (Pr[T+]), and relative risk (RR). These estimates may be described in probabilistic terms (see the Appendix at the end of this chapter).

Some caution should be exercised in generalizing results from one population to another population with different baseline disease prevalence rates, because the sensitivity and specificity of the genetic test depend on the baseline disease prevalence. The reader is urged to consult appropriate epidemiology or statistics texts for additional information about the derivation and interpretation of these terms (Hennekens and Buring 1987).

Occupational Beryllium Disease

The science of genetic susceptibility assessment is an immature but rapidly emerging field of study. One of the best delineated gene–environment exposure models is for chronic lung disease to the element beryllium, which is used in various high-strength industrial applications. In the following discussion, we examine existing information about a single-gene polymorphism associated with this disease. The discussion provides a basis for comparing the criteria of acceptance for disease susceptibility biomarkers used in clinical medicine and explains how they might be modified when considering use of a susceptibility biomarker in occupational disease prevention. We provide background information to support the genetic basis of the disease and detail how the analytic criteria are derived from the existing literature. The goals of this exercise are to explore the specific case of occupational beryllium disease and to provide an approach that can be generalized to other case studies.

Beryllium is a strong, lightweight metal with good electrical properties, resistance to rotational deformation, and a high heat capacity. It is passive to X-rays and emits neutrons when bombarded with alpha-particles. These physical and molecular properties make beryllium useful in various high-technology applications, including aerospace equipment, medical devices, ceramics, electronics, composite materials, and nuclear energy and weapons production.

Epidemiological studies conducted since the 1940s have associated beryllium exposure with many diseases, usually impaired respiratory function caused by reduced gas exchange between the lungs and blood. Recognition of the acute pulmonary effects of beryllium led to development of exposure guidelines in 1949 and development of industrial alternatives to beryllium that have greatly reduced the incidence of occupational beryllium disease (Eisenbud and Lisson 1983; WHO 1990). However, research conducted in the 1980s among beryllium workers in the ceramics and nuclear industries revealed chronic fibrotic scarring in their lungs, induced by beryllium and mediated by immune system activation. Microscopic analyses of tissues from patients with this chronic beryllium disease (CBD) have revealed the presence of pulmonary nodules, or *granulomas,* similar to those observed with sarcoidosis, another pulmonary granulomatous disease (WHO 1990), which suggests that the disease may have been misclassified in the past (Newman and Kreiss 1992). These granulomas are also similar to those formed on exposure to aluminum, zirconium, and titanium (Saltini and others 1998). Although beryllium diseases (including cancer and acute beryllium disease) are usually associated with respiratory dysfunction, beryllium exposure also can result in systemic disease involving the lymphatic system, heart, skin, liver, spleen, and other organs (Freiman and Hardy 1970; WHO 1990).

CBD is currently incurable and is often quite debilitating; it causes a steady decline in respiratory function. In individuals in whom death from other causes does not intervene, CBD often eventually progresses from respiratory impairment to death. As in sarcoidosis, the disease manifestation and success with treatment vary considerably between individuals. Medical treatment generally involves the administration of steroids and other immunosuppressive therapy with supportive respiratory care and disease monitoring.

As with other diseases that involve improper immune system function, the exposure of sensitive individuals to even minute doses of beryllium can elicit an immune response. CBD has been observed among individuals who were incidentally exposed, including office workers, residents near facilities that work with beryllium, and spouses of beryllium workers presumably exposed to beryllium on their spouses' clothing (Eisenbud and Lisson 1983; WHO 1990; Stange, Hilmas, and Furman 1996).

Genetic Basis of CBD

Investigations in animals have shown substantial differences in disease susceptibility to inhaled beryllium between animals of different strains within the same species (Barna and others 1981, 1984a, 1984b). These observations were the first to suggest that minor genetic differences could account for large differences in disease risk. Additional clues regarding the nature of the responsible genetic characteristics that contributed to CBD risk soon were identified through examination of the role of the immune system in disease progression and microscopic analysis of diseased tissues. These examinations suggested a potential immunological basis for disease that involved a specific processing of beryllium through immune cells in the lung. In particular, this processing is believed to involve cellular uptake of beryllium into scavenging cells (*macrophages*) and a resultant cellular interaction between macrophages and immune system–activating cells (*T-cells*). Recent research has largely focused on identifying genetic factors that help determine this response.

In 1993, Richeldi, Sorrentino, and Saltini reported results from a case–control investigation of genetic variants and their correlation with CBD among beryllium-exposed workers. Among tested gene variants, this group found that a single-gene variant, Glu-69, was highly associated with the presence of CBD among beryllium-exposed workers. Of CBD cases, 97% tested positive for this substitution (that is, Glu-69 test sensitivity is 97%), and roughly 73% of controls tested negative (that is, Glu-69 test specificity is 73%).

Depending on how genetics and environmental beryllium exposures interact, which is still largely unknown, the sensitivity and specificity of a genetic test based on the Glu-69 variant CBD may or may not be stable at different levels of beryllium exposure and CBD risk. Because the baseline CBD prevalence rate is assumed to be a function of the amount of beryllium exposure, the sensitivity and specificity are constant across different baseline prevalence rates under the specified model. At present there is no accepted dose–response model for beryllium.

In Figure 12-1, we used sensitivity and specificity rate estimates based on Richeldi, Sorrentino, and Saltini (1993) to calculate the predictive utility of the Glu-69 biomarker across a range of baseline CBD prevalence rates, assuming that the sensitivity and specificity were constant with respect to CBD prevalence. Although PPV ranges from 3.5% to 42% as baseline CBD prevalence changes from 1% to 17%, NPV is greater than 99.97% across all estimates of CBD prevalence. Such a high NPV indicates that few individuals without the gene are likely to develop disease with beryllium exposure. This characteristic may make Glu-69 useful as a screening test by which Glu-69–negative workers could be reasonably confident that they were not likely to develop CBD.

The predicted impact of occupational Glu-69 screening is dramatic. In the absence of the test and assuming 1% disease prevalence, 99% of those who are exposed would not develop disease. By using the Glu-69 test, one would expect that more than 99.92% of those who test negative would be disease-free with subsequent beryllium exposure (assuming a test sensitivity of 95% and a test specificity of 70%). However, examination of exposure and receptor factors (which also appear to play a role in disease prevalence) and confir-

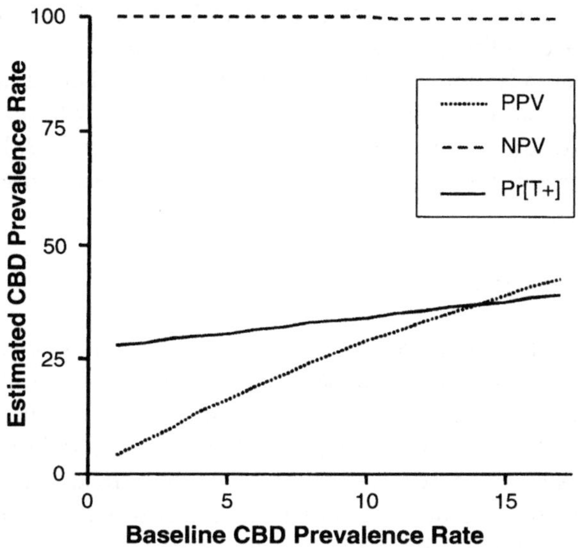

FIGURE 12-1. Estimated PPV, NPV, and Probability of Testing Positive for the Glu-69 Genetic Variant as a Function of Baseline CBD Prevalence Rate

Note: Pr[T+] = prevalence of the genetic biomarker in the population.

mation of the sensitivity and specificity of the polymorphism across worker populations are required before these results are generalized.

We can gain insight into the sources of variability underlying the estimated PPV through an analysis of the probabilities that are used in its estimation: test sensitivity, baseline disease prevalence, and prevalence of the polymorphism.

Baseline CBD Prevalence Rates

Compared with other occupational diseases, CBD occurs with an extremely high rate. Reported CBD prevalence rates range from 1–5% among all beryllium workers to 10–16% among high-risk workers (for example, machinists) (Kreiss and others 1993a, 1993b, 1997; Stange, Hilmas, and Furman 1996; Richeldi and others 1997).

The prevalence of beryllium sensitization, which is established through a test of the immune system's ability to recognize beryllium as a foreign particle, has been observed to be at least twice the CBD prevalence rate (Kreiss and others 1993a, 1997), and roughly 50% of sensitized individuals became CBD-positive in one longitudinal study at a rate of 8–10% per year. These rates are significantly affected by the highly variable latency in disease, which ranges from a few weeks to more than 30 years (Newman 1995). Such variability strongly influences estimates of the PPV, but NPV estimates are relatively insensitive to this parameter. Uncertainties remain regarding whether this gene is causally implicated in disease initiation. The linkages to other genes, which are almost certainly involved, and the contribution of exposure conditions to disease are yet to be determined (Newman 1993). Mechanistic roles for Glu-69 in CBD (and other diseases) have been suggested and are discussed below.

Glu-69 Test Sensitivity

The sensitivity of the Glu-69 test for predicting CBD has been estimated as 97% (Richeldi, Sorrentino, and Saltini 1993). However, estimates of PPV based on this case–control study depend on limited enrollment of CBD-positive (33 workers) and CBD-negative (44 workers) beryllium workers and therefore are subject to relatively large error. In a follow-up study of 125 workers, Richeldi and others (1997) reported a sensitivity rate of 83% (five of six beryllium workers with CBD tested positive for the polymorphism). Using the overall disease prevalence rate reported in that study (4.8%), we estimate a PPV of 12%, which is slightly lower than the value of 15.3% that is estimated using the 97% specificity rate reported by Richeldi, Sorrentino, and Saltini (1993) at the same baseline CBD prevalence rate. However, the 12% PPV estimate is based on few disease observations and therefore is also subject to relatively large binomial error.

To investigate the impact of uncertainty regarding test sensitivity, we evaluated the PPV of the Glu-69 assay as test sensitivity varied from 65% to 100% at baseline CBD prevalence rates ranging from 1% to 15% (Figure 12-2, top panel). As shown in Figure 12-2, the PPV is extremely sensitive to disease prevalence rate, and low baseline rates of disease (for example, 1%) effectively limit the PPV of the test despite extremely high test sensitivity. This dependency of the PPV on the baseline disease prevalence rate will effectively cap the PPV of all genetic biomarkers used to screen populations at low risk of occupational disease.

Prevalence of the Glu-69 Polymorphism

If the Glu-69 sensitivity and specificity rates are constant with respect to changes in baseline disease prevalence, the prevalence of the polymorphism among all workers can be estimated from the sensitivity, specificity, and baseline disease prevalence (Figure 12-1). Both baseline disease prevalence and test sensitivity influence estimates of Glu-69–positive polymorphism prevalence as shown in Figure 12-2 (bottom panel). As shown, the likelihood of

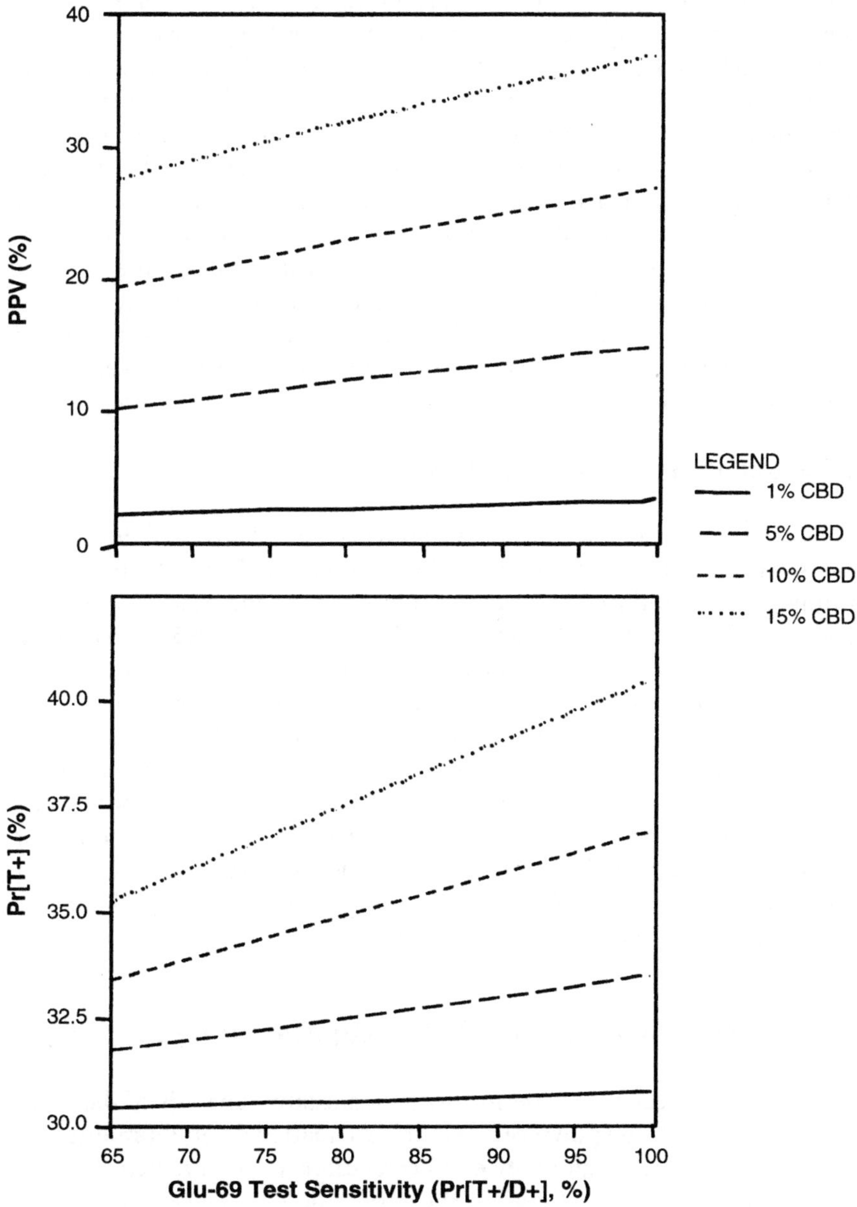

FIGURE 12-2. Estimated PPV (Top) of the Glu-69 Genetic Test and Probability of a Positive Test Result (Bottom) as a Function of Glu-69 Test Sensitivity at Four Baseline CBD Prevalence Rates

Note: Test specificity was assumed fixed at 70%.

testing positive for the Glu-69 polymorphism is positively correlated with both test sensitivity and the CBD prevalence rate in the population.

These analyses demonstrate that the relatively low PPV of the test is driven by a low disease prevalence rate (1–16%) and a relatively high fraction of individuals predicted to test positive for the test (28–38%, which is also a function of baseline disease prevalence). Because the sensitivity of the test is relatively high and the disease prevalence rate is not likely to increase, future research to reduce the number of false positives and improve the PPV should focus on establishing those exposure factors that contribute to disease and on reducing the number of false positives (that is, reducing the prevalence of polymorphism to agree with baseline CBD prevalence rates), perhaps by identifying linked genetic variants. Wang and others (1999) recently published such an effort.

Role of Glu-69 in Other Human Diseases

The presence of the Glu-69 variant also has been associated with lung disease among hard metal workers (Potolicchio and others 1997); systemic sclerosis, including disease induced by uranium mining (Rihs and others 1996); childhood acute lymphoblastic leukemia (Taylor and others 1995); celiac disease (Bugawan and others 1989); and pauciarticular juvenile rheumatoid arthritis (Begovich and others 1989). In addition, the Glu-69 polymorphism has been positively, but not consistently, associated with susceptibility to sarcoidosis (Maliarik and others 1998; Schurmann and others 1998). This latter association is interesting because the clinical manifestations and pulmonary pathology of sarcoidosis and CBD are so similar that before the routine use of a specific clinical test for identifying individuals who are sensitized to beryllium (the *lymphocyte proliferation test*), some cases of CBD probably were misclassified as sarcoidosis (Newman and Kreiss 1992).

Genetic Screening for Glu-69 in Occupational Health Policy

In the preceding section, we performed a quantitative analysis of the predictive value of the Glu-69 polymorphism for occupational beryllium disease. In this section, we discuss alternative interpretations of the predictive value of the test and means for estimating the economic or social costs associated with the use of this genetic information in occupational settings. We close with a discussion of possible industrial hygiene options for managing the disease and describe some of the implications of genetic testing. In this discussion, we explore the role of uncertainty in the quantitative information and the implications for policymaking.

Some researchers have argued that the poor specificity and low PPV preclude any practical applications of the Glu-69 susceptibility gene (Holtzman 1994; Newman 1998). For example, Marshall (1999) suggests that genetic testing for CBD susceptibility is currently hindered by poor disease predictivity (that is, PPV) and its potential use as a discriminatory screening tool. These fundamental issues will arise in the use of any genetic biomarker to be applied in occupational health programs, not only for beryllium susceptibility. Other factors that discourage the use of this marker in occupational health policy include a lack of clear understanding regarding the mechanistic role for the polymorphism in disease susceptibility; the likely multifactorial nature of disease incidence; limited epidemiological information to confirm and reproduce findings; and ethical, legal, and social concerns about genetic screening in general. In light of the absence of a suitable animal model and the complications of doing invasive bronchoscopy and biopsies in normal individuals, some of the desired information is essentially unattainable.

Our analysis of the predictive value of testing for the presence of the Glu-69 variant as a susceptibility biomarker has relevance in determining the value of this, and other, genetic tests in occupational screening programs. For example, even if the sensitivity and specificity of a given susceptibility biomarker were both 99%, the PPV of a given genetic test—including the Glu-69 test—would be only 50% if the baseline rate of disease prevalence was low (that is, 1%). Under ideal assumptions of baseline CBD prevalence (up to 17% as reported for machinists) and test sensitivity, the ability of the Glu-69 genetic test to positively identify individuals who will contract disease on exposure to beryllium does not exceed 42% and may be as low as 3–4% (Figure 12-2). Thus, a substantial fraction of work applicants who test positive for the Glu-69 variant might never get the disease with subsequent beryllium exposure yet might be denied jobs on the basis of their genetic status. In contrast, the ability to identify individuals who would be disease-free in the absence of the polymorphism exceeds 99.97% (Figure 12-2), indicating that few individuals without the gene are likely to develop disease after beryllium exposure. This characteristic may make Glu-69 useful as a screening test by which Glu-69–negative workers could be reasonably confident that they were not likely to develop CBD at current levels of beryllium exposure.

As we have shown, the ability of a genetic test to predict future disease is based on the fraction of the diseased population that carries the genetic marker (that is, the sensitivity), the fraction of the population that has the polymorphism, and the baseline disease prevalence rate. Compared with other occupational diseases, the values for CBD are quite high, suggesting that at least for the NPV, this test performs relatively well. Even as genetic epidemiological research identifies genetic polymorphism combinations that have very high sensitivity, the low disease prevalence that currently exists for occupational illnesses will ultimately limit the ability to predict disease among genetically susceptible individuals. Does this mean that developing relevant genetic susceptibility markers with low predictive value is all for naught? We don't think so. Rather, the informational improvement from use of a genetic test lies in the marginal gain in predictive value compared with not using the test. If one accepts this argument, then the practical utility of the test can be evaluated if one understands the implications of having a false-positive test outcome, the value of reducing disease prevalence, the implications of developing disease, and the industrial hygiene options available to reduce disease.

The Value of Reducing Occupational Beryllium Disease Prevalence

There are several ways to estimate the value of reducing disease among beryllium workers, including analyses based on the economic cost and the social value associated with disease. One approach is to use multiattribute utility models, which have been developed for use in a wide range of policy analyses, including benefit–cost models, social cost models, and utility models used for health policy (Pratt, Raiffa, and Schlaifer 1996). They usually are applied when a decision has multiple consequences and when decision criteria are measured on different scales. Although the models are powerful tools for summarizing complex information, their use is often controversial because they require explicit judgments regarding the valuation of significant health, social, and environmental impacts.

In one form of such a model, all inputs are valued in monetary terms. In a typical analysis, available decision options are compared, and the option with the highest net value is regarded as the preferred policy. Within this context, *value of information* (VOI) is typically used to describe either the change in net value associated with collecting and acting on additional information, or the reduction in opportunity loss resulting from the uncertainty reduction associated with additional information (Weinstein and Fineburg 1980).

Evaluation of Alternative Policy Options for Managing Occupational Beryllium Disease

We present results of an example VOI analysis of the first type (to describe the change in net value) for Glu-69 screening in occupational health. Specifically, we evaluated whether the use of genetic information could inform occupational health policymaking by examining three worker-screening options based on the Glu-69 test (Bartell and others 2000):

- required screening of workers for placement in beryllium-contaminated areas,

- increased screening for the presence of early disease among genetically susceptible beryllium-exposed workers, and
- voluntary genetic testing and informed decisionmaking by workers with potential beryllium exposure.

Model inputs are presented in Table 12-1, and the estimated postintervention CBD prevalence rates across the evaluated occupational health options and over a range of baseline CBD prevalence are presented in Table 12-2. Corresponding VOI estimates are provided in Table 12-3 (as value of CBD avoidance). VOI is presented as the total social savings (VOI > 0) or costs (VOI < 0) accrued per person tested for Glu-69 status across the various occupational health programs. Uncertainty analysis was conducted to evalu-

TABLE 12-1. Input Parameters for Glu-69 Value-of-Information Model

Parameter	*Description*	*Value*[a]	*Confidence*[b]
Sn	Sensitivity of Glu-69 for CBD prediction	β(33,2)	High
Sp	Specificity of Glu-69 for CBD prediction	β(32,14)	High
B_d	Statistical value of CBD ($million/case)	Various[c]	Low
$C_{genotype}$	Cost of Glu-69 testing ($/applicant)	T(100,150,500)	Medium
C_{LPT}	Present value of costs of increased LPT frequency ($/Glu-69+ worker)	U(700,2200)	Medium
$C_{counsel}$	Cost of genetic counseling ($/Glu-69+ applicant)	T(50,100,200)	Medium
w_{screen}	Removal probability for a Glu-69+ worker for required screening	100%	High
w_{LPT}	Removal probability for a Glu-69+ worker for increased LPT frequency	0%	Medium
$w_{volunteer}$	Removal probability for a Glu-69+ worker for voluntary screening	U(0%,100%)	Medium
e_{screen}	Risk reduction efficacy (for a Glu-69+ worker) of required screening	0%	High
e_{LPT}	Risk reduction efficacy (for a Glu-69+ worker) of increased LPT frequency	T(0%,5%, 50%)	Low
$e_{volunteer}$	Risk reduction efficacy (for a Glu-69+ worker) of voluntary screening	0%	Medium

Note: Glu-69+ = tested positive for Glu-69 polymorphism.

[a]β(*v*,*w*) indicates a beta distribution, U(*l*,*u*) indicates a uniform distribution, and T(*l*,*m*,*u*) indicates a triangular distribution. Values are assigned on the basis of empirical evidence, theoretical models, and judgment as described in Bartell and others 2000.

[b]Confidence indicates the assessors' level of comfort that the distribution shape and parameters assigned to the given parameter are appropriate given the existing state of knowledge.

[c]See Table 12-3.

Source: Bartell and others 2000.

TABLE 12-2. Predicted Postintervention CBD Risk by Baseline Prevalence and Intervention Type

Baseline prevalence (%)	*Intervention type*	*Postintervention CBD risk[a] (%)*
0.1	Required genetic screening	0.008 (0.001, 0.020)
0.1	Increased monitoring	0.08 (0.06, 0.10)
0.1	Voluntary genetic screening	0.06 (0.02, 0.10)
1	Required genetic screening	0.08 (0.02, 0.20)
1	Increased monitoring	0.8 (0.6, 1.0)
1	Voluntary genetic screening	0.6 (0.2, 1.0)
10	Required genetic screening	0.9 (0.2, 2.1)
10	Increased monitoring	8.3 (6.3, 9.7)
10	Voluntary genetic screening	6.5 (1.6, 9.7)

[a] First value indicates expected value of chronic beryllium disease (CBD) risk; numbers in parentheses indicates 90% simulation interval as defined by the 5th and 95th percentiles of Monte Carlo simulation results. These values depend on hypothetical models and subjective parameter estimates shown in Table 12-1.

Source: Bartell and others 2000.

ate the impact of variability and uncertainty in model assumptions, but only results for mean (expected) values are presented.

These analyses show that across a broad range of baseline CBD prevalence rates, any of the considered uses of the Glu-69 genetic susceptibility test would reduce CBD risk relative to no intervention (that is, no genetic screening; Table 12-2). Under most conditions (including low disease prevalence) and using a minimal valuation of health effects, the benefits of CBD reduction by use of Glu-69 genetic information outweigh the direct financial costs to the employer of genetic testing. We also find that the incremental gains from genetic screening are roughly proportional to the baseline CBD prevalence rate (Table 12-3).

Industrial Hygiene Options Available for Reducing CBD

We recognize that ethical, social, and legal concerns apply to genetic testing in occupational health screening programs. Although many issues must be resolved, we believe that these tests can be used in acceptable ways in predictive screening programs. One option is to offer confidential, elective, and independent third-party testing for all potential beryllium workers. With appropriate genetic counseling (both before and after testing), workers can be offered information about their genetic status and the uncertainties inherent in the genetic test. Such a model would not deny workers of potentially

TABLE 12-3. Predicted Glu-69 Value of Information by Baseline Prevalence, Intervention Type, and Value of CBD Avoidance

Baseline prevalence (%)	*Intervention type*	*Value of CBD avoidance*[a] *($)*			
		$12,200	*$240,000*	*$1,450,000*	*$16,300,000*
0.1	Required screening	−350 (−610, −180)	−140 (−400, 40)	970 (680, 1,200)	15,000 (13,000, 16,000)
0.1	Increased monitoring	−690 (−1,000, −410)	−650 (−990, −370)	−440 (−840, −40)	2,100 (−180, 5,400)
0.1	Voluntary screening	−330 (−560, −180)	−250 (−450, −70)	210 (−280, 870)	5,700 (220, 14,000)
1	Required screening	−250 (−510, −80)	1,800 (1,500, 2,100)	13,000 (11,000, 14,000)	150,000 (130,000, 160,000)
1	Increased monitoring	−680 (−1,000, −400)	−290 (−790, 270)	1,800 (−230, 4,700)	28,000 (4,800, 61,000)
1	Voluntary screening	−290 (−500, −150)	560 (−220, 1,700)	5,100 (120, 12,000)	60,000 (4,900, 140,000)
10	Required screening	710 (410, 950)	21,000 (19,000, 23,000)	130,000 (110,000, 140,000)	1,500,000 (1,300,000, 1,600,000)
10	Increased monitoring	−570 (−980, −180)	3,300 (10, 8,100)	24,000 (4,100, 52,000)	280,000 (54,000, 610,000)
10	Voluntary screening	70 (−320, 630)	8,000 (360, 20,000)	51,000 (3,800, 120,000)	570,000 (49,000, 1,400,000)

[a]First number indicates expected value in dollars per beryllium worker; numbers in parentheses indicate 90% simulation range as defined by the 5th and 95th percentiles of Monte Carlo simulation results. These values depend on hypothetical models and subjective parameter estimates shown in Table 12-1.

Source: Bartell and others 2000.

useful information that could help guide their decisionmaking yet would reduce the potential for imposed discrimination. From a public health perspective, disease reduction would be maximized by encouraging high participation in the screening program and by ensuring confidentiality and competitive alternative occupations for those who test positive.

Because an immune response underlies the development of CBD, this disease is unlike other toxicant-induced environmental diseases wherein increased exposure causes a proportional increase in response, and for which one may expect to establish a threshold below which no disease is expected to occur. Thus, it is not clear whether exposure reduction measures alone can be expected to reduce disease prevalence, particularly for individuals who are hypersensitive. Indeed, both the U.S. Department of Energy and the Occupational Safety and Health Administration are currently planning to reduce the current occupational standards. In addition, as early as 1972, the National Institute for Occupational Safety and Health had recommended exposure be reduced as low as feasible (Eisenbud 1982). This trend implies that a principal means for reducing disease prevalence would be to identify hypersensitive individuals before they are exposed rather than to rely solely on engineering solutions to reduce exposure. The argument holds even in the case in which genetic susceptibility and exposure factors combine to contribute to disease development, as appears to be the case with CBD.

Because the disease does not appear to occur in the absence of beryllium exposure, and because it occurs largely in occupational settings, disease incidence among workers can be prevented by avoiding occupational beryllium exposure. Such an exposure-free environment requires careful industrial hygiene and emissions controls to eliminate incidental exposures. In many older facilities, such comprehensive elimination appears unlikely because of the nature of previous work practices. Data comparing beryllium sensitization rates in U.S. Department of Energy construction and maintenance workers with those in production workers suggest a greater risk for the nonproduction workers, probably because of the presence of fugitive beryllium dust deposited in the largely inaccessible interstices of older buildings and equipment. Therefore, exposure reduction in production areas is inadequate protection for workers, particularly those involved in decontamination and demolition of such buildings. In these cases, exposure assessment itself is challenging because access to building interstices to characterize exposure risks often cannot occur until demolition is started.

Implications of Genetic Testing

Taking the results from the decision model at face value, the "best" health decision would be to require mandatory genetic screening because it results in both the greatest potential disease reduction and the greatest VOI. However, mandatory screening is probably not legal in the United States or, many would argue, ethical. Without legal guarantees, the use of even voluntary genetic testing could result in workplace discrimination, employer liability, and insurability problems. On the other hand, choosing to avoid using the available genetic information has legal and ethical implications because its use allows workers to make a more informed choice and potentially reduces the prevalence of a preventable disease among beryllium workers.

To resolve the ethical and legal conundrum of applying genetic testing in occupational health screening programs, we urge a dialogue to develop consensus among workers, employers, and regulators. Of particular interest would be answers to questions about who would perform testing, who would control information, who would have access to information, what would happen to positive individuals, and who would pay for testing and genetic counseling. An appealing option for consideration might be independent third-party testing with personal counseling for workers. In such a model, workers could make informed decisions about whether to be tested and then whether to work in a beryllium environment. Liability of the employer could be reduced. However, ensuring employment for individuals who choose to self-select out of beryllium occupations and minimizing the potential for employment and insurance discrimination would require stringent protection of confidentiality.

Conclusions

Scientific understanding regarding the role of exposure, toxicity, and individual susceptibility has provided a basis for establishing workplace regulations that have effectively reduced occupational disease during this century. As a result, it is uncommon to have an occupational disease with a baseline prevalence rate in excess of 1% among exposed workers. CBD is a modern-day exception to this observation.

We demonstrate that the use of genetic susceptibility information cannot be expected to provide PPVs above 50% even for biomarkers with 99% sensitivity and specificity with low population disease prevalence. This result suggests that criteria for accepting a genetic test in occupational health based on PPV would be unlikely to support the use of most genetic markers unless the test had extremely high sensitivity and specificity. So, when is it prudent to use such information in occupational health programs?

We argue that reliance on positive (or negative) predictive value alone is insufficient for developing occupational health recommendations for susceptibility biomarkers. Rather, the marginal improvement in predictive value from use of the biomarker must be weighed against constraining limitations of the social cost; seriousness of the disease; available alternatives for exposure reduction (or technology substitution); regulatory and legal acceptability; and potential socioeconomic harm, including occupational discrimination and loss of employability, privacy, and insurability. In other words, society may gain from small improvements in predictive value if the cost of illness is high and the consequences of testing are small. Because the predictive value of the genetic information in identifying populations at risk from occupational exposures is often highly uncertain, genetics-based decisions that are made to maximize worker health and socioeconomic benefits should consider this uncertainty.

Quantitative decision models that account for uncertainty in available information are useful tools for summarizing complex biological information regarding proposed genetic occupational health applications; however, few models have been developed and applied for specific genetic biomarkers. Although researchers have recommended that data on risks and benefits (including social, legal, and ethical factors) be collected, little guidance is available for determining whether the benefits outweigh the risks for a specific genetic testing program. One possible approach is to summarize and compare risks and benefits using multiattribute functions (Pratt, Raiffa, and Schlaifer 1996). Whereas multiattribute functions are controversial as decisionmaking systems, they may be used as informative tools for assisting decisions about genetic testing.

Although we examine the design and evaluation of alternative policies applied to a specific occupational disease, the analytic principles we present are generalizable. Regardless of the approach that is ultimately adopted in developing genetics-based occupational health programs, the researchers, decisionmakers, and interested parties should be informed of how to assess the clinical utility of a particular genetic test. We argue that reliance on absolute predictive value is a relatively uninformative metric for gauging the utility of a genetic marker. Rather, the marginal improvement in PPV and NPV should be used along with valuation metrics to establish the conditions under which a given test has positive social utility. Such information can then be used in a broader decision context to define future appropriate uses of genetic information in occupational health.

Acknowledgements

This project was supported in part by U.S. Department of Energy Cooperative Agreement DE-FC01-95EW55084 and the Center for the Study and Improvement of Regulation at Carnegie Mellon University and the University of Washington.

Appendix

Estimation of Specificity, Sensitivity, Positive Predictive Value, Negative Predictive Value, and Relative Risk

Quantitative evaluation of genetic biomarkers is accomplished by estimating their positive predictive value (PPV) and negative predictive value (NPV). These parameters are in turn estimated with knowledge of the sensitivity and specificity of the genetic biomarker, the prevalence of the biomarker in the population, and the prevalence of disease. Estimation of these parameters is given below.

Let variables *a, b, c,* and *d* reflect the number of study individuals classified in each of four categories:

	Disease present	Disease absent
Susceptibility gene present	a	b
Susceptibility gene absent	c	d

Point estimates of several genetic test characteristics can be obtained directly by using the following formulas (Hennekens and Buring 1987):

$$\text{sensitivity} = \Pr[T+ \mid D+] = \frac{a}{a+c}$$

$$\text{specificity} = \Pr[T- \mid D-] = \frac{d}{b+d}$$

If the baseline disease prevalence ($\Pr[D+]$) is known for the case–control study population, then the following point estimates also can be obtained (Bartell and others 2000; Hennekens and Buring 1987):

$$\begin{aligned}\Pr[T+] &= \Pr[T+ \mid D+]\Pr[D+] + (1-\Pr[T- \mid D-])(1-\Pr[D+]) \\ &= \frac{b}{b+d} + \frac{ad-bc}{(a+c)(b+d)}\Pr[D+]\end{aligned}$$

$$\text{PPV} = \Pr[D+ \mid T+] = \frac{\Pr[T+ \mid D+]\Pr[D+]}{\Pr[T+]} = \frac{a\Pr[D+]}{(a+c)\Pr[T+]}$$

$$\text{NPV} = \Pr[D- \mid T-] = \frac{\Pr[T- \mid D-](1-\Pr[D+])}{1-\Pr[T+]} = \frac{b(1-\Pr[D+])}{(b+d)(1-\Pr[T+])}$$

$$RR = \frac{\Pr[D+ \mid T+]}{1-\Pr[D- \mid T-]} = \frac{PPV}{1-NPV}$$

where Pr[T+] is prevalence of the genetic biomarker in the population, Pr[D+] is baseline disease prevalence rate, and RR is relative risk.

References

Badawi, A.F., S.J. Stern, N.P. Lang, and F.F. Kadlubar. 1996. Cytochrome P-450 and acetyltransferase expression as biomarkers of carcinogen-DNA adduct levels and human cancer susceptibility. *Progress in Clinical and Biological Research* 395: 109–140.

Barna, B.P., T. Chiang, S.G. Pillarisetti, and S.D. Deodhar. 1981. Immunologic studies of experimental beryllium lung disease in the guinea pig. *Clinical Immunology and Immunopathology* 20(3): 402–411.

Barna, B.P., S.D. Deodhar, T. Chiang, S. Gautam, and M. Edinger. 1984a. Experimental beryllium-induced lung disease. I. Differences in immunologic responses to beryllium compounds in strains 2 and 13 guinea pigs. *International Archives of Allergy and Applied Immunology* 73(1): 42–48.

Barna, B.P., S.D. Deodhar, S. Gautam, M. Edinger, T. Chiang, and J.T. McMahon. 1984b. Experimental beryllium-induced lung disease. II. Analyses of bronchial lavage cells in strains 2 and 13 guinea pigs. *International Archives of Allergy and Applied Immunology* 73(1): 49–55.

Barrett, J.C., H. Vainio, D. Peakall, and B.D. Goldstein. 1997. Twelfth meeting of the Scientific Group on Methodologies for the Safety Evaluation of Chemicals: Susceptibility to environmental hazards. *Environmental Health Perspectives* 105(Suppl. 4): 699–737.

Bartell, S.M., R.A. Ponce, T.K. Takaro, R.O. Zerbe, G.S. Omenn, and E.M. Faustman. 2000. Risk estimation and value-of-information analysis for three proposed genetic screening programs for chronic beryllium disease prevention. *Risk Analysis* 20(1): 87–99.

Begovich, A.B., T.L. Bugawan, B.S. Nepom, W. Klitz, G.T. Nepom, and H.A. Erlich. 1989. A specific HLA-DPbeta allele is associated with pauciarticular juvenile rheumatoid arthritis but not adult rheumatoid arthritis. *Proceedings of the National Academy of Sciences of the United States of America* 86(23): 9489–9493.

Bugawan, T.L., G. Furlong, J. Larrick, S. Auricchio, G.B. Ferrara, and H.A. Erlich. 1989. A combination of a particular HLA-DPbeta allele and an HLA-DQ heterodimer confers susceptibility to coeliac disease. *Nature* 339(6224): 470–473.

Caporaso, N. 1996. Genetic susceptibility and the common cancers. *Biomarkers* 1: 174–177.

Costa, L.G., W.F. Li, R.J. Richter, D.M. Shih, A. Lusis, and C.E. Furlong. 1999. The role of paraoxonase (PON1) in the detoxication of organophosphates and its human polymorphism. *Chemico-Biological Interactions* 119–120: 429–438.

Eisenbud, M. 1982. Origins of the standards for control of beryllium disease (1947–1949). *Environmental Research* 27(1): 79–88.

Eisenbud, M., and J. Lisson. 1983. Epidemiological aspects of beryllium-induced nonmalignant lung disease: A 30-year update. *Journal of Occupational Medicine* 25(3): 196–202.

Freiman, D.G., and H.L. Hardy. 1970. Beryllium disease. The relation of pulmonary pathology to clinical course and prognosis based on a study of 130 cases from the U.S. beryllium case registry. *Human Pathology* 1(1): 25–44.

Furlong, C.E., W.F. Li, R.J. Richter, D.M. Shih, A.J. Lusis, E. Alleva, and L.G. Costa. 2000. Genetic and temporal determinants of pesticide sensitivity: Role of paraoxonase (PON1). *Neurotoxicology* 21(1–2): 91–100.

Gochfeld, M. 1998. Susceptibility biomarkers in the workplace: Historical perspective. In *Biomarkers Medical and Workplace Applications*, edited by M.L. Mendelsohn, L.C. Mohr, and J. P. Peeters. Washington, DC: Joseph Henry Press, 3–22.

Hayes, R.B. 1995. Genetic susceptibility and occupational cancer. *La Medicina del Lavoro* 86(3): 206–213.

Hennekens, C.H., and J.E. Buring. 1987. *Epidemiology in Medicine.* Boston, MA: Little, Brown and Company.

Holtzman, N.A. 1994. Benefits and risks of emerging genetic technologies: The need for regulation. *Clinical Chemistry* 40(8): 1652–1657.

———. 1997. Genetic screening and public health. *American Journal of Public Health* 87(8): 1275–1277.

Kreiss, K., S. Wasserman, M.M. Mroz, and L.S. Newman. 1993a. Beryllium disease screening in the ceramics industry: Blood lymphocyte test performance and exposure-disease relations. *Journal of Occupational Medicine* 35(3): 267–274.

Kreiss, K., M.M. Mroz, B. Zhen, J.W. Martyny, and L.S. Newman. 1993b. Epidemiology of beryllium sensitization and disease in nuclear workers. *American Review of Respiratory Disease* 148(4 Pt 1): 985–991.

Kreiss, K., M.M. Mroz, B. Zhen, H. Wiedemann, and B. Barna. 1997. Risks of beryllium disease related to work processes at a metal, alloy, and oxide production plant. *Occupational and Environmental Medicine* 54(8): 605–612.

Maliarik, M.J., K.M. Chen, M.L. Major, R.G. Scheffer, J. Popovich, B.A. Rybicki, and M.C. Iannuzzi. 1998. Analysis of HLA-DPB1 polymorphisms in African-Americans with sarcoidosis. *American Journal of Respiratory and Critical Care Medicine* 158: 111–114.

Marshall, E. 1999. Beryllium screening raises ethical issues. *Science* 285(5425): 178–179.

Moran, J.L., D. Siegel, and D. Ross. 1999. A potential mechanism underlying the increased susceptibility of individuals with a polymorphism in NAD(P)H:quinone oxidoreductase 1 (NQO1) to benzene toxicity. *Proceedings of the National Academy of Sciences of the United States of America* 96(14): 8150–8155.

Newman, L.S. 1993. To Be^{2+} or not to Be^{2+}: Immunogenetics and occupational exposure. *Science* 262(5131): 197–198.

———. 1995. Metals. In *Occupational and Environmental Respiratory Disease*, edited by P. Harber, M. Schenker, and J. Balmes. St. Louis, MO: Mosby, 469–513.

———. 1998. Beryllium biomarkers: Application of immunologic, inflammatory, and genetic tools. In *Biomarkers: Medical and Workplace Applications*. Washington, DC: Joseph Henry Press, 285–300.

Newman, L.S., and K. Kreiss. 1992. Nonoccupational beryllium disease masquerading as sarcoidosis: Identification by blood lymphocyte proliferative response to beryllium. *American Review of Respiratory Disease* 145(5): 1212–1214.

Potolicchio, I., G. Mosconi, A. Forni, B. Nemery, P. Seghizzi, and R. Sorrentino. 1997. Susceptibility to hard metal lung disease is strongly associated with the presence of glutamate 69 in HLA-DPbeta chain. *European Journal of Immunology* 27(10): 2741–2743.

Pratt, J., H. Raiffa, and R. Schlaifer. 1996. *Statistical Decision Theory.* Cambridge, MA, MIT Press.

Richeldi, L., R. Sorrentino, and C. Saltini. 1993. HLA-DPB1 glutamate 69: A genetic marker of beryllium disease. *Science* 262(5131): 242–244.

Richeldi, L., K. Kreiss, M.M. Mroz, B. Zhen, P. Tartoni, and C. Saltini. 1997. Interaction of genetic and exposure factors in the prevalence of berylliosis. *American Journal of Industrial Medicine* 32(4): 337–340.

Rihs, H.P., K. Conrad, J. Mehlhorn, K. May-Taube, B. Welticke, K.H. Frank, and X. Baur. 1996. Molecular analysis of HLA-DPB1 alleles in idiopathic systemic sclerosis patients and uranium miners with systemic sclerosis. *International Archives of Allergy and Immunology* 109(3): 216–222.

Ross, D. 1996. Metabolic basis of benzene toxicity. *European Journal of Haematology Supplementum* 60: 111–118.

Saltini, C., M. Amicosante, A. Franchi, G. Lombardi, and L. Richeldi. 1998. Immunogenetic basis of environmental lung disease: Lessons from the berylliosis model. *The European Respiratory Journal* 12(6): 1463–1475.

Schurmann, M., G. Bein, D. Kirsten, M. Schlaak, J. Muller-Quernheim, and E. Schwinger. 1998. HLA-DQB1 and HLA-DPB1 genotypes in familial sarcoidosis. *Respiratory Medicine* 92(4): 649–652.

Semenza, J.C., and L.H. Weasel. 1997. Molecular epidemiology in environmental health: The potential of tumor suppressor gene p53 as a biomarker. *Environmental Health Perspectives* 105(Suppl 1): 155–163.

Smith, C.M., K.T. Kelsey, J.K. Wiencke, K. Leyden, S. Levin, and D.C. Christiani. 1994. Inherited glutathione-*S*-transferase deficiency is a risk factor for pulmonary asbestosis. *Cancer Epidemiology, Biomarkers, and Prevention* 3(6): 471–477.

Smith, M.T. 1999. Benzene, NQO1, and genetic susceptibility to cancer. *Proceedings of the National Academy of Sciences of the United States of America* 96(14): 7624–7626.

Stange, A.W., D.E. Hilmas, and F.J. Furman. 1996. Possible health risks from low level exposure to beryllium. *Toxicology* 111(1–3): 213–224.

Stellman, J.M. 1999. Women workers: The social construction of a special population. In *Occupational Medicine: State of the Art Reviews—Special Populations* (Vol. 14, No. 3), edited by H. Frumkin and G. Pransky. Philadelphia, PA: Hanley and Belfus, Inc., 559–580.

Taylor, G.M., M.D. Robinson, A. Binchy, J.M. Birch, R.F. Stevens, P.M. Jones, T. Carr, S. Dearden, and D.A. Gokhale. 1995. Preliminary evidence of an association between HLA-DPB1*0201 and childhood common acute lymphoblastic leukaemia supports an infectious aetiology. *Leukemia* 9(3): 440–443.

U.S. Congress, OTA (Office of Technology Assessment). 1983. *The Role of Genetic Testing in the Prevention of Occupational Disease.* Washington, DC: U.S. Congress, OTA.

———. 1990. *Genetic Monitoring and Screening in the Workplace.* Washington, DC: U.S. Congress, OTA.

Wang, Z., P.S. White, M. Petrovic, O.L. Tatum, L.S. Newman, L.A. Maier, and B.L. Marrone. 1999. Differential susceptibilities to chronic beryllium disease contributed by different Glu69 HLA-DPB1 and -DPA1 alleles. *Journal of Immunology* 163(3): 1647–1653.

Weinstein, M.C., and H.V. Fineburg. 1980. *Clinical Decision Analysis.* Philadelphia, PA: W.B. Saunders Company.

WHO (World Health Organization). 1990. *Environmental Health Criteria 106: Beryllium.* Geneva, Switzerland: WHO, International Programme on Chemical Safety.

Xu, X., J.K. Wiencke, T. Niu, M. Wang, H. Watanabe, K.T. Kelsey, and D.C. Christiani. 1998. Benzene exposure, glutathione-*S*-transferase theta homozygous deletion, and sister chromatid exchanges. *American Journal of Industrial Medicine* 33(2): 157–163.

13

International Technology Policy

Challenges in Regulating Ship Emissions

James J. Corbett and Paul S. Fischbeck

Until recently, most people concerned with air pollution impacts did not think much about oceangoing ships. Three general perceptions supported the assumption that ship emissions were not an international pollution concern. First, ships are very fuel-efficient and therefore would have low emissions. Second, even if ship emissions were not negligible, their effects would be small, because ships operate in remote ocean regions where they would have little impact. Third, even if ship air pollution impacts were noticeable (or large in some regions), the costs to reduce these emissions would be much higher than the costs of reducing emissions from other sources, such as cars and power plants.

However, research in the 1990s showed that these general perceptions held by industry, environmental experts, and policymakers were not correct. Although ship diesel engines are perhaps the most fuel-efficient combustion systems in transportation, paradoxically, their emissions are large as a direct consequence of this efficient combustion. International shipping uses approximately 2% of the world's fossil fuels consumed annually but accounts for about 14% of the nitrogen emissions (as NO_x) and 5% of the sulfur emissions (as SO_x) produced by human activity.

Geographically resolved emissions inventories (Corbett and Fischbeck 1997; Corbett and others 1999) have shown that ships in global trade operate 70% of the time within 400 kilometers (km) of shore (with 85% of ship traffic occurring along Northern Hemisphere trade routes). With long-range transport of air pollution, these emissions near the coastline contribute significantly to air quality problems in "downwind" regions. In fact, in many coastal and port regions along heavily traveled international trade routes, annual emissions from ships equal or exceed those of adjacent land-based sources.

In addition, ship emissions that occur in remote ocean areas contribute to global climate change. The mechanisms and magnitude of these global effects are still being researched, but we know, for example, that ship emissions increase the brightness and lifetime of clouds, that sulfur emissions from ships are comparable to biogenic sources of sulfur, and that emissions of nitrogen oxides (NO_x) and the resulting tropospheric ozone pollution can affect global climate.

In response to these findings, in 1997 the International Maritime Organization (IMO) adopted regulations to reduce NO_x and SO_x emissions from the main engines of new ships. In 1999, the U.S. Environmental Protection Agency (EPA) finalized additional regulations for new ships that apply IMO standards to large ships and require additional NO_x reductions for smaller U.S.-registered vessels.

These regulations address only new engines on ships and do not require any retrofit controls for the existing fleet. Unfortunately, with the long lifetime of ships (more than 20 years), these regulations will take decades to significantly reduce emissions in the oceangoing shipping fleet.

Figure 13-1 shows the projected NO_x emissions from international shipping, using accepted growth rates for trade and industry-average fleet turnover of existing vessels. Although IMO policy will reduce NO_x emissions from the uncontrolled emissions projections, substantial industry pollution reductions will take decades without international standards for existing engines. The realization that dealing only with new construction may not be enough has motivated several efforts to control shipboard emissions from existing ships with retrofit technologies. Depending on policy framework and implementation, if international standards required existing engines to achieve the same per-ship average NO_x reduction as IMO standards for new engines, the annual emissions from international shipping could drop and remain at or below current levels. In this chapter, we describe the problems and opportunities associated with such a policy.

NO_x Technologies Applicable to Existing Commercial Ship Engines

NO_x emissions from large ship diesel engines contribute to the same air quality problems as NO_x from automobile exhaust and power plants, which include direct health effects (such as lung irritation, increased asthma prob-

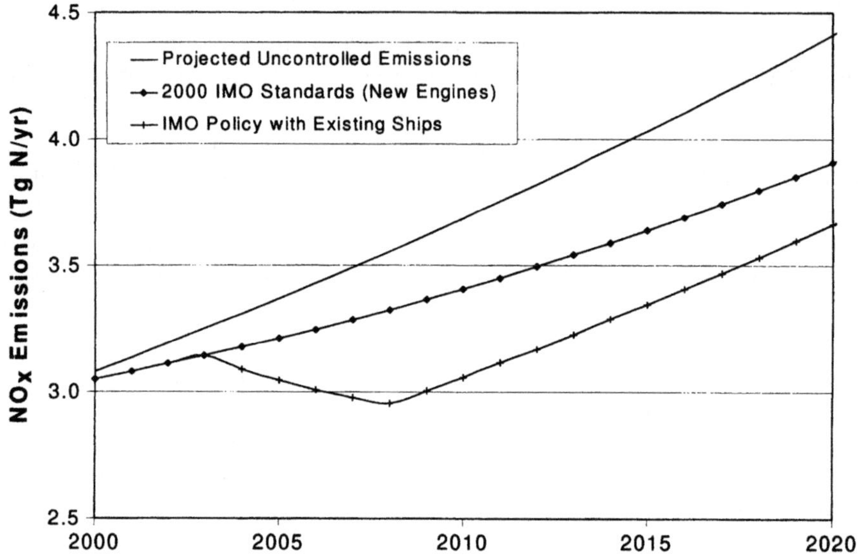

FIGURE 13-1. Projected NO_x Emissions under Various International Regulatory Policies
Note: TgN = teragrams (10^{12} grams) of nitrogen.

lems, and lower resistance to respiratory infections) and tropospheric ozone pollution (smog). One reason IMO and EPA regulated only new engines in their recent regulations is the uncertainty (in both performance and costs) associated with retrofit technologies for controlling NO_x emissions from existing ships. Each retrofit technology performs differently by design and operating conditions. From a ship operator's perspective, the life-cycle costs of these technologies are uncertain; technologies that are inexpensive to install often are expensive to operate because they degrade fuel efficiency. From the policymaker's perspective, cost-effectiveness measures that consider both the emissions reduction and the costs of the technology are doubly uncertain. Moreover, further complicating the analysis, the most cost-effective technology varies by policy scenario.

By considering these uncertainties, policymakers can evaluate the trade-off between regulations that are technology-based (for example, mandating a particular system for emissions reduction) or performance-based (for example, requiring a certain level of reduction or control). Policymakers also can consider whether local policies (for example, requiring emissions control in ports only) will result in behaviors that also meet regional and global objectives for pollution reduction. We consider how different a policy scenario appears from the ship operators' perspective when calculated as a life-cycle cost, and from the policymakers' perspective when considered as policy cost-effectiveness.

NO_x Pollution Mechanisms and Their Control

Combustion in a compression-ignition (diesel) engine results from autoignition at high temperatures and pressures; in spark-ignition engines, external energy from a spark begins the combustion process. Marine diesel engines (and similar land-based versions) operate at much lower engine speeds than most other diesels—as low as 122 maximum revolutions per minute (rpm) (MAN B&W 1997b).[1] Moreover, they have very large cylinder volumes, with diameters that routinely range from 20 centimeters (cm) to 90 cm and piston speeds as slow as 7 meters per second (m/s) (Heywood 1988; Harrington 1992). Over the past 25 years, the power output per cylinder has more than doubled, and fuel consumption has decreased by 25%—mainly through increased maximum firing pressure and better fuel injection control, supercharging, and scavenging (Heywood 1988). To be efficient, the fuel injected into large cylinders has to atomize and be well mixed with air before the long-duration power strokes. This combustion leads to high peak temperatures and long residence times, which cause marine engines to have typically higher NO_x emissions than other diesel engines (Carlton and others 1995; U.S. EPA 1997). Also, the fuel NO_x that results from nitrogen bound in the heavy residual fuel used by oceangoing ships can lead to higher NO_x emissions.

Technologies that reduce NO_x emissions can be divided into three groups (MAN B&W 1996): those that require engine modifications (in-engine controls), those that are implemented in the fuel or air system (pre-engine technologies), and those that are on the exhaust system (post-engine technologies). Nine specific technologies for NO_x control are considered here because of their application to existing marine engines or similar systems.[2] The six in-engine NO_x control strategies are aftercooler upgrades, engine derating, injection timing delay (retard), fuel system modifications to increase supply pressure, fuel injector upgrades, and exhaust gas recirculation (EGR). The two pre-engine technologies are water added to the combustion air, and water/fuel emulsions; the post-engine technology analyzed here is selective catalytic reduction (SCR).

In-Engine Technologies

Aftercooler Upgrades. An aftercooler, between the turbocharger and the engine, increases the intake air density and lowers the temperature of the charge air (Genovesi and Browning 1998). Both of these factors act to reduce the peak combustion temperature and can reduce NO_x emissions by about 10% (Sierra Research 1991; Venkatesh 1996). Nearly all medium-speed and slow-speed marine engines have turbochargers with aftercooling to optimize fuel economy (Alexandersson and others 1993), and overcooling can increase fuel consumption.

Engine Derating. NO_x emissions from diesel engines sometimes can be reduced by running the engine at less than design power or using lower-rated fuel injectors (Sierra Research 1991). These different settings can reduce NO_x by some 14% but increase fuel consumption.

Injection Timing Retard. Reducing the pressure at autoignition by retarding the timing of fuel injection will lower the peak flame temperature and reduce NO_x; however, it also increases fuel consumption (MAN B&W 1996; Hellmann 1997).

Fuel System Modifications to Increase Supply Pressure. Diesel engines operate with excess air because the fuel and air are not homogeneously mixed when autoignition occurs. One way of obtaining a better fuel–air mixture is to improve fuel atomization. This can be done either by upgrading the fuel injectors (discussed next) or by increasing the injection pressure; for many large marine engines, this means upgrading the common rail fuel system to accommodate a 25–50% increase in maximum injection pressures (Alexandersson and others 1993).

Fuel Injector Upgrades. Similar to increased fuel pressure, the objective of upgrading the fuel injectors is to improve atomization. The number and distribution of injector holes is limited partly by the strength of the injector tip and partly by manufacturing limits on hole size; another limitation is the ability for the spray pattern to avoid oxygen-deficient areas (Alexandersson and others 1993).

Exhaust Gas Recirculation. EGR entails recirculating some of the exhaust gases into the cylinder with the intake air. It increases the specific thermal capacity and decreases the temperature in the combustion space, and displaces oxygen that might otherwise participate in NO_x formation (Alexandersson and others 1993).

Pre-Engine Technologies

Water Added to the Combustion Air. Conceptually similar to increased aftercooling, injecting water into the intake air is another way to cool the charge air. Water injection is a proven method to reduce NO_x in gas turbines (Sierra Research 1991; Urbach and others 1997). Water injection into the air system is easier to implement than water/fuel emulsions but can cause corrosion of engine parts, and water quality is a greater concern (Sierra Research 1991; Venkatesh 1996).

Water/Fuel Emulsions. Adding water to the fuel is a proven technique to reduce NO_x, and it lends itself to application with residual fuels that often already contain some emulsified water and other blended fuels (MAN B&W 1996). A standard engine design permits the addition of about 20% water (as a percentage of fuel mass) at full load, although addition of more than 50% water has been tested (MAN B&W 1996).

Post-Engine Technologies

Selective Catalytic Reduction. SCR provides greater reductions in NO_x emissions than any of the technologies discussed earlier. It has been used on shoreside power plants and land-based industry combustion systems, and marine application of this technology has been the focus of considerable interest (Gibson and Groene 1991; Alexandersson and others 1993; Wartsila NSD 1994; Cooper and Peterson 1995; MAN B&W 1996, 1997a; Venkatesh 1996; Woodyard 1998). Catalytic reactions with ammonia or urea reduce the oxidized nitrogen to nitrogen gas (Alexandersson and others 1993). Although both types of catalyst systems have been installed on marine vessels, urea may be favored because it is nontoxic and biologically harmless and can be transported without problems (Woodyard 1998).

Table 13-1 is a summary of the technology discussion above, focused on the attributes most important in considering the potential NO_x control. Some technologies are better than others in terms of NO_x reductions, but trade-offs that result include reductions in fuel economy and increases in other pollutant emissions. The determination of average values for NO_x reduction and fuel penalty was based on a weighted average of reported values. Upper and lower bounds of NO_x reduction and fuel penalty parameters are provided to convey both variability in controls and uncertainty in technology performance. Average values quantifying the effects of each technology on other pollutants were developed similarly from the literature but with much less information available. These quantitative values are taken from the available literature; however, the emissions trade-off between NO_x reduction and increased emissions for other pollutants (especially particulate matter) is not always reported.

Cost and Pollution Reduction Assessment

Because of the different trade-offs and performance associated with each technology, cost considerations must be included. In this section, we describe an approach to estimating life-cycle costs of the nine NO_x control policies for existing ships. Because these technologies are currently available for marine

TABLE 13-1. Summary of NO_x Control Technology Performance Attributes

	Change in NO_x (%)			*Change in PM and other pollutants (%)*	*Change in fuel use (%)*		
Technology	*Average*	*Lower bound*	*Upper bound*		*Average*	*Lower bound*	*Upper bound*
Aftercooler upgrade	−10	−12	−10	+1	+2	+10	+1
Engine derating	−14	−9	−9	+10	+4	+4	+4
Fuel pressure increase	−14	−8	−8	+21	+2	+3	+1
Injector upgrade	−16	−6	−10	+21	+2	+3	+1
Injection timing retard	−19	−11	−9	+11	+4	+6	+3
Water in combustion air	−28	−12	−18	−1	+3	+3	+2
Exhaust gas recirculation	−34	−16	−15	+51	0	+2	0
Water/fuel emulsion	−42	−31	−22	−15	+2	+8	−2
Selective catalytic reduction	−81	−17	24	0	0	+2	0

Note: PM = particulate matter.

application, costs were taken directly from or derived from publicly available sources. The results, presented as net present values (NPVs), define a set of preferred technologies according to their NO_x reduction potential.

Cost Estimation Methodology

Two basic types of cost data are considered: fixed and annual. Fixed costs of NO_x control include completing the actual retrofit design, purchasing hardware and equipment, and installing the required parts. Annual costs result from absorbing fuel penalties associated with retrofit NO_x controls, purchasing other materials (such as ammonia or urea in SCR systems), and paying for other operating expenses (mostly labor) that result directly from the retrofit technology.

Data for these cost estimates are taken as reported where possible and derived otherwise. Our analysis uses new engine technology costs developed by EPA for the largest marine engines addressed by the proposed rule (U.S. EPA 1998a, 1998b), except where other technology cost data were considered to be more detailed.[3] The primary exception is that whereas EPA calculated only the incremental costs on top of a new engine without NO_x controls, we used the full price of equipment. Labor rates are assumed to be $500 per person-day, typical of labor rates (plus benefits) in public shipyards (Rivenbark 1996). Costs developed here are idealized to some degree, because they apply to a "typical" engine that propels a large ship operating internationally. However, these costs were discussed and generally validated through discussions with representatives of Wartsila NSD Corporation (personal communication from C. Broman and O. Koivisto to J.J. Corbett, Feb. 17, 1999).

To determine how much it would cost to design the retrofit system, we applied an approach similar to the one used by EPA to estimate R&D costs for a new engine (U.S. EPA 1998a, 1998b). Both manufacturer costs to design a general solution to the problem and specific implementation costs for various kinds of ships in the fleet are included. Total manufacturer costs were calculated and then distributed across all ships to determine a per-ship cost. A 10-year implementation schedule was used to estimate the number of retrofits that would occur in the first year. Similar to the EPA method, our approach places retrofit design costs up front and makes the estimate larger (more conservative). Retrofit design costs range from $0 for options such as derating and injection timing retard (which require only operational modifications) to $3,300 per ship for fuel system modifications for the case where an engine manufacturer will have to spread the cost of centralized research (as much as $1,500,000) over the expected number of engines to be retrofitted to recover costs.

Hardware costs for engine derating assumed that new injectors were needed that would limit the maximum engine power to the lower rating.

Injection timing retard has no hardware costs—only labor costs to retime the engine. Additional water distillation capacity was assumed in the costs for humidifying combustion air, with distillation costs provided through an industry quote (personal communication from Paul Choules, MECO, to J.J. Corbett, Feb. 5, 1999) and other costs taken from a report for EPA (Venkatesh 1996). The same increased distillation capacity was assumed for water/fuel emulsification, with equipment costs for emulsification provided by a Swedish Transport Research Board report (Alexandersson and others 1993). Costs for SCR technology onboard ship were taken from the same report.

To estimate the annual increase in fuel costs caused by implementing the retrofit technologies, average daily fuel usage was converted to an annual basis and multiplied by the percent increase in fuel usage shown in Table 13-1 and by a representative fuel price (MGN 1999): $66 per ton of residual fuel, $130 per ton of marine distillate oil. Other annual costs were taken from the various sources that provided technology costs.

Fixed, Annual, and NPV Costs

Table 13-2 presents the fixed and annual costs estimated for each NO_x reduction technology considered. Several observations can be made at this point. First, hardware and installation costs vary considerably but generally increase with increased NO_x reduction. The exceptions are the costs for injection timing retard and EGR. Second, retrofit design costs are generally similar, and relatively small, across technologies. Third, annual maintenance and operating costs appear small in general compared with increased annual fuel costs, with the exception of SCR. This difference is expected because SCR does not increase fuel consumption but does involve significant consumption of either urea or ammonia, as discussed earlier.

The NPV of these estimates was calculated using a 15% interest rate and adopting a very conservative approach, a life span of 23 years after retrofit. This estimate is equivalent to assuming that once upgraded, a ship would be in service for as many more years as a new ship would be. However, with a 15% discount rate, this assumption has less effect on the NPV costs than a lower discount rate would. The NPV costs for each technology are presented in Table 13-2.

The data in Table 13-2 include the average NO_x reductions. However, these NO_x reduction values may vary across the ranges shown in Table 13-1, as a result of either uncertainty in technology effectiveness or intentional variations in engineering design. For example, the percentage of water in water/fuel emulsions may be adjusted to achieve different levels of NO_x control. Additionally, whereas annual costs should be relatively robust, because they depend on the increased annual fuel usage more than any other factor,

TABLE 13-2. Summary of Expected NO_x Reductions and Fixed, Annual, and NPV Costs for NO_x Control Technologies Applicable to Existing Engines

		Fixed costs (thousand \$)		*Annual costs (thousand \$)*		
Control technology	*Change in NO_x (%)*	*Retrofit design (R&D-like)*	*Hardware and installation*	*Increased maintenance and operating costs*	*Increased fuel costs*	*NPV cost[a] (thousand \$)*
Aftercooler upgrade	–10	0.7	11.0	0.0	27.0	184
Engine derating	–14	0.0	34.0	1.5	54.0	386
Fuel pressure increase	–14	2.1	34.0	0.0	29.0	220
Injector upgrade	–16	3.3	38.0	0.0	24.0	192
Injection timing retard	–19	0.0	0.3	3.0	54.0	363
Water in combustion air	–28	1.8	130.0	2.6	34.0	365
Exhaust gas recirculation	–34	1.1	2.5	0.0	2,600.0	16,900
Water/fuel emulsion	–42	1.8	117.0	1.0	31.0	325
Selective catalytic reduction	–81	1.8	283.0	30.0	0.0	475

Notes: Some differences from previous numbers may appear due to rounding of calculations. NPV = net present value.

[a] At 15% interest over 23 years.

the NPV cost estimates in Table 13-2 may be sensitive to errors in estimated fixed costs. Specifically, U.S. EPA (1998a, 1998b) reports that after-market engine equipment may cost three times more than the same equipment when provided in a new engine package.

To illustrate both of these uncertainties, Figure 13-2 presents the NO_x control technologies according to their relative costs and pollution control. All costs shown assume that control technologies are operated at all times. Later, we explore the potential impact of policy scenarios where the operation of NO_x control equipment is more limited.

Relative Performance of NO_x Control Technologies

As seen in Figure 13-2, the point estimates developed in this chapter suggest a clear progression of preferred technologies that increase NO_x reductions at the least cost. The least-cost control curve would include aftercooler upgrade, fuel injector upgrade, water/fuel emulsion, and SCR. The technologies that simply appear too expensive include engine derating and EGR. Notice that

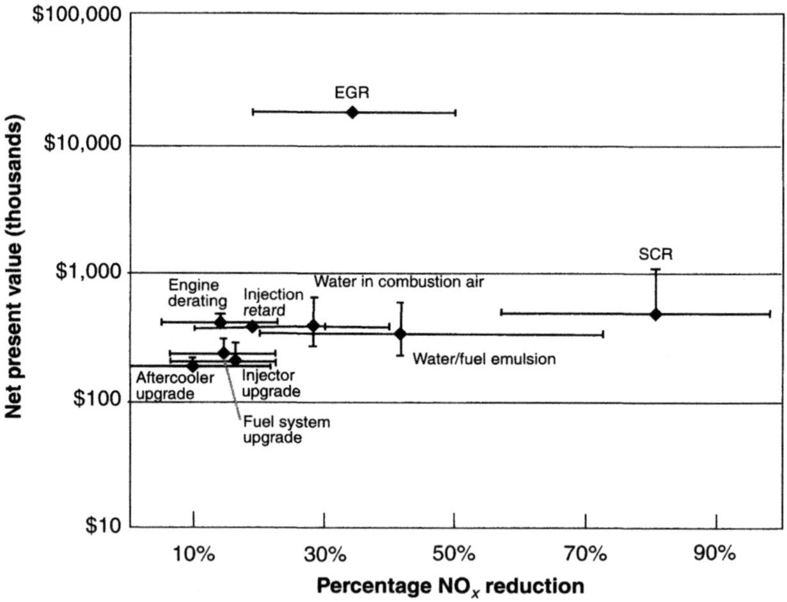

FIGURE 13-2. NPV of Control Costs (Fixed + Annual) and NO_x Control for Existing Engines

Notes: The interest rate used in the calculation was 15%. Emissions control is assumed during 100% of operations. Horizontal bars indicate the range of reported NO_x reductions; vertical bars indicate the range of NPV costs. EGR = exhaust gas recirculation; SCR = selective catalytic reduction.

the technologies with lower average NO_x reductions overlap NO_x reduction ranges. This fact may imply that when costs are similar, an operator may select the technology with greater average NO_x control, even if it exceeds the minimum required control. It could ensure greater confidence in the amount of NO_x reduction and some insurance against future regulatory changes. The most interesting technologies in this regard are water in combustion air and water/fuel emulsions. With such a wide range of options available for reducing NO_x emissions, the ability to control the technology performance during operation may be desirable for an operator.

Impact of Technology Cost on the Ship Operator

The costs involved in controlling NO_x will not be negligible, especially to the ship operator. Table 13-3 lists the type and importance of general operating costs for an international bulk carrier; fuel makes up the largest proportion of total operating cost (Wartsila NSD 1997). To demonstrate how significant different NO_x control costs would be, they can be compared with normal operating costs. For example, using the costs associated with SCR operated during 100% of main engine operations (annualized cost of about $86,000 and an estimated annual fuel cost of $1.34 million), NO_x controls require approximately 6% of annual fuel costs,[4] or less than 4% of total operating costs. Depending on how quickly shipping rates can adjust to account for variations in the incurred costs, operators at the margin could be most affected by aggressive regulatory policies that mandate retrofit technologies.

The impact on the price of goods is expected to be lower. Freight rates are between 4% and 13% of the import price of low-value bulk commodities; the worldwide average is about 9% (UNCTAD 1995). For high-value goods, such as manufactured items and containerized cargoes, freight rates make up a much smaller percentage of the final commodity price. If shipping contributes about 9% to the cost of goods and we assume that in the long run, 100% of operator costs will be passed on to the consumer, then the 4% change in operating costs could eventually increase the costs of goods by about 0.3%.

TABLE 13-3. Operational Costs for a Bulk Carrier

Operational cost	*Amount (%)*
Total	100
Fuel	59
Crew	24
Repair and maintenance	7
Other	9

Source: Wartsila NSD 1997.

Fuel price increases also will increase the fuel penalties associated with NO_x control technologies. However, they would change proportionally, motivating research into minimizing the fuel-related costs of NO_x control.

Cost-Effectiveness Comparison with Other NO_x Control Efforts

The cost-effectiveness of these retrofit NO_x controls can be compared with NO_x control strategies for new ships and for other combustion sources (both mobile and stationary). Regulators define cost-effectiveness as the annualized cost of control divided by the annual mass of emissions reduced (Koeberlein and others 1997). Policymakers use cost-effectiveness to compare and justify control technologies and regulations.

To determine the annualized cost of each technology, NPV costs were spread over 23 years, the maximum expected life of a retrofitted ship. Reducing the expected lifetime to 10 years increased the annualized costs for capital-intensive technologies (for example, 10-year annualized costs increase by 26% for SCR, by 25% for water/fuel emulsion and water in combustion air, and by 14% for aftercooler upgrade).[5]

By using estimated annual NO_x emissions reduction per ship,[6] cost-effectiveness for each technology can be determined. Despite its high costs, the most cost-effective technology is SCR; the significant NO_x reduction potential (denominator) offsets its higher technology cost (numerator). For comparison purposes, Figure 13-3 shows the minimum, median, and maximum cost-effectiveness values currently accepted by the California Air Resources Board (CARB) for stationary engines (Koeberlein and others 1997).

These values are for stationary internal combustion engines (which are most similar to marine diesel engines) and range between \$140 and \$25,000 per ton NO_x removed; CARB also has reported cost-effectiveness for mobile sources, including those with diesel engines (Koeberlein and others 1997). EPA estimated the cost-effectiveness of regulatory standards for new marine diesel engines (Tier 2 and Tier 3 standards that take effect in 2004 and 2008, respectively) and provided four other recent U.S. EPA mobile source regulations that required NO_x reductions (U.S. EPA 1998a, 1998b). These cost-effectiveness estimates are summarized in Table 13-4.

Can Control Technologies Enable Existing Engines to Meet Proposed NO_x Limits?

Two kinds of policy strategies could be used to reduce emissions from existing ships: mandate that ships install a particular technology under certain operating conditions, or mandate an emissions standard that ship engines

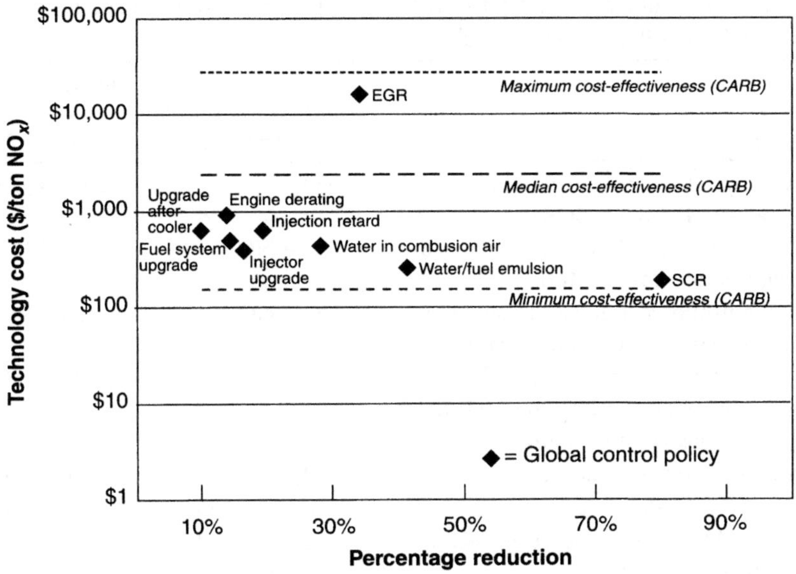

FIGURE 13-3. Cost-Effectiveness of NO_x Control Technologies

must meet. We have shown that at least nine technologies could reduce NO_x from existing ship engines and have estimated the various NO_x reductions and costs for each technology. However, the question of whether and which of these technologies could comply with proposed standards (as written for new engines) was not directly addressed; it is taken up here. An analysis using emissions data from actual ship engines currently in service is used in a technology policy "thought experiment" to explore whether the two policy strategies have different outcomes.

IMO (1998) and U.S. EPA (1999) regulations apply to a composite emission factor obtained by applying a duty cycle to emissions taken at various loads;[7] Table 13-5 summarizes these standards. EPA calculates that meeting IMO standards will require, on average, a 17% NO_x reduction for new engines (U.S. EPA 1998a, 1998b). However, existing engines may require greater NO_x control to meet new engine standards. This increase can be expected because manufacturers can design a new engine to meet the standard, but for an existing engine, they may have to select one of the off-the-shelf technologies that ensure compliance.[8]

The E3 duty cycle[9] for heavy-duty marine engines was used because it is considered to represent typical overall engine loads for exhaust emission measurement (ISO 1996; Markle and Brown 1996). It was applied to each of the engines tested by Lloyd's to convert raw observations of emissions at var-

TABLE 13-4. Summary of Cost-Effectiveness for NO_x Control from Existing Ships (This Work) and Other Recent NO_x Control Programs

NO_x rule	*Pollutants considered in calculations*	*Cost-effectiveness (\$/ton NO_x removed)*
This study, all 9 control technologies (low = SCR, high = EGR)	NO_x	200–16,000
U.S. EPA proposed Tier 2 for new marine engines (including IMO)	HC + NO_x	76–738
U.S. EPA proposed Tier 2 and Tier 3 for new marine engines (excluding IMO)	HC + NO_x	30–280
Clean fuel fleet program (heavy-duty)	NO_x	1,300–1,500
2.5 g/hp-h NMHC + NO_x standards for highway heavy-duty engines	NMHC + NO_x	100–600
Locomotive engine standards	NO_x	160–250
Nonroad Tier 2 standards	NMHC + NO_x	480–540
Various CARB mobile source emissions control measures	Both THC + NO_x and NO_x only	180–38,000
Various CARB stationary IC engine measures	NO_x	140–25,000

Notes: SCR = selective catalytic reduction; EGR = exhaust gas recirculation; EPA = Environmental Protection Agency; IMO = International Maritime Organization; HC = hydrocarbon; g/hp-h = grams/horsepower-hour; NMHC = nonmethane hydrocarbons; CARB = California Air Resources Board; IC = internal combustion (engine).

ious loads to average emissions factors for each vessel. NO_x concentrations (in parts per million) from 61 engines on 59 vessels were measured at various loads during the Marine Exhaust Emissions Research Programme and reported by Lloyd's Register (1990). Figure 13-4 presents the estimated composite emissions for each engine in Lloyd's data after applying the E3 duty cycle. Without any retrofit technologies, approximately 36% of the slow-speed engines and approximately 40% of the medium-speed engines met IMO standards for new engines. Table 13-6 summarizes the percentage of engines that met proposed NO_x standards.

To get an idea of the benefit of NO_x control for existing engines, the average reductions estimated for each of the technologies described earlier were applied to the uncontrolled E3 duty cycle emissions in Figure 13-4. E3 composite emissions were directly multiplied by average NO_x reductions, assuming that the NO_x control technology would affect all E3 modes equally, or in such a way as to provide the average NO_x control overall. Only two technologies achieved NO_x reductions great enough to achieve 100% IMO compliance for the engines tested by Lloyd's: water/fuel emulsion and SCR. Only SCR

TABLE 13-5. IMO and EPA Marine Diesel Engine NO_x Standards

Policy body	*Effective year*	*Regulation applied*	*NO_x (g/kwh)*
IMO	2000	$n > 2{,}000$ rpm	9.8
		$130 \leq n \leq 2{,}000$ rpm	$45^{-n-0.2}$
		$n < 130$ rpm	17
EPA	2004	Tier 2 limits (HC + NO_x)	7.0
	2008	Tier 3 limits (HC + NO_x)	5.0

Notes: n = maximum rated engine speed at each engine speed shown; rpm = revolutions per minute.

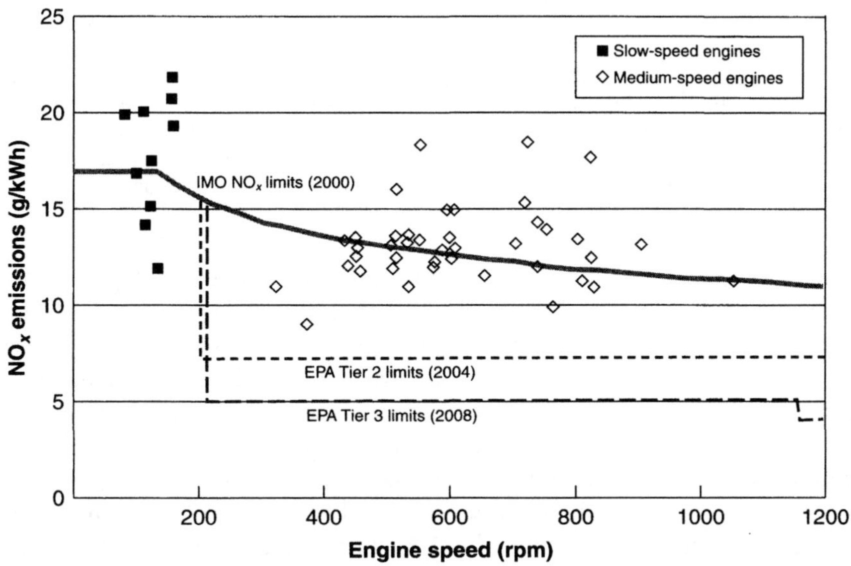

FIGURE 13-4. Uncontrolled Ship NO_x Emissions and Proposed Emission Limits

Notes: Data were estimated after applying the E3 duty cycle. g/kWh = grams per kilowatt-hour; rpm = revolutions per minute.

Source: Lloyd's Register 1990.

achieved NO_x reductions that were great enough for all engines to meet U.S. EPA Tier 2 and Tier 3 limits. These results are shown in Figures 13-5 and 13-6.

Therefore, if regulatory bodies wanted to ensure compliance through technology-specific policies, they would have to require SCR on all existing ships. However, this type of policy would overcontrol the emissions of some low-polluting ships. Requiring SCR for ships that are already compliant or that could be compliant with relatively inexpensive technologies would be economically inefficient. Table 13-6 shows the percentage of ships that could

TABLE 13-6. Summary of Compliance Percentages for Existing Engines under Proposed Standards for NO_x Control for Least-Cost Technologies Providing >17% Reduction

	Ships requiring each technology to meet standards[a] (%)				
Standard	*None*	*Injection timing retard*	*Water/fuel emulsion*	*SCR*	*Performance-based weighted costs (thousand $)*
IMO					
Slow-speed engines	36	55	9	0	229
Medium-speed engines	40	48	12	0	442
U.S. EPA[b]					
Tier 2 (2004)	0	0	35	65	423
Tier 3 (2008)	0	0	0	100	475

[a] Under E3 Duty Cycle, these percentages refer to the number of engines in Lloyd's data that met the NO_x limits shown (Lloyd's Register 1990).

[b] EPA standards apply to engines with smaller cylinder volumes than typical slow-speed engines; therefore, these estimates are applied only to the medium-speed engine data.

achieve at least 17% NO_x reductions by the least-cost technologies.[10] If we multiply the percentage of ships in each column by the cost of each technology from Table 13-2, then the technology-based approach costs between $33,000 and $246,000 more per ship than a flexible, performance-based approach. However, even though different ships may be able to meet emissions standards with different technologies, there are economic reasons for a shipping company to adopt a common solution across its fleet.

Only IMO limits apply to all international ships regardless of engine size, according to the equations in Table 13-5, which makes NO_x emissions dependent on the maximum rpm rating for each engine. EPA regulations (U.S. EPA 1999) apply to engines defined by cylinder sizes, not rpm ratings. However, for the purposes of this analysis, we include an assessment of the potential for existing oceangoing ships to comply with U.S. EPA Tier 2 and Tier 3 standards.

Likely Policy Scenarios for Existing Engines

In the analysis so far, we have considered only a global policy in which all existing ships are required to meet new emissions regulations all the time. Next, we consider two other possible policies: regional control and port control. If regional- or port-level policies are adopted, not all the operating costs previously outlined would be incurred. To minimize significant annual costs

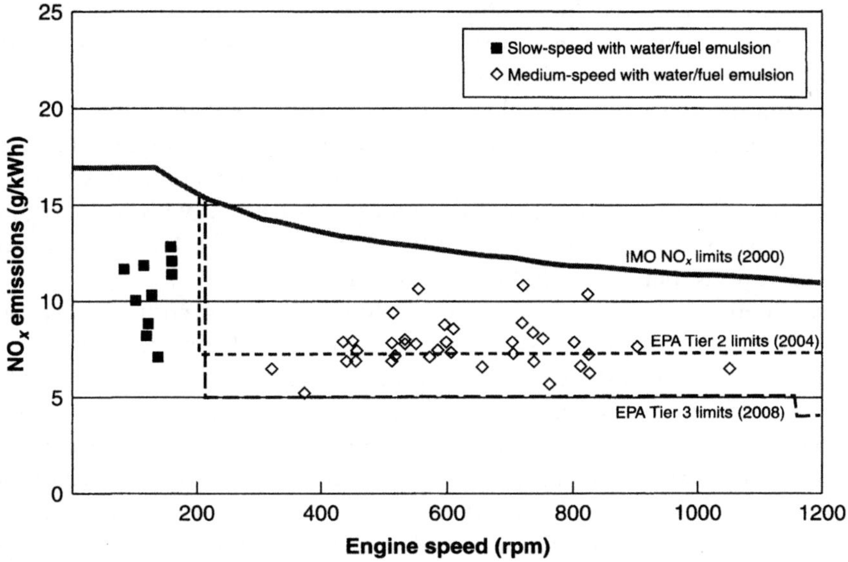

FIGURE 13-5. Ship NO_x Emissions and Limits Controlled with Water/Fuel Emulsion Technology

Note: Data were estimated after applying the E3 duty cycle.

Source: Lloyd's Register 1990.

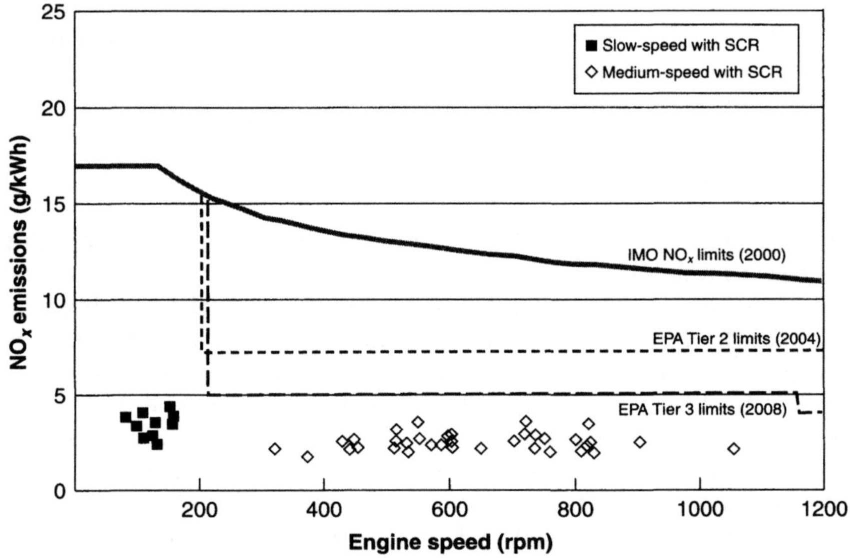

FIGURE 13-6. Ship NO_x Emissions and Limits Controlled with SCR Technology

Note: Data were estimated after applying the E3 duty cycle.

Source: Lloyd's Register 1990.

(for example, fuel penalties), ship operators that complied with emission requirements while in the controlled area could bypass or disconnect the NO_x control device when in uncontrolled regions.

A regional emissions control policy could be justified, according to scientific evidence (Benkovitz and others 1994) that suggests that emissions from ships can spread at least 400 km (with a mean of 900 km and a maximum of 1,700 km). Conceptually, this kind of policy is consistent with other regulations for ships, which describe "pollution control areas" intended to protect coastal regions that can extend many miles from shore.[11]

Port-control emissions policies would limit emissions within a port itself or within a series of collocated ports and could extend control requirements to some reasonable distance considered within the port's jurisdiction. For the purposes of our analysis, port control extends 50 miles from the port, a distance consistent with the port studies conducted in southern California. Port calls assume that a ship will spend a total of two days in the port region per visit: half a day transiting inbound, one day in port, and half a day transiting outbound of the 50-mile region. No downward or upward adjustments are made to account for decreased load while in port or increased load if cargo pumps, refrigeration units, and so forth, are needed. These types of policies have precedence in terms of ship pollution, where the U.S. Coast Guard defines bilge water discharge limits for all ships within 12 nautical miles (U.S. Coast Guard and Department of Transportation 1998a, 1998b). Because requirements regarding international trade, freedom of navigation, and national ports of entry do not permit ports to establish independent standards, port-control scenarios for NO_x must be done through a federal or international mechanism.

Costs to the Shipper

The costs per ship to reduce NO_x emissions by the nominal control achievable by each technology are presented for the three policy scenarios in Figure 13-7. Costs change as a direct function of annual costs avoided during operations outside of the emissions control zone. In general, the port-control case accrues the least cost per ship. This case defines the lower boundary of preferred technologies, and the global scenario with emissions control 100% of the time defines the upper boundary of preferred technologies. Although the costs to the shipper vary under different policy scenarios, the rank order of least-cost technologies does not change, except that injection timing retard (with almost zero fixed costs) replaces injector upgrade under all regional- and port-policy cases.

However, some technologies that might be considered contenders in the global analysis cannot compete under other policies. Specifically, engine der-

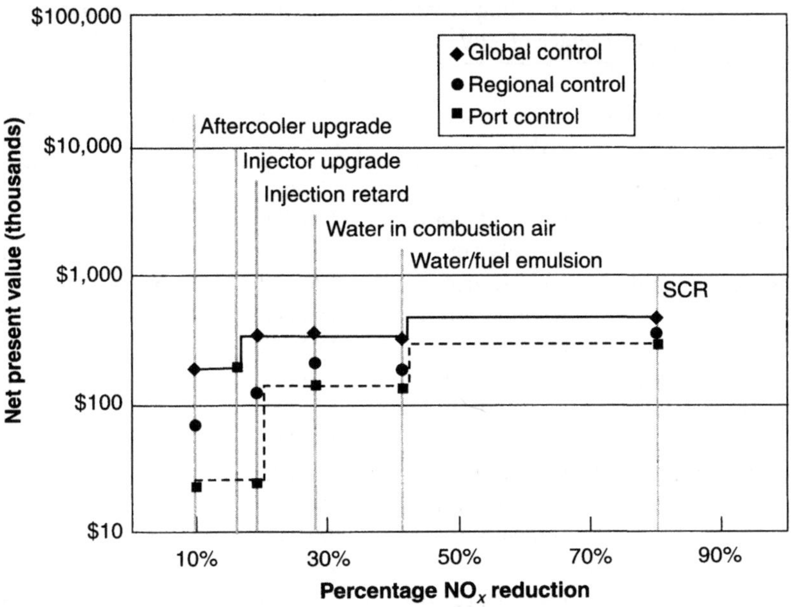

FIGURE 13-7. NPV of Control Costs (Fixed + Annual) and NO_x Control for Existing Engines under Various Cases for Each Policy Scenario

Note: The interest rate used in the calculation was 15%.

ating, fuel system upgrade, and injector upgrade are "permanent" retrofits in the sense that the ship operator cannot bypass them, so resulting annual costs (primarily from fuel penalties) cannot be avoided under less than global control. Also, EGR cannot be competitive under any policy scenario because of the requirement that it use marine distillate oil for the equipment to work.

Comparison of Cost-Effectiveness Estimates under Different Policy Scenarios

Table 13-7 shows the cost-effectiveness measures for various combinations of technology policy. Note that the NPV costs and cost-effectiveness of the different technologies vary considerably. Using an effectiveness cutoff of $738 per ton NO_x removed (EPA cost estimate for their new regulations [Table 13-4]), only water/fuel emulsion and SCR are acceptable across all three policy scenarios.[12] Moreover, there is a significant trade-off between cost to shippers and cost-effectiveness (the more effective technologies are more expensive). This trade-off is shown in Figure 13-8, where the combination of technologies and policies reveals a convex curve defining a "frontier" that maximizes

TABLE 13-7. Summary of Existing Engine Technologies and Policy Scenarios Showing Annual Costs as a Percentage of NPV

NO_x control technology (change in NO_x)	*Policy scenario*	*NPV costs, 10 years at i = 15% ($)*	*Annual costs in NPV (%)*	*Cost-effectiveness ($/ton NO_x)*
Aftercooler upgrade (–10%)	Port control	21,000	43	920
	Regional control[a]	103,000	89	640
	Global control[a]	146,000	92	620
Fuel system upgrade (–14%)	Port control	180,000	80	2,420
	Regional control	180,000	80	770
	Global control[a]	180,000	80	520
Unit injection upgrade (–16%)	Port control	160,000	74	1,877
	Regional control[a]	160,000	74	600
	Global control[a]	160,000	74	410
Injection timing retard (–19%)	Port control	19,000	99	790
	Regional control[a]	194,000	100	620
	Global control[a]	285,000	100	620
Water in combustion air (–28%)	Port control	146,000	8	1,100
	Regional control[a]	257,000	48	560
	Global control[a]	315,000	58	470
Water/fuel emulsion –42%)	Port control[a]	130,000	8	670
	Regional control[a]	229,000	48	340
	Global control[a]	281,000	58	280
Selective catalytic reduction (–81%)	Port control[a]	295,000	3	670
	Regional control[a]	386,000	26	300
	Global control[a]	434,000	34	230

Notes: NPV = net present value; *i* = interest rate used in the calculation.

[a] More cost-effective than U.S. EPA Tier 2 standards for new engines (see Table 13-4).

policy cost-effectiveness at the least cost to the ship operator. The frontier of dominating technology policy combinations includes (in order of increasing cost to the ship operator) injection timing retard under a port-level policy, injector upgrade under a global policy, water/fuel emulsion under a regional or global policy, and SCR under a global policy.

The reason for this trade-off is that more NO_x is removed during greater periods of operation, and technology costs (particularly capital costs) are distributed over this larger reduction. These trade-offs are important if policymakers intend to allow ship operators to select among several NO_x control alternatives rather than mandate one or two. Devising a regulatory framework that accounts for this trade-off would require negotiations between ship operators and regulators.

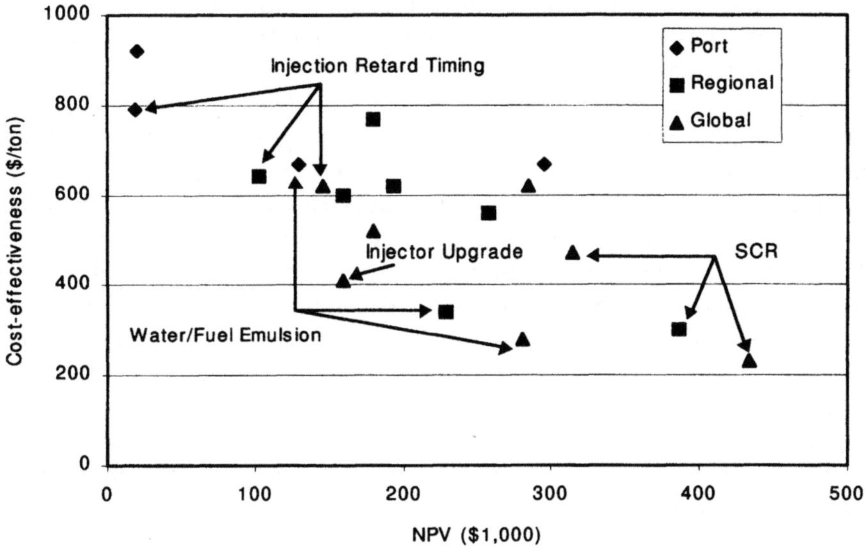

FIGURE 13-8. Trade-Offs between Cost to Ship Operators and Policy Cost-Effectiveness under Global, Regional, and Port Policy Scenarios

Incentives

The apparent dominance of technology choices under port and global policies shown in Figure 13-8 does not mean that market incentives would be ineffective under regional policy scenarios. For example, Sweden established differentiated dues that entered into force January 1, 1998 (National Swedish Maritime Administration 1996; Wood-Thomas 1999).[13] By reducing NO_x emissions to less than 12 grams per kilowatt-hour, another $0.20 per gross registered metric ton can be saved. Before 2000, Sweden also offered an investment subsidy for the installation of SCR equipment for NO_x control that equaled 40% of investment costs; a 30% subsidy is available for ships that install SCR before 2002. The European Commission is in the process of evaluating the market implications for maritime transportation and trade in the region (Swedish Port and Stevedores Association 1999; Wood-Thomas 1999). However, the incentives are considered to be neutral for competition because they are applied to all vessels and not limited to vessels registered in Sweden (Wood-Thomas 1999).

Market incentives also may take the form of emissions trading schemes, in which a portion of the reductions in emissions from one NO_x source can be sold to facilitate permitting or expansion by another source. For ships to opt into an emissions trading scheme, some conditions will need to be met. Actual ship emissions occurring within the eligible region on a per-ship basis must be accurately measured or estimated. This task could involve continuous or predictive emissions monitoring (Cooper and Andreasson 1999). The current market for NO_x allowances ranges from \$2,000 to \$5,500 per ton of NO_x emissions (ICF Resources 1995; Cantor Fitzgerald 1999; Farrell and others 1999; Natsource 1999), values clearly higher than the costs per ton of NO_x emitted shown for ships in Table 13-7. If these prices hold, ship operators might be able to realize significant economic return from installing emissions control systems—and incentive to install technologies with greater NO_x reduction such as SCR would be strong. Although this mechanism would be most easily tailored to vessels that operate within the emissions trading region 100% of the time (for example, ferry vessels), the opportunity for a cargo ship operating along a coastline (or between ports with separate NO_x trading programs) to participate in different trading regions could be possible.

Concerns from the Shipping Industry Perspective

As discussed earlier, NO_x controls would equal approximately 6% of annual fuel costs, or less than 4% of total operating costs. If the retrofit policy schedule is lengthy, then the fleet would include both newly built vessels with new engines and older vessels with retrofitted engines during the transition. The increased operating costs for the retrofitted ships (apparent from the analysis presented here) could make them less competitive unless subsidies or increased shipping rates were put in place. Operators at the margin would be most affected by aggressive policies in this regard. Most likely, a ship operator would simply delay a retrofit until the last opportunity within the regulatory framework. This situation could all change if the emissions trading program became feasible. With the current NO_x trading price, rapid adoption of retrofit technologies could be possible. In any event, the pass-through price to consumers appears to be modest (about 0.3% the costs of goods).

Concerns from the Policymakers' Perspective

Policymakers want to implement a strategy that is capable of achieving affordable, meaningful reductions when and where they are needed. Achieving such a goal would require global, or at least multinational, policy coordination—especially for international shipping. Individual nations concerned with emissions from shipping traffic off their coasts would not be able to achieve local or regional objectives by regulating their own international ships, because most ships are registered under flags of convenience. Perhaps, as in Sweden, strategies could be implemented through market mechanisms if incentives are not biased in favor of home nation's flag (nontariff trade barrier) and not biased against the home nation's flag. This problem may be why EPA waived U.S. flag ships operating in foreign trade from its proposed rules.

Other policymaker concerns are more traditional and fall into the areas of compliance verification, enforcement, and monitoring. Written logs and computerized engine data logs can satisfy some of these concerns. For example, automatic engine monitoring systems will record when injection timing retard on modern engines is done (personal communication from C. Broman and O. Koivisto to J.J. Corbett, Feb. 17, 1999). In fact, verification and monitoring would not be difficult for the technologies that appear to be most feasible in terms of flexible technology, cost, and policy attributes. The more difficult concern would be enforcement, because for international shipping, it involves nongovernmental bodies (such as IMO), bilateral and multilateral international treaties, national territories, and regional impacts. These concerns will likely persist into the discussions of policy mechanisms and technical details.

Summary

Because of a series of misconceptions, international shipping has managed to avoid emissions control regulations. This situation changed recently, but currently proposed policies regulate only new vessels. Current policies are unlikely to have meaningful impact in the short-to-near term, because of the long life of ships. In this chapter, we have shown how several current technologies can reduce NO_x emissions from existing ships. The technologies vary on many dimensions, forcing policymakers to make difficult trade-off decisions. Technologies that are most cost-effective in reducing NO_x emissions would be expensive to install and operate.

The difficulties of establishing a viable regulatory policy are complicated not only by technological and economic uncertainties but also by international politics, verification requirements, and enforcement issues. Identifying possible technological solutions is only the first step in this process. Setting the geographic scope of the policy probably would have a direct impact on which technologies are likely to be used and on the amount of pollution controlled. For example, the sequential establishment of independently regulated port-control zones could lead to some suboptimal decisions. On the other hand, complementary regulatory standards implemented by a few dominant ports in world trade could encourage industry-wide behavior very similar to a shipping industry response to a single international standard. In either case, understanding the uncertainties can assist all stakeholders in the policy-making process.

Notes

1. Compare this speed with high-speed diesel engines in trucks, which typically operate at higher than 2,100 rpm, and medium-speed diesel engines in power generation, which often operate at 500–1,200 rpm.

2. Other engineering control strategies exist, some of which are proven to reduce NO_x in new marine engines and others that are being actively researched (NAVSEA 1992, 1994; Alexandersson and others 1993; NIAG 1996; Venkatesh 1996; U.S. DOE 1997; U.S. EPA 1998a, 1998b; Wartsila NSD 1998). They include direct in-cylinder water injection, in-cylinder ceramic coatings, nonthermal plasma systems, and variable geometry turbocharging. Because most of these technologies require substantial engine redesign, they are infeasible for retrofitting on existing propulsion systems and are not considered as options for achieving significant NO_x reductions in the existing fleet within the next couple of decades.

3. U.S. Environmental Protection Agency costs were used for four technologies: aftercooler upgrades, fuel system modifications to increase supply pressure, fuel injector upgrades, and exhaust gas recirculation.

4. Annualized costs are calculated from net present value costs over 10 years at 15% for this analysis, resulting in more conservative, higher annual costs than if a 23-year time period were used.

5. Absolute annual cost differences were less than $700 in all cases. For simplicity, this analysis ignores the transient costs that may occur during the implementation of NO_x control and the effects of fleet growth or emissions increase over time. However, this difference using a 10-year period does not change significantly any of the insights that result from the longer annualization period.

6. A regional perspective also can be used, but in previous inventory work (Corbett and Fischbeck 1997; Corbett and others 1999), we found that the differences between these approaches are relatively small and that cost-effectiveness on a per-ship basis provides the highest costs.

7. A *duty cycle* is a time-weighted load profile, where the percentage of time operating within set power/throttle settings is multiplied by the energy or emissions at those settings. For example, it represents the amount of time spent at idle, low loads, medium loads, and full loads. International Organization for Standardization (ISO) standard duty cycles have been defined for marine engines and prescribed by U.S. Environmental Protection Agency regulations for emissions testing purposes.

8. Engine manufactures are currently offering new engines that they claim achieve 50% NO_x reductions from uncontrolled levels (Wartsila NSD 1997, 1998).

9. The E3 duty cycle is one standard duty cycle applied to marine vessels.

10. Injection timing retard is used here because the performance of the less costly injector upgrade did not achieve a 17% NO_x reduction on average. This result ignores

the likelihood that technological innovation would occur to enable injector upgrade performance from our estimate of a 16% reduction.

11. For example, the International Maritime Organization's regulations for oily water discharge from crude-oil tankers under MARPOL 73/78 prohibit discharges in special areas or within 50 nautical miles of nearest land (Harrington 1992; U.S. Coast Guard and Department of Transportation 1998a, 1998b). Other international regulations define limits for large regions such as exclusive economic zones, territorial seas, and contiguous zones (Bernaerts 1988).

12. This cost-effectiveness value is estimated by the U.S. Environmental Protection Agency for new engines meeting Tier 2 standards (see Table 13-4).

13. These dues in 1999 were 5 Swedish krona (SKr; about $0.60) per gross registered metric ton (GRT) for a ship with no pollution control, but they drop to SKr 4.10 (about $0.50) per GRT for vessels that operate on 0.5% sulfur and 1% sulfur for ferries and cargo vessels, respectively.

References

Alexandersson, A., E. Flodstrom, and others. 1993. *Exhaust Gas Emissions from Sea Transportation.* Gothenburg, Sweden, Mariterm AB, Swedish Transportation Research Board.

Benkovitz, C.M., C.M. Berkowitz, and others. 1994. Sulfate over the North Atlantic and Adjacent Continental Regions: Evaluation for October and November 1986 Using a Three-Dimensional Model Driven by Observation-Derived Meteorology. *Journal of Geophysical Research* 99(D10): 20,725–20,756.

Bernaerts, A. 1988. *Bernaerts' Guide to the 1982 United Nations Convention on the Law of the Sea.* Coulsdon, Surrey, U.K., Fairplay Publications, Ltd.

Cantor Fitzgerald. 1999. *NO_x Budget Allowance Bid/Offer Bulletin.* http://www.cantor.com/ebs/mb012699.htm.

Carlton, J.S., S.D. Danton, and others. 1995. *Marine Exhaust Emissions Research Programme.* London, U.K., Lloyd's Register Engineering Services.

Cooper, D., and K. Peterson. 1995. *Emission Measurements from a Urea-Based SCR/OXI Catalytic NO_x/HC Exhaust Gas Treatment System on Board a Diesel Powered Passenger Ferry—Operation after 12,000 Hours of Service.* Gothenburg, Sweden, Swedish Environmental Research Institute (IVL).

Cooper, D.A., and K. Andreasson. 1999. Predictive NO_x Emission Monitoring on Board a Passenger Ferry. *Atmospheric Environment* 33(28): 4637.

Corbett, J.J., and P.S. Fischbeck. 1997. Emissions from Ships. *Science* 278(5339): 823–824.

Corbett, J.J., P.S. Fischbeck, and others. 1999. Global Nitrogen and Sulfur Emissions Inventories for Oceangoing Ships. *Journal of Geophysical Research* 104(D3): 3457–3470.

Farrell, A., R. Carter, and others. 1999. The NO_x Budget: Market-Based Control of Tropospheric Ozone in the Northeastern United States. *Resource and Energy Economics* 21(2): 103–124.

Genovesi, K., and L. Browning. 1998. *Incremental Cost Estimates for Marine Diesel Technology Improvements.* Ann Arbor, MI, U.S. Environmental Protection Agency, Office of Mobile Sources.

Gibson, J., and O. Groene. 1991. Selective Catalytic Reduction on Marine Diesel Engines. *Automotive Engineering* 99(10): 18–22.

Harrington, R.L. (ed.). 1992. *Marine Engineering.* Jersey City, NJ, Society of Naval Architects and Marine Engineers.

Hellmann, J.O. 1997. Emission Control: Two Stroke Low-Speed Diesel Engines. Presented at the General Meeting of the Institution of Diesel and Gas Turbine Engineers, London, U.K.

Heywood, J.B. 1988. *Internal Combustion Engine Fundamentals.* New York, NY, McGraw-Hill, Inc.

ICF Resources. 1995. Estimated Effects of Alternative NO_x Cap and Trading Schemes in the Northeast Ozone Transport Region, Fairfax, VA, ICF Kaiser.

IMO (International Maritime Organization). 1998. *Annex VI of MARPOL 73/78 and NO_x Technical Code.* London, U.K., IMO.

ISO (International Organization for Standardization). 1996. *Reciprocating Internal Combustion Engines—Exhaust Emission Measurement—Part 4: Test Cycles for Different Engine Applications.* Geneva, Switzerland, ISO.

Koeberlein, D., B. Riddell, and others. 1997. *Sources and Control of Oxides of Nitrogen Emissions.* Sacramento, CA, California Environmental Protection Agency, Air Resources Board (CARB).

Lloyd's Register. 1990. *Marine Exhaust Emissions Research Programme: Steady State Operation* (including Slow Speed Addendum). London, U.K., Lloyd's Register of Shipping.

MAN B&W. 1996. *Emission Control: Two Stroke Low-Speed Diesel Engines.* Copenhagen, Denmark, MAN B&W Diesel A/S.

———. 1997a. *Emission Control: Two-Stroke Low-Speed Diesel Engines.* Copenhagen, Denmark, MAN B&W Diesel A/S.

———. 1997b. *How to Deal with Emission Control.* Copenhagen, Denmark, MAN B&W Diesel A/S.

Markle, S.P., and A.J. Brown. 1996. Naval Ship Engine Exhaust Emission Characterization. *Naval Engineers Journal* 108(5): 37–47.

MGN. 1999. *MGN Newsletter and HK Shipping News International.* Warren, RI, Maritime Information Systems, Inc.

National Swedish Maritime Administration. 1996. *Summary of a Proposal to the Swedish Government Concerning Environmentally Differentiated Shipping Dues.* Norrkoping, Sweden, Sjöfartsverket, National Swedish Maritime Administration.

Natsource. 1999. *AirTrends.* Natsource, Emissions Brokerage Desk, New York, NY, 2(12): 4. http://www.natsource.com/images/UplImages/Jan991.pdf (accessed March 2001).

NAVSEA (Naval Sea Systems Command). 1992. *Internal Combustion (Gas Turbine and Diesel) Engine Exhaust Emission Study.* Pittsburgh, PA, NAVSEA and Westinghouse Machinery Technology Division.

———. 1994. *U.S. Navy Marine Diesel Engine and Gas Turbine Exhaust Emissions.* Washington, DC, NAVSEA 03X31.

NIAG (NATO Industrial Advisory Group). 1996. *Prefeasibility Study on a NATO Environmentally Sound Ship of the 21st Century.* Brussels, Belgium, NIAG.

Rivenbark, T. 1996. *Financial and Operating Statements for Norfolk Naval Shipyard: Facts Derived from Year Ending FY 1996.* Norfolk, VA, Norfolk Naval Shipyard.

Sierra Research. 1991. *Regulatory Strategies for Reducing Emissions from Marine Vessels in California Waters.* Sacramento, CA, California Air Resources Board.

Swedish Port and Stevedores Association. 1999. *Differential Charging for Environmental Purposes.* Stockholm, Sweden, Swedish Port and Stevedores Association.

UNCTAD (U.N. Conference on Trade and Development). 1995. *Review of Maritime Transport 1994.* New York, NY, and Geneva, Switzerland, UNCTAD.

Urbach, H.B., D.T. Knauss, and others. 1997. The Reduction of NO_x Emissions from Marine Power Plants. Presented at Air and Waste Management Association's 90th Annual Meeting and Exhibition, Toronto, Ontario, Canada.

U.S. Coast Guard and Department of Transportation. 1998a. Vessels Carrying Oil, Noxious Liquid Substances, Garbage, Municipal or Commercial Waste, and Ballast Water. *Code of Federal Regulations* Title 33, Part 151.

———. 1998b. Oil or Hazardous Material Pollution Prevention Regulations for Vessels. *Code of Federal Regulations* Title 33, Part 155.

U.S. DOE (Department of Energy). 1997. *Abstracts of Papers Presented at the 1997 Diesel Engine Emissions Reduction Workshop, San Diego, CA.* Washington, DC, U.S. DOE.

U.S. EPA (Environmental Protection Agency). 1997. *AP-42: Compilation of Air Pollutant Emission Factors.* Research Triangle Park, NC, U.S. Environmental Protection Agency.

———. 1998a. Control of Emissions of Air Pollution from New CI Marine Engines at or above 37 kw; Proposed Rule. *Code of Federal Regulations* Title 40, Part 94.

———. 1998b. *Draft Regulatory Impact Analysis: Control of Emissions from Compression-Ignition Marine Engines.* Ann Arbor, MI, U.S. EPA Office of Air and Radiation, Office of Mobile Sources, Engine Programs and Compliance Division.

———. 1999. Control of Emissions of Air Pollution from New CI Marine Engines at or above 37 kW, Final Rule. *Code of Federal Regulations* Title 40, Parts 89, 92, and 94.

Venkatesh, S. 1996. *Reduction of NO_x and PM from Navy Diesel Engines: A Feasibility Analysis.* Mountain View, CA, Acurex Environmental Corp.

Wartsila NSD. 1994. *Wartsila Diesel Group Customer Journal: Marine News.* Annapolis, MD, Wartsila NSD Corporation.

———. 1997. *Wartsila 32: The Performer.* Annapolis, MD, Wartsila NSD Corporation.

———. 1998. *Wartsila 32: Technology Review.* Annapolis, MD, Wartsila NSD Corporation.

Wood-Thomas, B. 1999. *Swedish Incentives Reducing Ship Emissions.* Washington, DC, U.S. Environmental Protection Agency.

Woodyard, D. 1998. Designers Clean Up Diesel Exhausts. *Jane's Speed at Sea* June: 46–48.

14

Private Eyes in the Sky

Implications of Remote Sensing Technology for Enforcing Environmental Regulation

MOLLY K. MACAULEY AND TIMOTHY J. BRENNAN

An important potential benefit of emerging telecommunications technology involves the enhanced ability to detect environmental pollution by using remote sensing devices. Detailed information can be collected by existing and proposed systems of satellites, aircraft, and automated ground monitors and transmitted through telecommunications networks back to a central location for analysis and evaluation. Data collected by commercial civilian satellite systems can be particularly precise, with sophisticated spatial, spectral, and temporal resolution.

The environmental benefits of these technologies are obvious. They are capable of not only measuring air and water emissions, soil contaminants, and other indicators but also of helping to illustrate the "big picture" in assessing the ecological context of these factors, that is, showing the effects of pollution on humans and the environment.[1] In addition, emerging remote sensing technologies can improve the understanding of and in many cases reduce the range of uncertainty regarding the extent of environmental degradation (Judd 1994). The new technologies also may reduce the costs of compliance with and enforcement of domestic regulations and international agreements concerning the environment. Such finely tuned monitoring has several implications for the way that the U.S. Environmental Protection Agency (EPA) and other decisionmakers regulate. For instance, the data may permit closer monitoring of information that is self-reported by regulated parties and play a key role in adjudicating environmental lawsuits and disputes.

In this chapter, we focus on enforcement properties of space-based remote satellite sensing (RSS) technologies with respect to their implications for how regulatory policy is created and implemented.[2] These issues include the extent to which RSS is likely to reduce significantly the costs of monitoring, compliance, and enforcement for regulated parties, regulators, or both; whether the technologies affect incentives for regulated parties to self-report accurately;[3] and other ways RSS may change the structure and implementation of regulation. We assess RSS with respect to the environmental conditions it can sense; the timeliness, accuracy, and reliability of RSS; the accessibility of RSS data to regulators, regulated parties, and third parties; the implications of the ability of remote sensing satellites to observe without being observed; and, to a limited extent given the uncertainty and scarcity of the data, the cost-effectiveness of RSS compared with alternative sources of information. In our view, the growing role of new RSS technologies in environmental regulatory enforcement makes these economic, political, and legal questions increasingly important.

This chapter is organized as follows. In the next section, we describe RSS technologies. Then, we review economic analyses of environmental protection and regulatory enforcement, to indicate which factors are most important in deciding how RSS is likely to influence these activities. In the last two sections, we combine a review of the regulatory models and our assessment of RSS technologies to draw inferences about future policy directions for enforcing environmental regulations.

Remote Satellite Systems: Technology Overview

We first describe remote sensing technology and then discuss remote satellite systems, including both current technologies and technologies expected to be available in the near future. We focus on those aspects of the technologies that are most relevant to environmental enforcement.

Background[4]

Generally speaking, *remote sensing* is the act of observing an object at some distance removed from the object. Two of the most frequently cited definitions are offered in textbooks on remote sensing. According to Lillesand and Kiefer (1987),

> Remote sensing is the science and art of obtaining information about an object, area, or phenomenon through the analysis of data acquired by a device that is not in contact with the object, area, or phenomenon under investigation.

More specifically addressing the application of remote sensing to studying our planet, Campbell's (1987) definition of *remote sensing* is

> the science of deriving information about the earth's land and water areas from images acquired at a distance. It usually relies on measurement of electromagnetic energy reflected or emitted from the features of interest.

According to either definition, remote sensing can include a host of technologies, including aircraft-based observations, ground-based devices (such as sensors on a smokestack), and space-based observations.

Introduction to RSS

Our focus is space-based RSS observations. There are two general types of RSS observations: electrooptical and radar. The optical systems use either photographic or electromechanical sensing devices, and until recently, the bulk of space-based imagery has been the province of these electrooptical remote sensors. Radar imaging systems began with experimental sensors in the late 1970s and now include operating sensors, launched in 1995. Electrooptical and radar technologies may be situated on aircraft, spacecraft, or on the ground, but our discussion centers on space-based technologies.

RSS technologies usually are described on the basis of their spatial resolution. A conventional rule of thumb for judging spatial resolution is that at 10 meters, an airport runway and its navigation markings are discernible; at 5 meters, aircraft and large objects being loaded or unloaded can be seen; and at 1 meter, automobiles can be identified.

Photographs are the most typical and traditional means of remote sensing. They can be two- or three-dimensional (for three-dimensional or stereo photos, the camera obtains overlapping frames that are then exposed in sequence) and black and white (panchromatic), natural color, or infrared. The spatial resolution or sharpness of the photo depends on the camera lens, film, and distance from the lens to the object of interest. The spatial resolution of commercial satellites now ranges from one to three meters.

Electromechanical devices are detectors that sense the radiation that is naturally reflected from the surface of an object. Such devices include line scanners and charge-coupled devices (CCDs). Line scanners are mirrors that collect radiation from the surface of an object as the mirrors sweep across it. CCDs consist of arrays of thousands of tiny light-sensitive detectors that measure radiation directed onto the arrays through a system of lenses. The devices record the information in either analog or digital mode; digital data are particularly useful in that they can be readily processed, enhanced, and analyzed by computers. One advantage of electromechanical devices that facilitates the analysis of the collected data is that the orderly array of lines and image elements, or pixels, that they produce and their collection of different wavebands of radiation from the same portion of the surface means that the data from all wavebands of the radiation are exactly registered to each other.

Passive microwave imaging uses antennas to scan radiation from the microwave region of the spectrum. The resolution of an image from such a system is approximately the wavelength divided by the antenna diameter. Historically, because there has been a physical limit to the size of a portable antenna mounted on a spacecraft and because microwave wavelengths are millimeters to centimeters long, the spatial resolution at orbital altitudes has been quite coarse (on the order of 10–30 kilometers). Recent technological advances, together with public policies that now permit the use of these technologies for non-military-related applications, have greatly increased the spatial and spectral resolution available. Regardless of these advances, however, a disadvantage of passive microwave imaging for some applications is that it is affected by weather (clouds obscure the imaging) and cannot operate at night.

Active microwave or radar imaging directs energy towards the surface; energy reflected from the surface then travels back to the sensor. The time taken for the energy to travel back is a very accurate measure of distance between the sensor and the object. Space-based radar is useful for producing three-dimensional information; data about the height of the ocean surface, for instance, can be the basis for deductions about bathymetry and the gravitational field or about wind speeds. Radar imaging is unaffected by weather and can operate night or day. However, a major problem with radar is that the beam of energy spreads out with distance. For acceptable resolution, the antenna must be quite large.

Synthetic-aperture radar (SAR) solves this problem by exploiting the Doppler effect that shifts the frequency of returned energy. The Doppler history of the returned energy is recorded and after additional processing yields data of resolution equal to or greater than that obtained from real-aperture radar. SAR permits a short antenna to mimic one up to several hundred meters long. Even with SAR, however, the data rates and processing requirements can be quite high, taxing the capability of both the spacecraft and the data communication downlinks. Advances in data compression and bandwidth utilization techniques have begun to alleviate these problems.

Hyperspectral imaging is the most revolutionary of emerging RSS technologies. It senses energy in much higher regions of the electromagnetic spectrum and in much finer gradations than most RSS systems to date. The energy is in the form of “spectral fingerprints” in the infrared and shortwave light spectra—ranges invisible to humans. These data will complement the imagery from existing systems and enable the collection of data about a wider range of terrain, vegetation, and atmospheric conditions to produce ultrahigh-resolution information. Whereas a standard sensor with fewer than 10 bands, or ranges, of spectrum can differentiate between gross classes of vegetation, for instance, a hyperspectral imager can operate across 300 or even 400 wavelengths to distinguish wheat from alfalfa and identify healthy and unhealthy vegetative growth.

Existing RSS Systems

Although the RSS systems used to enforce environmental regulations can be and have been implemented as ground- and aircraft-based sensors, space-based platforms for this purpose are a relatively recent development. Space-based remote sensing in the civilian sector began in 1963 with programs under the aegis of the National Aeronautics and Space Administration (NASA) and the National Oceanographic and Atmospheric Administration (NOAA). The programs focused largely on experimenting with how to build and launch the spacecraft and sensors; record, transmit, and process the data; and interpret the data. Environmental applications of these early RSS systems typically were to record land use; monitor the health and yields of some agricultural crops and forest stands; and identify geologic formations useful in oil, gas, and nonfuel minerals exploration.

Since the advent of these experimental programs in the 1960s and 1970s, additional civilian remote sensing spacecraft have been launched by the U.S. private sector and by the U.S. government and foreign governments (see Table 14-1).[5] The earliest systems operated at large-scale resolution (spatial, spectral, and temporal), but newer systems have the potential to be used for observing details such as chemical interactions, air particulates, and other small-scale phenomena. The systems have been less useful for phenomena that are best observed in continuous time. Rather, the niche for RSS to date has been observation at more discrete time intervals. Aircraft (including remotely piloted aircraft, or drones), small rockets, balloons, and in situ sensors have been the preferred technologies for measurements on a much smaller scale, and real-time continuous monitoring has been the province of in situ sensors.[6]

At the present time, civilian remote sensing systems include the U.S. government's Landsat system; commercially owned and operated U.S. systems; and government systems operated in France, Russia, India, and Israel. Russia, Canada, Japan, and a joint venture by several European nations operate SAR systems. In most of these cases, the systems are directly owned by or finan-

TABLE 14-1. Existing and Planned Remote Sensing Satellites

Country	*Owner*	*Program*	*Launch date*	*Instrument type*	*Resolution (m)* P	M	R
Germany	G	Rapid Eye AG	2005 or later				
France	G	Spot 5B	2004	P, M	5	10	
United States	G	EOS AM-2	2004	P, M	10	30	
United States	C	Space Imaging	2003–04	P, M	0.5	2	
United States	C	GDE	2002		1		
United States	C	RDL	2002	R			1
United States	C	Boeing	2002	M		10	
Canada	G	Radarsat 2	2002	R			9
United States	C	Orbimage	2000	P, M	1	4	
United States	G,C[a]	Orbimage	2000	P, M, H			
United States	G	Space Imaging	2000	P, M	1	3	
United States	C	Earth Watch	2000	P, M	1	4	
United States	C	Space Imaging	2000	P, M	3	15	
United States	G	Terre (formerely EO-1)	1999	H			
France	G	Spot 5A	1999	P, M	5	10	
India	G	IRS-1 D	1999	P, M	10	20	
Korea	G	KOMSAT	1998	P, M	10	10	
United States, Japan	G	EOS AM-1	1999	M	15	15	
United States	G	Landsat-7	1998	P, M	15	30	
ESA	G	ENVISAT	1998	R			30
United States	C	Orbimage	1997	M	4		
France	G	Spot 4	1997	P, M	10	20	
Russia	G	Almaz 2	1996	R			5
Japan	G	ADEOS	1996	P, M	8	16	
China, Brazil	G	CBERS	1996	P, M	20	20	
Israel	G	EROS	1996		2		
Canada	G	Radarsat 1	1995	R			9
India	G	IRS-1 C	1995	P, M	10	20	
China, Brazil	G	CBERS	1995	P, M	20	20	
Russia	G	Resours-02	1995	M		27	

Notes: This list is not comprehensive. G = government funded; C = commercially funded; P = panchromatic; M = multispectral; R = radar; H = hyperspectral.

[a] The spacecraft is commercially owned and operated, but the hyperspectral instrument is owned by the U.S. government.

Source: *Aerospace America*, various issues; *Aviation Week and Space Technology*, various issues; www.licensing.noaa.gov.

cially backed by governments. For example, Landsat is owned by the U.S. government but operated by a private company, the Earth Observation Satellite Company (EOSAT). Radarsat International, Inc., of Vancouver, British Columbia, Canada, owns the rights to the imagery produced by the Canadian Radarsat system. Radarsat was the first spacecraft designed for the commercial radar imagery market.

Emerging RSS Technologies

Only in the past few years have RSS technologies begun to advance sufficiently, at low enough cost and with approval from national security agencies for the civilian use of high-resolution imagery, to make space-based sensing useful for a wide variety of environmental regulatory enforcement activities.[7] New RSS technologies promise such improved resolution that many experts claim the technologies will revolutionize the process of environmental regulation.

As of summer 2000, 11 U.S. companies, representing small to large corporations, had been licensed to deploy and operate commercial high-resolution satellites. Their offerings are largely data collected by electrooptical sensors, but two systems scheduled for launch in the next few years include SAR and hyperspectral capabilities. These commercially owned and operated systems join the some dozen or so government civilian systems described above.

The first commercial satellite, known as Ikonos and owned and operated by Space Imaging, Inc. (a subsidiary of the Lockheed Martin Corp.), was successfully launched in September 1999. Ikonos collects black-and-white images with one-meter resolution and color images with three-meter resolution. It circles Earth 14 times a day, imaging a given area on the Earth every three days at high resolution and every day at low resolution. The geographic coordinates for taking the imagery are entered into a computerized ordering system. Once collected by the satellite, data are sent to customers largely by way of the Internet, which has become the primary means of disseminating data; because the imagery is digital, it can be processed and shipped to a customer within a day. Ikonos' imagery has won widespread praise for its clarity and detail. To expand its product line to include radar data, Space Imaging has become a distributorship for imagery from Canada's Radarsat spacecraft and the European Space Agency's Radar Satellite. Following a similar business model, including forming international partnerships but offering different scales of resolution, Orbital Imaging Corporation and Earthwatch, Inc., planned to launch satellites in 2001, and 12 more systems (including radar and hyperspecral techniques) are scheduled by various companies during the next five years. In all cases, customers for the data are wide-ranging but include natural resource and environmental managers in government and industry, land use and urban planners, in-car navigation systems, and real estate markets.

Figure 14-1 illustrates the typical network that companies use to acquire and distribute data. During acquisition, the imagery is stored on-board the satellites. It is then downloaded to ground stations and sent to a master archive, where it may be processed and stored. Customers buy the images, receiving them via the Internet or other delivery services. Customers include licensed or franchised distributors, value-added firms that use the imagery as input into another product (say, a map, consulting study, or other product), and actual end-use customers who use the data for their own activities. Under some companies' arrangements, data may be sent directly to customers from the ground stations rather than only from the master archive, reaching customers just a few hours after being collected by the spacecraft.

In general, the companies have contracted with international firms that host the ground stations and in some cases serve as the regional distributors of the data. For example, Lockheed Martin has entered into joint partner-

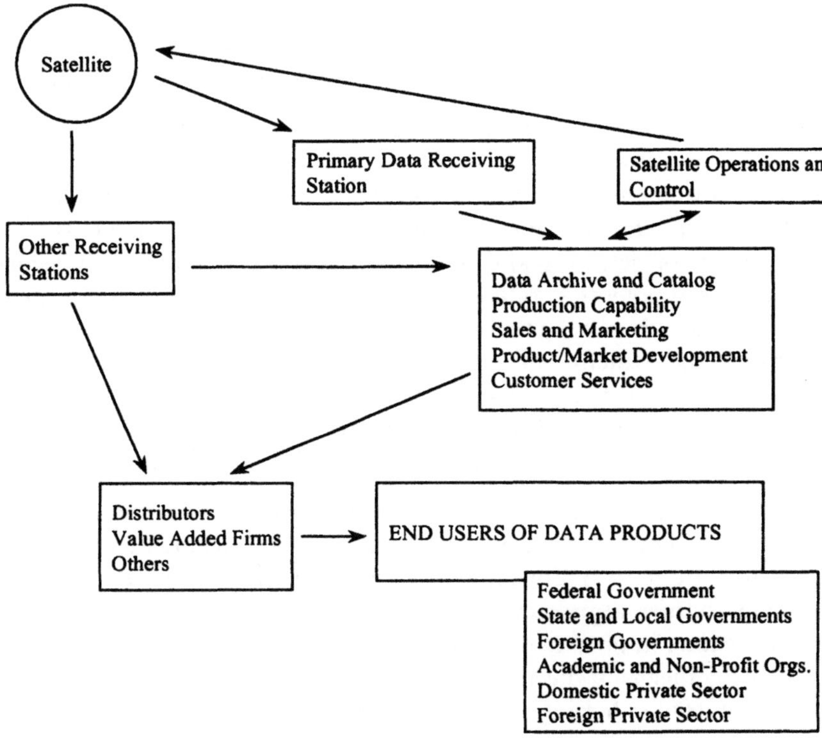

FIGURE 14-1. Architecture of the Commercial Remote Sensing Data Market

ships with domestic and overseas firms, including Hitachi Ltd. and Nuova Telespazio s.p.a., which will be licensed distributors of data from the spacecraft. Property rights to the data reside with Lockheed Martin. In some cases, customers also have the opportunity to task or schedule the satellite's acquisition of imagery, subject, of course, to its fixed operating parameters (such as its orbit and altitude).

Using RSS to Enforce Environmental Regulations: Economic Considerations

Various considerations may affect policies regarding what environmental practices we should regulate, where to place our effort in enforcing those regulations, and how much we should penalize or punish the violators we detect. Among those considerations are matters of environmental ethics, constitutional law, and equity issues associated with balancing the severity of the punishment with the consequences of the crime. Whereas we acknowledge the importance of these considerations, the criteria we use to evaluate the specific issue of how RSS technology could or should affect the design and enforcement of environmental regulation is primarily economic.

As we apply it here, the economic perspective embodies several principles:

- The purpose of environmental regulation is to reduce the level of pollution or ecosystem degradation to the point where the benefits of any additional environmental improvements equals the additional costs, in lost production or extra abatement, from achieving those improvements.
- The purpose of detecting and punishing those who violate such regulations is to lead potential polluters to account for the costs of environmental degradation in making their production and abatement decisions.
- Achieving the benefits of enforcing environmental regulations itself involves various costs associated with identifying and measuring potential violations, the accuracy of those measurements, the resolution of disputes regarding liability, and the imposition of specific punishments.

Our objective in this chapter is to review the underlying economic analyses of environmental protection and regulatory enforcement, to indicate which factors are most important in deciding how RSS can best achieve the goals of environmental protection and pollution management.[8] In doing so, we believe that an important consideration in achieving these goals is to find the least costly means of meeting environmental policy targets and to ensure that the benefits of enforcing regulation exceed those minimized costs.

In this section, we highlight four sets of results from the extensive literature on the economics of environmental externalities with respect to their relevance to four characteristics of RSS technology in its application to environmental monitoring and regulatory enforcement:

- quantity and quality of the information supplied by RSS,
- access to RSS information (say, by regulators, regulated parties, and third parties),
- cost of RSS information compared with information from other sources, and
- strategic effects of RSS detection and concealment (that is, the "concealed" nature of the act of remote sensing from space).

Macauley and Brennan (1998) present a more detailed literature review and discussion of these topics.

Information (Quantity and Quality)

As early as Pigou's (1932) seminal description of what later became defined as "externalities," the literature has emphasized that effective environmental policy requires that we have good information regarding the damages that emissions may cause, whether the approach be Coasian negotiation, command and control, or marketlike incentives. In addition, Spulber (1985) raises the point that ideal policies should be designed around emissions rather than output, and Weitzman (1974) describes how uncertainty about the quantity of emissions may influence the effectiveness of quantity-based methods of regulating (such as those in the EPA sulfur dioxide emissions trading program) and improve on tax-based means of pollution control. Finally, economic costs and other values bring up the need to minimize the punishment of regulated parties who do not violate regulations, making accuracy an important consideration in assessing enforcement technologies.

Access (by Regulators, Regulated Parties, and Third Parties)

With respect to access to information, Coase (1960) implies that one should investigate how observation technologies may help private parties enforce agreements to reduce emissions and perhaps reduce transaction costs to the point where private bargaining might substitute for public policy. Extending the literature on the economics of law enforcement (for example, Becker 1968) to the case of environmental policy, we know that in general, technologies that make data available to third parties can facilitate more efficient enforcement of environmental regulations, consistent with environmental statutes and tort law. Moreover, methods by which supervisors and the public can better monitor emissions may lead to more effective monitoring of efforts by enforcement agents, which in turn can improve their performance and reduce the need for unorthodox compensation systems to induce greater effort (for example, see Binder 1995).[9]

In addition, enforcement of environmental regulations is showing increasing reliance on third-party enforcement. Current federal environmental law grants citizens the right to sue parties who they allege have violated environmental emissions or permitting statutes, where other governmental authorities have not begun similar actions.[10] Rights to sue under common tort law are also preserved. States and other governmental jurisdictions are allowed to institute and enforce environmental standards more strict than those of the federal government.[11] Technologies that permit third parties to monitor environmental regulations may be able to help environmental authorities improve compliance and reduce not only their own enforcement costs but also the costs to all parties able to secure compliance (see Florini 1995).

Also, with regard to access, an enhanced ability to audit compliance and reduce the costs of self-reporting can improve the performance of self-reporting schemes, a ubiquitous and potentially cost-saving means of meeting enforcement goals, particularly by making credible commitments to inspect regulated parties who report no violations. However, the more widespread self-reporting is, the less valuable improvements in detection accuracy and cost reductions are, because self-reporting substitutes for inspections and monitoring.

Finally, the ubiquity of specific compliance rather than general deterrence in the environmental enforcement system increases the importance of ensuring that all sources that compete with each other in product markets are treated equally.

Cost

Conclusions in the literature on the costs of information used in environmental regulation and compliance enforcement (Harford 1991; Harrington 1992) also are relevant to our discussion of RSS. Technologies that can reduce the cost and improve the precision of observing emissions will make both tax-based and quantity-based environmental regulation more effective and reduce the need to rely on theoretically less efficient command-and-control methods. To the extent that the goal of the regulator is deterrence at the lowest cost, the first step is to determine whether fines can substitute for enforcement effort, to produce expected penalties sufficient to meet those goals.

Nevertheless, a host of factors limits the size of the fines that can be tied to violations, warranting attention to enforcement effect and, consequently, monitoring costs. These factors include bankruptcy constraints on people and corporations. In the environmental context, this situation may mean that, all else equal, proportionally more effort should be devoted to uncovering violations by smaller firms rather than larger ones, because the former are more limited in their vulnerability to fines. In any of these cases, the fact that positive effort has to be devoted to enforcement means that enforcement costs and efforts to employ new technology become important. Along these lines, it becomes important to consider how monitoring costs may vary according to the location of the source, the chemical and optical properties of the specific pollutant, and the variation in time over how that pollutant is emitted. For example, some monitoring technologies may be the least costly means for measuring emissions in remote areas, whereas others may be more cost-effective for those close to urban areas.

Detection and Concealment

Detection accuracy also matters. The U.S. systems of legal and regulatory processes and appeals are predicated in part on the belief that, certainly in their absence, there is some chance of error in that violators will go undetected and the innocent will be punished. Increasing detection accuracy may decrease the cost of detecting violators of environmental rules, for instance.

Violators are not likely to wait passively to see whether they will face punishment for their violations, however, and they may engage in activities to reduce the chance of being caught (for instance, surreptitious dumping of hazardous waste). In addition, an important characteristic of many pollutants—particularly those associated with industrial production—is whether they are emitted as produced or whether they can be stored in such a way to control the rate at which they are released into the environment (Russell, Harrington, and Vaughan 1986).[12] A rough generalization may be that gaseous or atmospheric pollutants are in the first category, whereas liquid and solid wastes are in the second. The source of variability in the production of pollutants is that they tend to match production runs, which can vary within the standard workday; across periods of the day, week, or year; and with demand.[13] If they cannot be stored, the emissions will match these production variations; however, if they can be stored, the polluter can reduce this variation by smoothing out emissions levels.

This distinction can affect technology for monitoring in two quite different ways. First, suppose that the enforcer can be reasonably assured that stored emissions are emitted on a relatively constant basis. In that case, it would take relatively few measurements to get an accurate reading of the polluter's emissions. A more variable emissions pattern would require more measurements—continuous measurement, in the limit—to get an equally accurate reading. This variability would suggest that technologies with low variable cost of additional measurement of the same source are more useful when emissions cannot be stored.

Another related factor is marginal deterrence (Posner 1992). If a potential violator is thinking about whether to commit a more serious or less serious violation, then the penalty structure should discourage the former more than the latter. One way to do this is to make sure that the less serious violations have less serious punishments. Alternatively, enforcers could achieve marginal deterrence by not increasing punishments for more severe violations but putting more effort into detecting them. In either case, to the extent that considerations of marginal deterrence put a ceiling on punishment levels or warrant differential detection efforts, enforcers will have a need to consider the costs and accuracy of alternative means of detection as well as the possibility of concealment.

Another aspect of marginal deterrence looks at violators that have been or expect to be detected. In the environmental context, one example might be that if a firm expects to get the maximum punishment for dumping one ton of a toxic chemical into a river, the firm has no motive not to dump another hundred tons into the river. The solution is to limit the amount of punishment for lesser violations in order to preserve the marginal incentives necessary to deter subsequent, more serious violations by that violator. Limiting punishments, however, gives enforcers the incentive to put resources into detection and thus using technologies that reduce the cost of detection.

An additional observation related both to the cost of information and to detection is that the relative costs of different monitoring technologies are likely to depend on the location of the source of emissions, their chemical and optical content, and variation over time in how those emissions enter the environment. Finally, reductions in the cost of repeated measurements can improve accuracy when emissions are subject to random variation from fluctuations in environmental, production, and market factors.

Implications of RSS Technologies for Environmental Enforcement

Nature of the Information to Be Supplied

RSS technologies offer opportunities to increase and improve the usefulness of remote sensing for enforcing environmental regulation. The types of environmental monitoring—for different media, such as air and water, and for different activities, such as land use or water quality monitoring—to which RSS is best suited varies widely among RSS systems, given that each differs in its spatial, spectral, and temporal resolution and geographic coverage. Systems may operate in the spatial domain of 1 meter or better, 3 meters, or around 5 or 10 meters; some systems collect finer gradations of spectral fingerprints than others do.

The systems also differ in their frequency of coverage and geographic coverage, with implications for monitoring activities. For example, some proposed satellite systems provide observations of given points on Earth as many as 14 times a day, but existing systems make observations only every couple of days. Frequency of coverage is related to geographic coverage, in that they are jointly determined by the spacecraft's orbital and operating parameters. Individual companies proposing to operate multiple spacecraft plan to coordinate their coverage parameters such that some of the spacecraft will orbit close to the poles and cross a given location at the equator every 60 hours, whereas others will orbit in a higher inclination, frequently revisiting the most densely developed land masses. Taken together, the spacecraft are intended to provide fairly comprehensive coverage. Similarly, radar data obtained during periods of high cloud cover or at night can complement electrooptical data obtained during cloud-free daylight.

Experts agree that it is not yet clear just how effective new RSS technology will be in all of its potential applications. There is agreement, however, that the systems now operating or planned for operation in the next few years will be able to supply significant detail about the following types of phenomena and activities:[14]

- water: watershed elevation, slope, and aspect; hydrography; detection and measurement of pollution and sediment plumes; assessment of land use impacts on water quality; water clarity measurement and tracking; measurement of sea level; location and status of remote lakes and streams; point sources of pollution
- air: climatic conditions; human and physical geography of surrounding air quality management areas; some detection, identification, and measurement of chemical content of atmosphere (for example, carbon dioxide, sulfur, aerosols, volcanic residue, and water vapor); some measurement of airborne particulates; ozone profiling; air clarity measurement; some point sources of pollution
- land: land use and land cover (cultivation and other agricultural uses, abandonment, urbanization, and urban infrastructure such as utilities and telecommunications siting and routing, monitoring of land use in corridors of rights-of-way, desert encroachment, integrity and area of forest canopy, and vegetation species diversification and distribution); precise location of sites; changes in land at a site over time; environmental profiling and planning; and geolocation of oil spills, other hazardous releases, flooding, fire, and other events
- wetlands: detection, mapping, and evaluation of extent and ecological health; historical studies
- waste: land use and cover in residential and commercial areas, topography, wind patterns, siting parameters, and flood plains; pre- and postimpact

comparisons; damage assessment; optimal siting of cleanup crews and safe disposal locations; and detection of old sites

Effects of the New Technologies on Environmental Regulatory Enforcement

In this section, we evaluate the particular advantages of high-resolution RSS information for environmental regulatory enforcement. We invoke as a framework conclusions from our earlier discussion of the economics of environmental regulation; thus, we assess the technologies with respect to the quantity and quality of the information they provide, accessibility of the data, cost of the data, and implications of RSS as a technology that can observe without being observed. Table 14-2 summarizes our discussion.

Information (Quantity and Quality)

The view afforded from space, as distinguished from aircraft or in situ measurement, is geographically broad. For this reason, the space systems supply information relevant to one of the "So what?" questions of environmental regulation: what is the contextual impact of activities—on the surrounding environment, the public in the area, and so forth? High-resolution systems provide particularly detailed synoptic information. In addition, as archives of such data are acquired, the data will offer an historical contextual perspective useful in "benchmarking" before-and-after changes (for example, when an industry is required to restore an environmental resource to its original status). The question also can be answered by using satellite data in conjunction with other data, such as census data, to see at a glance a potentially affected population, its demographic, housing, and other characteristics as well as other information in a spatial dimension.[15]

RSS thus offers to increase the information base with which to enforce regulation. It may do this in several dimensions: by providing more information, by providing more accurate information, and by providing more information over time.

Data Access

For numerous reasons, space-based RSS information may be more readily available to regulators, regulated parties, and third parties than aerial or in situ information. Once a spacecraft is in orbit, data can be collected routinely during the operating lifetime of the satellite. The spacecraft data also can be obtained in near-real time via distribution to ground stations. Commercial

TABLE 14-2. Implications of New RSS Technologies for Enforcing Environmental Regulations

Factors in regulation and enforcement	*Advantageous?*	*Implications for enforcement*
Information (quantity, quality)		
Geographic coverage	++	Permits more accurate assessment of damages, including context
Synoptic view	++	Permits improved targeting of enforcement activity
Accuracy	+	Informs choice between quantity- and tax-based control
Real-time coverage of some activities	+	Helps minimize punishment of nonviolators
Continuous monitoring	–	
Data access		
To regulators	++	Facilitates private enforcement by regulated parties
To regulated parties	++	Facilitates role of third parties in enforcement
To third parties	++	Reduces need for inspections
In real time	+	
Cost		
Difficult-to-access processes or sites	++	Likely to be cost-effective compared with aerial or in situ observation
Nonobtrusive observation	++	Permits some repeated measurements
Concealed sensing	++	Lowers enforcement costs Boosts deterrence effects
Sensor concealment	++	Increases cost of concealment by observed party Increases likelihood of detection Improves effectiveness of enforcement

Notes: ++ = significant benefit; + = some benefit; – = no particular advantage.

RSS systems all use high-speed telecommunications links in transmitting data from the spacecraft to the ground and, for customers without ground stations, use other electronic delivery mechanisms in sending information from ground stations to customers. The Internet, for instance, figures prominently as a data distribution mechanism in commercial RSS systems. In contrast to aircraft or in situ inspections, which usually are privately contracted for or conducted by regulators, access to RSS data is much more public—open to any party able to pay for the data. In short, all parties are likely to have access to the satellite data through ordinary market mechanisms.[16]

This accessibility to data may be one of the most significant attributes of RSS for enforcement. For example, RSS data availability may facilitate the enforcement of Coasian agreements by reducing transaction costs to the point where private bargaining may substitute for public policy. Alternatively, the efforts of enforcement agents may be cost-effectively bolstered, and actions by third parties—citizens' groups, environmental concerns, and others—also may be informed by RSS data.[17]

In addition, regulated parties may find that their own use of RSS data can improve their self-reporting efforts. As we noted earlier, we know that credible self-reporting can substitute for inspections and monitoring. Moreover, if the accessibility of RSS data encourages credible use (because other parties can acquire and verify the data), then RSS may further reduce the need for inspections.

Another implication is that data availability can increase the opportunity for even-handedness in monitoring and enforcement. The openness of data access can make it easier for all sources competing in product markets to be treated equally with respect to enforcement. Accessibility per se of course does not guarantee that RSS data will be used with such even-handedness, but it does improve the means to do so.

Cost

The new systems are more cost-effective than other means of collecting data when information is sought about processes or locales that are difficult to access (say, by in situ visits), require a synoptic view, or require nonobtrusive (such as a pristine habitat) or concealed sensing (such as random, unannounced site inspections). The cost of aerial photography is roughly $250 per hour. Mapping 1 million miles (say, the waterways in Louisiana) would cost on the order of $1,500,000; the cost of space-based RSS for the task would be about $140,000 for the data, processing, and analysis.[18] This ratio of around a 10-to-1 cost savings is aligned with other estimates of 5 to 1 or better for the use of RSS compared with aircraft measurements.

In general, the costs of using RSS have been declining for several reasons related to technological change. One is the decreasing cost of small spacecraft and launch services; another is declining costs and increasing capability of software (such as geographic information systems) and hardware (computers and processing).[19] Another example is the use of hyperspectral imaging to identify concentrations of leachate at Superfund sites. At Leadville, Colorado, EPA was able to save an estimated $2 million in identifying and establishing cleanup priorities of the most damaged areas of the site. The imaging technique sorted through 1,200 waste piles in 45 seconds (however, data analysis took 10 months).[20]

Another cost-related factor is that because the space systems indeed have such broad coverage, they collect information—at close to zero marginal cost—about features that may not have been of interest or noticeable at the time of the review but may become important at a later date. In contrast, because most information collected by aircraft or in situ has a marginal cost greater than zero, seemingly unimportant data may deliberately not be collected by such technologies in an attempt to reduce collection costs (for example, the aircraft shortens or reorients its flight path, recording sensors are turned off, or ground inspectors limit their spot checks [in situ monitoring]).

To the extent that space-based RSS indeed is cost-effective, there are several implications for policy. One is that reducing the cost of repeated measurements can improve accuracy and the likelihood of detecting violations of environmental regulations. Repeated measurements may also facilitate detection when polluting activities are subject to random variation from fluctuations in climatic and other factors, or when emissions can be stored for purposes of eluding detection. A related implication is that lower monitoring costs can facilitate enforcement and, in turn, boost any deterrence effects of that enforcement. This approach to deterrence can be an alternative to fines, which are limited in that they may not be high enough to be feasibly tied to violations (for example, due to bankruptcy constraints or marginal deterrence). In addition, more cost-effective information may permit greater use of economic incentive-based approaches to regulation (and enforcement) rather than command-and-control approaches, which are theoretically less efficient.

Strategic Effects from Sensor Concealment

Space-based RSS technologies have a unique attribute, that is, some ability to observe without awareness of the observed party. The orbital parameters of the systems are public knowledge through the licensing process the systems must undertake. Therefore, it is not inconceivable that someone could obtain these parameters, calculate when the spacecraft would pass overhead, and take actions to conceal activity during these times (although it is not clear how much it would cost parties to take such action).[21]

The possibility to observe without being seen (assuming the observed party does not take averting action) has several policy-related results. For instance, among three recommendations Russell (1990) makes in an analysis of environmental monitoring and enforcement policy, two relate to concealment. He advances the desirability of unannounced and warrantless inspections, arguing that "there should be no anticipation of privacy in the generation and discharge of pollutants." Whereas this recommendation need not require that the observation technology be hidden from view of the observed party, RSS offers the opportunity to do this, and to do it cost-effectively. Russell also argues for the desirability of remote sensing methods "that are capable of measuring stack emissions from outside the factory or utility gates." The usefulness of RSS's ability to observe without being observed is that it can increase both the likelihood of detection and the cost of concealment by the observed party, thus improving the effectiveness of enforcement activities.[22]

Conclusions

Our review of economic models of environmental protection and regulatory enforcement has highlighted several attributes that are particularly likely to benefit from new enforcement technologies such as RSS. These attributes include the quantity and quality of information supplied by the new technologies; the accessibility of the information to regulators, regulated parties, and third parties; the cost of the information; and whether the process of information collection can be concealed from the observer.

RSS technologies are likely to influence all of these attributes and, in general, improve the efficacy of enforcement. We make the following observations in this regard.

- Although the extent to which emerging RSS technologies will detect specific emissions in all media (such as some kinds of gas emissions) is not clear, high-resolution data can provide a context for understanding the damages associated with environmental pollution. In turn, this information confers at least three benefits for environmental enforcement: it can help assess damages more accurately; improve targeting of enforcement and remediation; and provide better information for deciding whether to use quantity- or tax-based means of pollution control. In addition, more accurate information such as that offered by high-resolution RSS may minimize the punishment of regulated parties who do not violate regulations.
- The accessibility of RSS data to all parties—regulators, regulated parties, and third parties—may facilitate private enforcement as a substitute for public policy by reducing transaction costs. Regulated parties may also use RSS data for their own self-reporting efforts, which if credible, can substitute for inspections and monitoring. Credibility of the data may be heightened by its public accessibility. Accessibility also increases the opportunity for even-handedness in monitoring and enforcement, in treating equally all sources that compete with each other in product markets.
- To the extent that RSS data represent a more cost-effective source of information than other data sources, repeated measurements might be economically taken using RSS, thereby improving accuracy and the likelihood of

detection of violations of environmental regulations. Repeated measurements also may improve detection of polluting activities subject to random fluctuations caused by climate or other factors. Lower monitoring costs also can boost any deterrence effects of enforcement by making enforcement easier, and this approach may be an alternative to fines. In addition, more cost-effective information may permit greater use of economic incentive-based approaches to regulation (and enforcement) rather than command-and-control approaches, which are theoretically less efficient.

- Finally, the opportunity for RSS to observe without being observed can increase both the likelihood of detection and the costs of concealment of activities on the part of observed parties, thus improving the effectiveness of enforcement actions. However, the operating parameters of RSS are a matter of public record and therefore make it possible for observed parties to take measures to conceal their actions when an observation spacecraft is expected to be overhead.

Acknowledgements

This research was jointly funded by the U.S. Environmental Protection Agency, Office of Policy Planning and Evaluation (Cooperative Agreement CX824429-01-0) and by Resources for the Future.

Responsibility for opinions and errors rests with the authors.

Notes

1. For example, automated ground monitors combined with aircraft photos are being used to facilitate the monitoring of water quality (*Geo Info Systems* 1995). As another example, ground-based remote sensing of auto tailpipe emissions has been conducted in pilot projects in several urban areas (Brooks 1992; Hughes Environmental Systems 1992). For general discussion of using satellite data for monitoring international environmental agreements, see Berkowitz 1992.

2. We thus extend the literature on monitoring and enforcement (e.g., see Russell, Harrington, and Vaughan 1986; Harrington and Macauley 1995).

3. The relevant literature on self-reporting is informative but sparse; see Harford 1987, Malik 1993, and Kaplow and Shavell 1994.

4. Much of this section is based on Drury 1990.

5. In addition, some RSS imagery collected by national security satellites has been released for the purpose of environmental analysis. See, for example, Gertz 1995, Lawler 1995, and the discussion in Berkowitz 1993.

6. Leary (1992) offers good discussion of the relative advantages of drones, balloons, and other techniques. There are two additional sources of RSS information on which we do not focus in this study: weather satellites, and sensors flown on the U.S.

space shuttle. Weather satellites provide data about atmospheric conditions, ocean temperatures, and ocean turbidity, for example. Sensors flown on the space shuttle include electrooptical and synthetic-aperture radar devices.

7. Two policies have facilitated commercial development of high-resolution space-based remote satellite sensing (RSS): the 1992 Land Remote Sensing Policy Act and Presidential Directive-23 (PD-23), issued in 1994. The 1992 act encourages commercial development and clarifies the licensing process for the satellites. PD-23 clarifies national security policy toward the commercial RSS industry, generally permitting it to operate with minimal restrictions.

8. Observation technologies also are crucial in enforcing agreements among governments to control emissions. For a discussion of the implications of space-based remote satellite sensing for both the design of international environmental treaties and potential threats to national sovereignty, see Macauley 1992.

9. Binder (1995) concludes that to the extent that these "large numbers" of affected parties fit into a few relatively well-organized, homogenous groups, representatives of their organizations may be able to negotiate agreements at relatively low cost through an "environmental dispute resolution" process.

10. 42 U.S. 7604 (1994), (a)–(b), (f).

11. 42 U.S. 7416 (1994).

12. As with almost all binary "either/or" classifications, it would be more technically more accurate to refer to a continuum of possibilities, depending on how much it would cost to store the pollutant. Russell, Harrington, and Vaughan (1986) provide a useful brief survey of the economics of enforcement in environmental contexts.

13. Depending on the pollutant, of course, the external cost of the emissions and hence the importance of monitoring and enforcement may vary over time as well.

14. Much of this discussion is based on NASA and U.S. EPA 1995.

15. The use of space-based remote satellite sensing (RSS) data as a base map is not new, and other data, such as aerial photos, also can be used as the base over which to lay the census data. However, our point is that the high-resolution RSS data are likely to offer an even better base map.

16. In cases where data costs may be prohibitive (say, for a community group), one might imagine that grants or other price breaks for data might be made available just as they are for other goods and services for public, charitable, and other social benefit.

17. Florini (1995) emphasizes the use of space-based remote satellite sensing (RSS) data in improving the role of third-party monitoring and enforcement. Some observers have questioned whether third parties and others are likely to be able to interpret RSS data. Low-resolution data are difficult to interpret; high-resolution images are much easier to understand. In addition, there is a growing consulting industry supplying photo-interpretation expertise. The bigger questions may be whether regulators and courts accept RSS data as proof of compliance.

18. This example is from NASA and U.S. EPA 1995.

19. Another important development is the availability of the global positioning spacecraft system (GPS), a U.S. Department of Defense network that provides precise location and timing signals for all types of space- and ground-based activities. For discussion, see Pace and others 1996.

20. See Robbins 1999.

21. To be sure, some activities are not readily concealed, but the concealment effort is likely to be largest when the penalties for noncompliance are most egregious.

22. We are not arguing that the enforcement authorities should necessarily conceal their activities. Such concealment would add to concerns now being expressed about individuals' rights to privacy (see Brennan and Macauley 1995), and this possible cost would need to be balanced by advantages of concealment.

References

Becker, Gary. 1968. Crime and Punishment: An Economic Approach. *Journal of Political Economy* 76: 169–217.

Berkowitz, Bruce D. 1992. *Use of Intelligence Data for Environmental Monitoring: Summary of Meetings Held by the Council on Foreign Relations.* May. Mimeo. Washington, DC.

———. 1993. *The Use of Intelligence Resources for Environmental Monitoring: A Critical Issues Report by the Council on Foreign Relations.* February. Mimeo. Washington, DC.

Binder, G. 1995. Resolving Environmental Disputes. *Conservation Issues* 6(2): 1–10.

Brennan, Timothy, and Molly Macauley. 1995. Remote Sensing Satellites and Privacy: A Framework for Policy Assessment. *Journal of Law, Computers, and Artificial Intelligence* 4: 233–248.

Brooks, R. 1992. California Report Recommends Drive-By Sensor. *Ward's Engine Update* December 1.

Campbell, James B. 1987. *Introduction to Remote Sensing.* New York: Guilford Press, 1987.

Coase, Ronald. 1960. The Problem of Social Cost. *Journal of Law and Economics* 3: 1–40.

Drury, S.A. 1990. *A Guide to Remote Sensing.* New York: Oxford University Press.

Florini, A. 1995. *Technologies for Verifying Compliance with Environmental Standards.* December 1. Mimeo. Environmental Defense: Washington, DC.

Geo Info Systems. 1995. Kansas City Builds a GIS to Defray Costs of Clean Water Act Compliance. *Geo Info Systems* June.

Gertz, Bill. 1995. Gore Touts Use of Spy Photos for Fighting Global Pollution. *Washington Times,* February 25, A5.

Harford, Jon. 1987. Self-Reporting of Pollution and the Firm's Behavior under Imperfectly Enforceable Regulations. *Journal of Environmental Economics and Management* 24(3): 241–257.

———. 1991. Measurement Error and State-Dependent Pollution Control Enforcement. *Journal of Environmental Economics and Management* 21(1): 67–81.

Harrington, Winston. 1992. Enforcement Leverage When Penalties Are Restricted. In *The Economics of the Environment,* edited by Wallace Oates. Brookfield, VT: Edward Elgar.

Harrington, Winston, and Molly Macauley. 1995. *The Value of Information and the Cost of Advocacy.* June. Mimeo. Washington, DC: Resources for the Future.

Hughes Environmental Systems, Inc. 1992. *Feasibility Study for the Use of Remote Sensors to Detect High Emitting Vehicles.* Final Report Prepared for the South Coast Air Quality Management District (California). December 22. El Segundo, CA: Hughes Environmental Systems, Inc.

Judd, D. 1994. Risk Assessment, Uncertainty Analysis, and Other Fancy Lies. *Earth Observation Magazine* 5: 6.

Kaplow, Louis, and Steven Shavell. 1994. Optimal Law Enforcement with Self-Reporting Behavior. *Journal of Political Economy* 102(3): 583–606.

Lawler, A. 1995. Spy Photos Come in from the Cold. *Science* 267(March 3): 1260.

Leary, W.E. 1992. Designers Plan Drones to Probe Atmosphere. *The New York Times,* March 31, c1 and c10.

Lillesand, T.M., and R.W. Kiefer. 1987. *Remote Sensing and Image Interpretation.* New York: John Wiley & Sons.

Macauley, Molly. 1992. Collective Goods and National Sovereignty: Conflicting Values in Global Information Acquisition. *Proceedings of the Conference on Space Monitoring of Global Change.* San Diego, CA: University of California Institute on Global Conflict and Cooperation, 31–47.

Macauley, Molly, and Timothy Brennan. 1998. Enforcing Environmental Regulation: Implications of Remote Sensing Technology. RFF Discussion Paper 98-33. Washington, DC: Resources for the Future.

Malik, Arun S. 1993. Self-Reporting and the Design of Policies for Regulating Stochastic Pollution. *Journal of Environmental Economics and Management* 24(3): 241–257.

NASA (National Aeronautics and Space Administration) and U.S. EPA (Environmental Protection Agency). 1995. *Perspectives from Afar: How Remote Sensing and Spatial Information Technologies Can Enable the EPA to Measure Progress towards Its Environmental Goals.* July 25. Washington, DC: NASA.

Pace, Scott, and others. 1996. *The Global Positioning System.* Washington, DC: RAND Corporation.

Pigou, Arthur C. 1932. *The Economics of Welfare.* London, U.K.: Macmillan.

Posner, Richard. 1992. *Economic Analysis of Law.* Boston. MA: Little, Brown.

Robbins, J. 1999. High-Tech Camera Sees What Eyes Cannot. *The New York Times,* September 14.

Russell, Clifford S. 1990. Monitoring and Enforcement. In *Public Policies for Environmental Protection,* edited by Paul Portney. Washington, DC: Resources for the Future.

Russell, Clifford S., Winston Harrington, and W. J. Vaughan. 1986. *Enforcing Pollution Control Laws.* Washington, DC: Resources for the Future.

Spulber, Daniel. 1985. Effluent Regulation and Long Run Optimality. *Journal of Environmental Economics and Management* 12(2): 103–116.

Weitzman, Martin. 1974. Prices vs. Quantities. *Review of Economic Studies* 41: 477–491.

PART IV

Evaluating Design and Performance

There are many perspectives on regulatory design and performance. Theoreticians develop models such as those that compare market-based instruments with various standards; survey researchers elicit responses from experts or the public; policy analysts of all stripes assess regulations, programs, and policies. This variety is well expressed by the lack of specificity in the Government Performance and Results Act, recent government legislation that requires federal agencies to evaluate their own performance. The chapters in this section investigate the details of regulatory performance in the context of workplace safety, water quality trading, and the Clean Air Act. Two of them (Chapters 16 and 19) also summarize tools of more general applicability.

In Chapter 15, Shultz and Fischbeck investigate issues of workplace safety. Designing inspection programs differently may create the opportunity to reduce accidents at the same or lower cost. Whereas the case is specific to offshore platforms, the approach can be generalized to apply to inspection and compliance activities throughout the environmental, health, and safety system. Elements of this chapter hark back to the technology focus of Chapter 14 and the challenge in observing regulated activity. In Chapter 16, Matthews and Lave continue with a focus on workplace safety, but their analysis cuts across all industries. Using a variation of an external impact and cost model that takes into account direct and indirect effects, they find that the national social cost of injuries exceeds that for fatalities along with interesting variations across industries. Few might suspect that the system supporting the "eating and drinking establishments" industry is one of the most dangerous, even before considering alcohol and driving impacts.

In Chapter 17, Schultz and Small turn to the environmental concerns of water and air quality, investigating the potential for water pollution trading. Trading has become accepted as environmentally effective and cost-effective in air pollution regulation. As researchers and regulators seek to transfer those lessons to water pollution, various challenges arise. Schultz and Small detail some of the challenges in measuring pollution loads in a watershed and in defining emissions trading ratios between locations to improve regulatory performance. Chapter 18 shifts to air quality: Matthews reviews a U.S. Environmental Protection Agency (EPA) report on the subject, investigates the robustness of the EPA results by using a different valuation approach, and then analyzes the relative contribution of mobile and stationary source pollution to the net benefits of the Clean Air Act. In Chapter 19, Farrow and others describe FERET (the computer program that is provided along with this book) as a computerized template to assist analysts and stakeholders in quantifying the impacts, benefits, and costs of air and safety regulation. As an illustration of the use of the template, they study a companion EPA report to the one discussed in Chapter 18 and conclude that the selection of health concentration–response functions is critical to evaluating performance but that the selection of alternative methods to calculate the value of a statistical life is not critical.

These chapters raise numerous questions for research and discussion. While reading, you might consider questions such as these:

- What are the regulatory gains and losses from targeting workplaces with different accident and fatality rates?
- Should analysts consider only direct impacts or impacts that occur throughout a chain of economic activity?
- What principles and objectives should guide the detailed design of a water-quality trading program?
- Can analysts distinguish the impact of alternative regulatory designs and programs on measures of performance such as net present value? What are some of the difficulties?
- What role can tools like extended input–output analysis and computerized benefit–cost analysis play in regulatory development and stakeholder discussion? What role should these tools play?

15

Workplace Accident and Compliance Monitoring

The Case of Offshore Platform Inspections

John R. Shultz and Paul S. Fischbeck

Offshore platforms are often very large, expensive, and complicated structures. Dozens of workers live, for weeks at a time, up to more than 100 miles from shore. If an accident occurs, it can be very difficult to evacuate people who are injured. If a spill occurs, it might take many hours for boats with repair equipment or oil slick booms to reach the platform. Difficulty in reaching the facility, lack of easy egress, potential cost to the owner, danger to the environment, and danger to the health and safety of the workers make it important to minimize the likelihood of an accident or spill. As can be seen in Figure 15-1, some platforms appear to be accidents waiting to happen.

The 4,000 platforms in the Gulf of Mexico account for 14% of all oil and 25% of all natural gas consumed in the United States and therefore are very important sources of energy. The primary government agency charged with maintaining safe operations on offshore platforms is the Minerals Management Service (MMS), which is part of the U.S. Department of Interior. MMS is required by the U.S. Congress to inspect every platform annually, regardless of its size, previous accident history, or location.

In this chapter, we present two models (statistical and expert judgement) that can accurately predict which platforms are likely to have an accident or spill. The models are developed using a set of risk factors associated with the platforms' physical characteristics (for example, size, number of components, and age) and operational characteristics (for example, number of prior accidents and spills, inspection records). The data set covers the years 1986–1995. Data are averaged on a five-year basis (for example, the average num-

FIGURE 15-1. Offshore Oil Platform in the Gulf of Mexico

ber of accidents on a platform is used as an input to the models). Model forecasts are performed for 1992, 1993, 1994, and 1995 and compared with the actual number of accidents or spills that occurred during those years. The models rank the platforms in terms of risk, and usually, 50% of the incidents occur in the top 10% of the ranks.

The models in this chapter illustrate how analytical techniques can be successfully applied to a complicated policy problem—the forecasting of the risk of accidents and spills at offshore platforms. However, MMS faces a challenge in applying this knowledge to increase the safety of offshore platforms. It is the government's responsibility to determine the appropriate strategies to minimize risk, given the constraints of limited financial and personnel resources. How to implement an inspection program or other control measure to decrease the risk to people and the environment must be carefully thought out. A discussion of the policy implications and how the ranking might be used is presented at the end of this chapter.

First, we explain why this study was important. Next, we present an overview of the data that form the basis of the study and explain some of the problems encountered in analyzing the information. Then, we present and compare two models (the logistic regression and the expert judgement model) and their results. To conclude, we explain how MMS can use this information to reduce risk and suggest what additional work is necessary to increase the safety of offshore facilities.

Motivation for Study[1]

The potential impact of catastrophic failure of an offshore platform is quite serious in terms of both lost lives and damage to the environment. Over the 10-year period from 1986 to 1995, approximately 800 accidents or spills occurred at the roughly 4,000 MMS-inspected platforms on the outer continental shelf (OCS).[2] Fortunately, most of the accidents or spills were not serious. The worst incident in the United States, in terms of loss of life, resulted in seven deaths (MMS 1990). However, accidents on platforms in other parts of the world have been more disastrous. For example, two catastrophic platform failures occurred in the North Atlantic. In 1980, the Alexander L. Kielland, a Norwegian platform, capsized and sunk, resulting in 123 human deaths and hundreds of millions of dollars in damage. In 1988, a fire and explosion at the Piper Alpha facility, a U.K. platform, resulted in 167 deaths and more than a billion dollars in damage and lost revenue. Several other total-loss events have occurred elsewhere in the world, but these two events were the worst, and they are responsible for the worldwide realization of the dangers associated with these facilities and the need for better safety procedures (Pitblado and Tuney 1996, 99).

To decrease the likelihood of accidents and spills, many nations—including the United States—have instituted inspection programs of offshore facilities in the hope that problems will be caught during the inspection process, before they culminate in a major incident. The inspection process serves the purpose of risk reduction in two ways: operators must manage the platforms in a conscientious manner to avoid risk of a platform shutdown or civil or criminal penalties, and information gathered about a platform's operational history might be useful in predicting the likelihood of future problems. Overall, the safety record for U.S. platforms has been good, and this good safety record has been attributed to the government's aggressive inspection program[3] (Paragon Engineering 1996, 8):

> The record of safety on the Outer Continental Shelf (OCS) has been good. In terms of injuries and fatalities, OCS drilling and production operations are comparable to other hazardous activities onshore, such as mining and construction. In terms of environmental impact, oil pollution from offshore operations contributes less than any other significant course to the release of hydrocarbons into the marine environment. U.S. offshore industry spillage volumes and the amount spilled compared to total production has been reduced. Thus, MMS and the offshore industry are not faced with the problem of correcting a manifestly poor safety record. The United States has succeeded under its present inspection program in averting the kinds of catastrophic disasters that have befallen the offshore operations of many other nations. Although the evidence of a direct connection is lacking, certainly the activities and vigilance of the federal government have been a factor. However, an increase in the margin of safety on the OCS can be achieved by improving the link between the MMS inspection program

and safety performance of the industry. The committee's recommendations are intended to accomplish that end.

Data Description

Early efforts at risk assessment of platforms using operational or historical information were unsuccessful, primarily for lack of properly organized data. In 1985, the situation was as follows (U.S. Congress, OTA 1985):

> There is currently no single comprehensive source of statistics on U.S. offshore accidents, and there are no reliable injury and fatality rate statistics for offshore operations beyond those compiled by the International Association of Drilling Contractors (IADC) for individual workplace accidents in offshore drilling. The lack of data makes it difficult to evaluate the level of safety achieved by oil and gas operators, safety-related equipment, and Federal regulation. It also makes it difficult to assess the effects on safety when changes are introduced.

With respect to record keeping, the situation today is improving over what existed in 1985. MMS is now the lead organization for maintaining records of accidents and spills at offshore facilities and has published several good data summary papers (MMS 1992, 1995, 1998). Additionally, the information about accidents and spills is now kept in one database, along with inspection reports instead of being spread over several databases at separate locations.[4] However, some gaps in the reported information remain, perhaps the most serious being that the platform or structure identification number is not listed in all reports of accidents or spills. When facilities are not clearly identified in the data, it is very difficult to draw any conclusions regarding the operation of individual facilities.

To develop the models in this research, we consulted two sources of information: databases maintained by MMS that contained historical information on platform inspections and accidents, and expert opinion surveys of platform inspectors. The risk factors in this research can broadly be defined as physical characteristics (for example, platform's distance to shore and age), operational characteristics (for example, platform's inspection history), and histories of accidents and spills.

Definition of Terms

The terms *accident* and *spill* are defined as binary events. For consistency, only one accident or spill is recorded for a platform for any 24-hour period.

Multiple related events are coded as a single event, but accidents and spills can and do occur independently from each other. According to these criteria, the number of accidents and spills has averaged about 100 per year; approximately 2–3% of the platforms have an accident or spill every year.

In our study, no attempt was made to address all risks associated with platform operation. Instead, the data were analyzed to look for trends that could be related to specific risk factors to find predictive relationships that could be applied across all 4,000 platforms.

MMS inspects each platform in the Gulf of Mexico at least once per year. If a problem is noted during an inspection, the inspector issues a violation notice to the platform operator, called an incident of noncompliance (INC).

Risk factor refers to terms associated with the physical design or operation of offshore platforms. Thousands of potential platform characteristics could have been chosen as risk factors and used as inputs to the models. The database fields were reviewed by MMS employees for the likelihood of each field being related to an increased risk of having an accident or spill.[5] From this screening effort, a subset of approximately 100 risk factors was chosen. Table 15-1 lists some of the risk factor codes and definitions.

Risk ranking refers to a ranking of the platforms based on the results of either the statistical or the expert judgement model. The risk ranking of any particular platform varies from year to year because of changing conditions. The platforms are ranked from most likely to have an accident or spill to least likely to have an accident or spill on the basis of two criteria: the probability estimate (which varies from zero to one) as derived in a logistic regression model, and the "risk index" as determined in the expert judgement model. *Risk index* refers to the raw output of the expert judgement model. This model is based on a survey of 59 platform inspectors and engineers.

To help the reader understand the information presented in this chapter, the number of accidents that occurred in 1995 was chosen as an example. Therefore, whenever it is necessary to explain how to interpret a graph, or

TABLE 15-1. Partial List of Risk Factors Associated with Each Platform

Risk factor	*Code*	*Description*
1	#ACC	Average number of accidents during the past 5 years
2	#INCs	Average number of poor inspections during the past 5 years
3	#EXP	Average number of explosions during the past 5 years
4	AGE	Platform age
⋮	⋮	⋮
100	#COMP	Average number of components on the platform during the past 5 years

perform a calculation, the sample set of data is the number of accidents that occurred in 1995.

Data Statistics

When performing an analysis of a large data set, one of the first things the analyst should do is summarize the statistics of each risk factor individually, before constructing multivariate regression models or performing other statistical analyses. A first-order analysis of the data in this study indicated that there was a significant number of errors in the databases. It is difficult to determine exactly how many errors there were, because to do so would necessitate the review of millions of individual entries. However, for some fields, a quick check was possible. For example, the distance from a platform to shore can be verified by geographically plotting each platform's location using its recorded latitude and longitude and then computing the distance by using a geographical information system.

Several percent of the platforms had this type of error. These errors were most likely caused by incorrect data entry but could have been caused by the fact that platforms are sometimes moved to new locations in the Gulf of Mexico and the databases are not updated. The reader should keep in mind that the results of the modeling effort are only as good as the data used to construct the models, and in some cases, the data are somewhat suspect.

Statistical Model

The risk factors were used to construct a multivariable logistic regression model to predict accidents or spills. The multivariable model uses time series data to forecast the probability of an accident or spill in the future. Predictions were performed for all platforms from 1992 to 1995, and platforms were ranked by the model prediction.

Risk Factors, Multicollinearity, and Principal Components

Multicollinearity will always exist in real data because the values of the independent variables cannot be chosen from the outset to be independent (as in a designed experiment). This means that there will always be some confounding effects between variables that are not easily separated. Principal components are groupings of risk factors based on their degree of correlation. The idea behind principal component analysis is to reduce the dimensionality of the data set and correct for multicollinearity. The logistic regression model uses a set of rotated principal components as the independent

TABLE 15-2. Four Risk Factor Groupings Used in the Regression Analysis

Component no.	*Description*	*Code*[a]
1	Inspection performance: the number of INCs received, the types of INCs received, and the type of enforcement action performed following each INC	INCs
2	Complexity: measure of size, activity, and complexity (for example, major complexes and numbers of components)	COMP
3	Age and experience: platform age and the experience levels of the operator and company	AGE
4	Accident and spill history: number and types of accidents and spills	ACC/SPILL

Note: INC = incident of noncompliance.

[a] Used in regression tables.

variables (Table 15-2). Note that each component is actually a collection of time series data.[6]

Logistic Regression Model

The goal of this time series analysis is to forecast what is likely to happen in the future, given what has happened in the past. A simplified version of the model used in this study that restricts probabilities between zero and one is

$$\text{probability (accident or spill)} = \frac{1}{1 + e^{-\beta_0 + \beta_1 \times \text{INCs} + \beta_2 \times \text{COMP} + \beta_3 \times \text{AGE/EXP} + \beta_4 \times \text{ACC/SPILL}}}$$

where β_1, β_2, β_3, and β_4 are the parameter estimates as determined by the model and β_0 is a constant; see Tables 15-1 and 15-2 for other abbreviations. If the parameter estimates are shown to be not significant, then the associated principal component is not significant in forecasting accidents or spills. Table 15-3 summarizes the significance of the parameter estimates derived in the regressions.

Statistical Model: Platform Risk Ranking

Figure 15-2 is central to understanding the model's predictions and platform rankings. Platforms are sorted by the probability of having an accident or spill, from most likely to least likely, thus providing the risk ranking of platforms. The vertical axis is the cumulative percentage of accidents that actually occurred in 1995. The baseline rate is the result that would be expected if

TABLE 15-3. Results of the Regression Model

	Parameter significance level				*Total*			
Code	*1991*	*1992*	*1993*	*1994*	***	**	*	*NS*
Regression to predict accidents								
INCs	***	***	**	***	3	1	0	0
COMP	***	***	***	***	4	0	0	0
AGE	*	NS	**	NS	0	1	1	2
ACC/SPILL	NS	***	***	NS	2	0	0	2
Regression to predict spills								
INCs	***	***	*	***	3	0	1	0
COMP	***	***	*	***	3	0	1	0
AGE	NS	NS	**	NS	0	1	0	3
ACC/SPILL	NS	***	***	NS	2	0	0	2

Notes: Breakpoint levels of significance: ***$p < 0.01$, **$p < 0.05$, *$p < 0.10$; NS (not significant) indicates $p > 0.10$.

accidents were randomly distributed (that is, if the model had no predictive value). (If the models were perfect, the curve would shoot up quickly to 100% level on the vertical axis, correctly identifying all the accidents and spills and not making errors.)

Figure 15-2 should be used as follows:

1. On the vertical axis (% of accidents), find 50%.
2. Move horizontally to the bold jagged line (logistic regression model prediction).
3. Read down to the horizontal axis. The value is about 12%; this means that 50% of the accidents that happened in 1995 occurred in the top 12% of the platforms as ranked by the model. If the accidents were randomly distributed, one would expect to get 50% of the accidents in 50% of the platforms. (A similar ranking for the "average" expert judgement model, discussed later, shows that 50% of the accidents occurred in the top 16% of ranked platforms.)
4. The area under the curve in Figure 15-2 is analogous to the model's accuracy, although the proper calculation of the area is more complicated than simply integrating the area under this curve.

Expert Judgement Model

A multiattribute expert judgement model was built from data collected in a survey. As part of the data reduction process to design the final survey, sev-

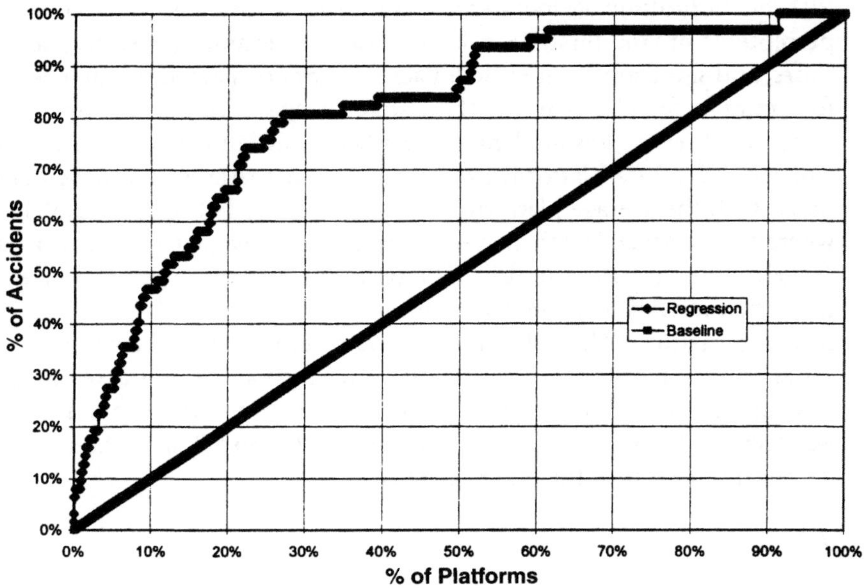

FIGURE 15-2. Risk Ranking Based on Regression Model, 1995

eral preliminary workshops were conducted with MMS personnel. The goal of holding the workshops was to "weed out" duplicated or hard-to-define risk factors. This process made the selection of the risk factors used in the final survey an interactive process. The data were evaluated and summarized, the experts were consulted, the experts made recommendations, and then the list of risk factors was revised to incorporate those views. A final survey was conducted that incorporated the expert's views on various risk factors and the influence that those factors might have on the likelihood of an accident or spill.

The expert judgement model was constructed by using the results of a 1998 survey of platform inspectors. The 59 respondents included 47 inspectors, 11 engineers, and 1 supervisor. (There are 56 total offshore platform inspectors in the Gulf of Mexico region.) All the respondents were involved in the inspection of offshore platforms, and all were government employees.

Survey Questions

The survey contained three sections:

1. Respondent data: This section was designed to gather descriptive data on the respondents to determine whether there was any relationship between their personal characteristics or experience and their perceptions of risk.

2. Risk quantification: Respondents were asked to describe and rate their perceptions of risk. First, 25 risk *categories* (for example, age, distance to shore, and slot count) were listed. Each category contained multiple *levels*; for example, for platform age, there were six age levels (for example, 0–5 years, 6–10 years). Respondents were asked to assign a *score* to each level (for example, "If a platform had this level, how risky would it be compared with the average platform?"). Scores were coded on a scale of 1–5, where 1 was "Much less than average risk," 3 was "Average risk," and 5 was "Much more than average risk." The respondents then were asked to estimate their level of *confidence* in their scores.
3. Risk comparisons: The risk categories were compared, and each category assigned a "weight" on a scale of 0–100. Each respondent was asked to give a weight of 100 to the category considered most important in predicting whether a platform would have an accident or spill and then assign decreasing weights to the other categories.

Risk Index

Using the survey results, a risk index was computed and the platforms ranked. This procedure is shown in the equation

$$\text{Risk index} = \text{Weight}_{\text{Age}} \times \text{Score}_{\text{Age}} + \text{Weight}_{\#\text{COMP}} \times \text{Score}_{\#\text{COMP}} + \\ \ldots + \text{Other risk factors}$$

In Table 15-4, a specific platform (no. 20,874) and a composite expert (that is, an average expert) were chosen to illustrate how the calculation is performed.

TABLE 15-4. Example of the Expert Judgement Model Calculation

				"Average" from survey		
Code	*Value*	*Category*	*Level*	*Score*	*Weight*	*Risk index*
COMPLEX_ID_NUM	20,874	Platform number				
AGE	53	Platform age	>25	4.48	0.62	2.78
#COMP	45	Average number of components	41–50	3.76	0.45	1.69
⋮	⋮	⋮	⋮	⋮	⋮	⋮
Total index						26.75

Platform Risk Ranking

The risk ranking that results from the expert model, when plotted as in Figure 15-3, is interpreted in the same manner as Figure 15-2. The difference is that the ranking for the expert judgement model is based on the magnitude of the risk index, not on a probability estimate.

Pooled Models

If the model results for all the experts are "pooled" together, a chart can be constructed to show where an individual platform would have typically ranked for 1992–1995. Figure 15-4 shows the results from "pooling" the average expert model to predict accidents from 1992 to 1995. Platform 20,874 is shown as an example of how to apply the results of the "pooled" model. This platform is ranked on the borderline of the top 10% for platforms likely to have an accident or spill in 1995. Figure 15-4 shows this platform's "typical" rank and how many accidents have "typically" occurred on platforms with a risk index of 26.75 or greater for the four-year period 1992–1995.

Overall, platform 20,874 would historically rank in the top 8.5% of the most risky platforms. Also, platforms that have a risk index of 26.75 or higher accounted for about 35% of the accidents that occurred during the

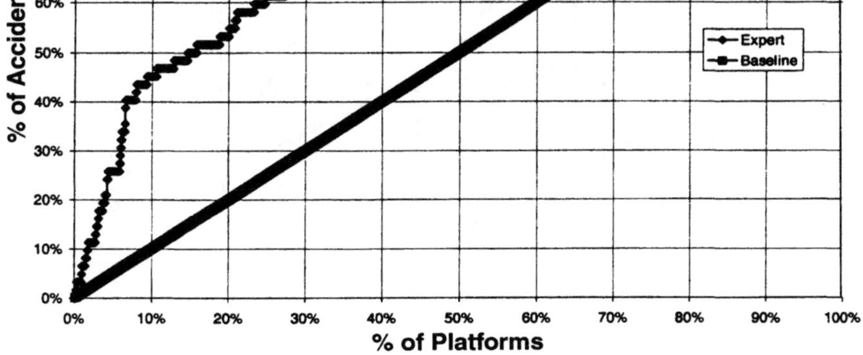

FIGURE 15-3. Risk Ranking Based on Average Expert Judgement Model, 1995

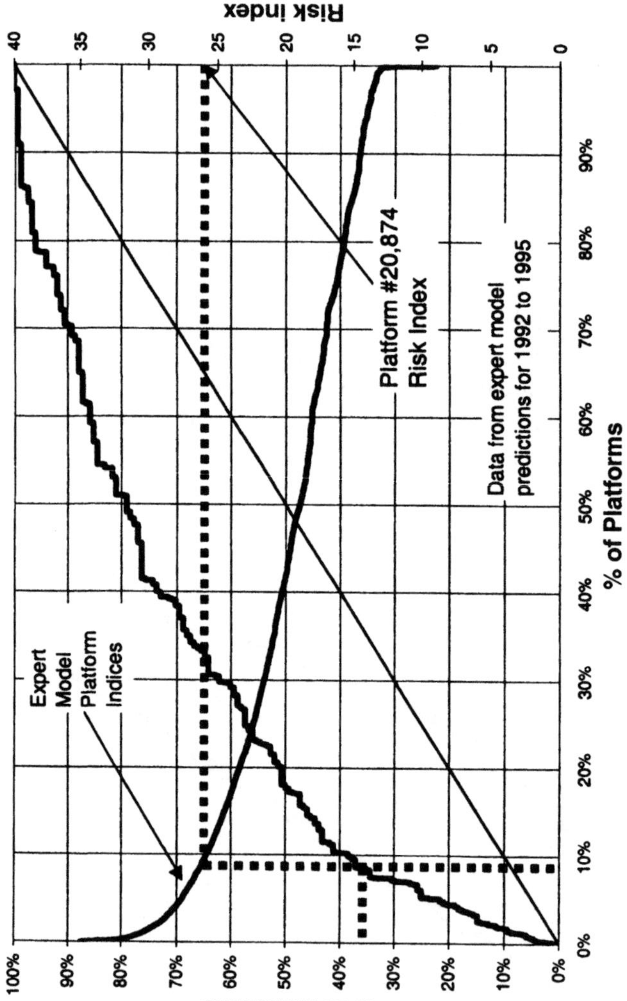

FIGURE 15-4. Risk Ranking Comparison Based on Pooled Expert Model

four-year period. The pooled average expert model in Figure 15-4 provides an easy, straightforward way of evaluating where any platform would rank based on the expert model results for 1992–1995.

Model Comparisons

It is important to note that no model is ever the "right" model under all situations. As Milton Friedemen states, "All models are wrong, but some are more useful than others. No method dominates all others over all situations, i.e., there is usually no uniformly 'best' method. Each model has a set of situations where it works best" (quoted in Langaas 1995, 36).

Figures 15-2 and 15-3 suggest that both the regression model and the average expert judgement models are much better than a random model. The logistic regression model appears to outperform the average expert models in ranking the platform (at least for the number of accidents that actually occurred in 1995).

We can do better than this simple visual comparison by applying techniques developed by Swets (1988): essentially, the true-positive proportion is plotted against the false-positive proportion, and the area under the curve represents an unbiased estimate of the accuracy of the model. Actual analysis of these models using the Swets approach is beyond the scope of this chapter. However, some of the differences in the various models can be highlighted by looking at which platforms each model predicted would have an accident or spill in the next year. By taking each model's list of the 10% of platforms most likely to have an accident or spill and then counting how many of them actually did have a problem, the models can be compared. Figure 15-5 shows the results of such an analysis. The "best" expert does not get as many correct in the top 10% as one might assume.

Comparing the Significance of Components: Expert to Logistic Model

Table 15-5 shows that both the experts and the logistic regressions consider measures of complexity the most important predictor of risk. However, "age and experience" is ranked number two by the experts but ranked last (four out of four) by the logistic regression. Therefore, the two models rely on different components to come up with the model predictions.

Other Model Accuracy Comparisons

At the beginning of this study, many people assumed that more experienced inspectors would be better than less experienced inspectors at estimating

TABLE 15-5. Important Factors for the Regression and Expert Judgement Models

	Rank of importance	
Code	*Logistic regression*	*Expert*
COMP	1	1
INCs	2	3
ACC/SPILL	3	4
AGE	4	2

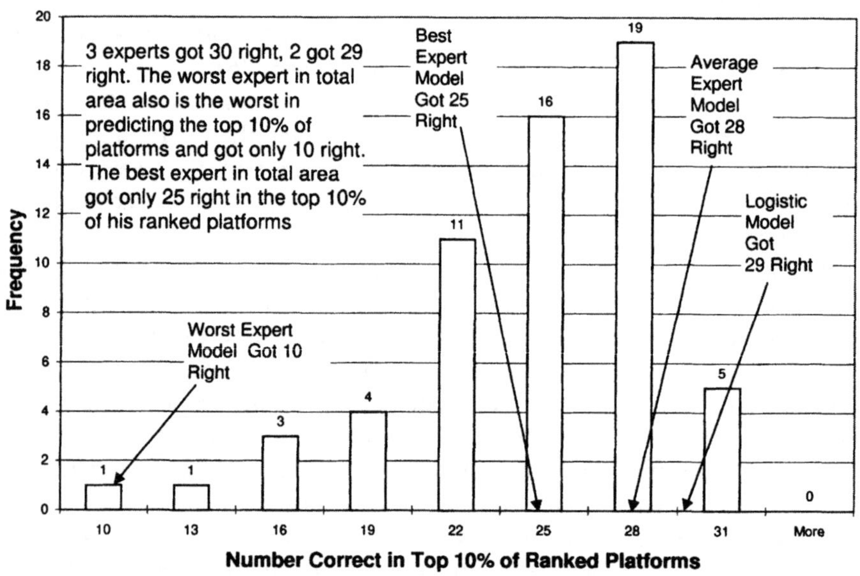

FIGURE 15-5. Comparison of Expert and Regression Models: Number Correct in the Top 10% of Ranked Platforms (62 Accidents, 3,380 Platforms in 1995)

risk. However, the data show that this assumption was incorrect. There is no statistically significant relationship between the experts' accuracy and their level of experience. In fact, there is a slight (not significant) negative correlation; that is, more experienced inspectors have slightly less accurate models than less experienced inspectors.

MMS management was curious whether the accuracy of the models differed according to the location of the inspectors. If there were a difference, MMS might be able to increase the accuracy of those inspectors by increasing the level of training in that district. However, no statistically significant difference was found based on the district in which the inspectors worked.

A check was done to see whether the inspectors that were good at predicting the worst platforms also had the most accurate model overall. The inspectors were ranked based on their model accuracy and ranked based on the number of correct predictions in the top 10% of ranked platforms. Surprisingly, the results indicated that some inspectors were very good at identifying the worst platforms but not so good for the remaining platforms.

In addition to comparing the accuracy between the logistic and average expert models, an interesting comparison is the platform rankings as determined by each method. This is done by comparing the platform's rankings as predicted by both the expert and logistic models for 1995. Although the models were similar in prediction area, the rank of the platforms was not as close as one might have assumed. The correlation of ranks was about 0.5, which is less than expected and indicated that the ranks of the platforms differed by model.

Model Comparison Summary

Pareto found that in most processes, about 80% of the problems are caused by only around 20% of the actors (see Wadsworth 1990, 219). As the models and statistics in our study show, the likelihood of an accident or spill at an offshore platform is very nearly a textbook example of a Pareto process. A relatively small number of platforms are responsible for most accidents and spills. Furthermore, both the statistical and expert judgement models are good at predicting which 20% are most likely to have those accidents or spills.

The results are summarized as follows:

1. The experience of the inspector and the number of years of offshore experience have no impact on the accuracy of their predictions.
2. The experts are all reasonably good at ranking platforms.
3. The variation between experts was less than was expected at the outset of the study.
4. The model accuracy between the districts is not statistically significantly different.
5. Some experts were very good at identifying the very worst platforms (top 10% of ranked platforms), but the same experts do not have the best mental model overall.
6. The expert who had the best overall model was not as good as many other experts at picking the top 10% of platforms.
7. Even the worst expert performed better than chance.
8. The logistic model was better than the best expert.
9. The average expert model was very good at predicting the very worst platforms (top 10% of ranked platforms).

10. The average expert model offers a method of ranking platforms in terms of perceived risk that compares very favorably with the logistic models in terms of accuracy and is much easier to calculate.

Policy Implications

In this chapter, we have shown that the risk of accidents and spills on offshore platforms can indeed be forecast with good accuracy using historical data and expert opinion. The issue facing policymakers is how to use this information to reduce the likelihood of accidents and spills. MMS faces several barriers if it attempts to implement a quantitative, risk-minimizing, inspection program on the basis of our study.

One barrier is that Congress has mandated that each platform be inspected once each year, regardless of size, complexity, and history of accidents and spills. MMS realizes that this mandate may not be an efficient use of the agency's limited inspection resources and would like to change the frequency of inspections so that "safe" platforms are visited less often and "risky" platforms more often than once a year. To circumvent this legislative restriction, MMS has devised what it calls the "mother/cluster" concept, whereby platforms are grouped into clusters. (Often, many small unmanned satellite wells simply pump hydrocarbons to a large manned processing platform.) These clusters are then defined as one "platform." MMS would satisfy the congressional requirement by inspecting each platform cluster once a year. This method would provide inspectors with greater freedom to allocate their time to the more important platforms within each cluster.

A second barrier is that currently, MMS simply does not have the technical expertise or budget to devise and implement a sophisticated, mathematically based, dynamic inspection protocol. It is conceivable that with some additional study, MMS could use the results of our risk ranking along with information about the cost (both time and money) and effectiveness of inspections at each facility to develop a constrained optimization method to minimize the total risk to individuals and the environment. However, few of the quantitative analytic assets of MMS, and the Department of the Interior as a whole, are being used to explore this opportunity. Therefore, the inspection scheduling will remain under the district manager's purview. Fortunately, the robustness of the expert models seems to imply that reasonable schedules will be used.

A critical building block of any rigorous risk-based inspection protocol is a strong commitment from MMS to data integrity and data acquisition. The results of this study have a limited "shelf life" and can become outdated as new platforms come on line and existing platforms age. Data must continue to be collected. In addition, the expert models highlighted data fields that the inspectors felt were important in predicting accidents and spills but were not being collected. These fields should be added to the database.

Finally, MMS can look at the modeling results and say, "Well, it looks like the expert's judgements are pretty good, let's just do what they think." Traditionally, this is how inspections have been scheduled. However, this argument is tricky to make, especially in a political environment. Overlooking (or downplaying) the statistical analysis of data could prove imprudent if a significant accident or spill were to occur. The insights provided by the logistic models were different from that of the expert models. These insights (if they led to a reduction of risk) could easily outweigh any associated cost. Measuring the budgetary cost of a new safety program or data collection effort is usually much easier than estimating the expected loss of a spill or accident. In the political world, justifying a cost-saving decision after an accident is very difficult. After a spill, it would be extremely hard for MMS to explain why the risky platforms were not properly identified and inspected and the corrective actions not taken.

How Can the Ranking Be Used to Decrease Risk?

MMS has many control strategy options, but they tend to fall into three basic categories: lower the threshold for when to issue an INC during inspections and thereby increase the number of INCs, inspect some facilities more frequently, or increase the penalty for an INC. Unfortunately, the efficacy of these strategies is difficult to determine with the current data. The old statistical adage that "correlation is not causation" must be remembered. This kind of modeling does not determine which factors cause accidents and spills, only that they happen together. For example, platforms with more INCs tend to have more accidents. This certainly does not mean that INCs cause accidents. In fact, INCs may decrease the likelihood of an accident.

Is Lowering the Threshold for INCs a Worthwhile Approach?

It is very difficult to determine the impact of lowering the threshold of when to issue an INC given the current data set. To the best of our knowledge, the same inspection standards (INC thresholds) were used across all geographic regions, and these standards did not change during the analysis period. Without a comparison set, it is not possible to predict whether a lower standard and the resulting increase in INCs would reduce risks. A dramatic lowering of the threshold could lead to "frivolous" INCs that hinder platform operations. To improve safety on the OCS, the risk ranking must be used in a cooperative way by the inspectors and the platforms' owners and operators.

However, if inspectors change their method of issuing INCs, the platform risk profiles will change. It could invalidate the risk-ranking models presented here and require the construction of new models that could prove difficult to calibrate.

Would Inspecting "At-Risk" Platforms More Frequently Lower the Likelihood of an Accident?

It is certainly logical that greater attention at higher-ranked facilities probably would result in fewer accidents. However, this result can be confirmed only with further analysis. Increased attention could have several results: the number and types of INCs issued could change (that is, greater scrutiny could lead to more careful inspections and more INCs), platform management could change, and insurance rates could increase. Any or all of these results could lead to reduced risk. However, increased inspection of high-risk platforms is not without cost to MMS. If inspectors were pulled off of some platforms to spend more time on others, there could be a shift in where the accidents and spills occur.

In What Other Ways Can the Risk Ranking Help Policymakers?

First, management should feel reassured that the inspectors have a good intuitive sense regarding which risks are likely to cause an accident or spill. The expert models are good at ranking platforms. Second, the data and models provide a means of answering criticisms of how and why performance measures are picked. Third, management now has some concrete examples of how improved procedures for data acquisition and storage can benefit OCS operations. These findings should help silence critics and budget-cutters. The research highlights the need to fix the problems with data acquisition and processing that MMS has experienced.

Should the Government Keep This Information Confidential?

The risk ranking should be considered an internal MMS management tool, and rankings should be published only in broad bands (for example, as "top 25%," "top 50%"). Individual platform scores should not be made public. As stated earlier, random errors are associated with the underlying data. Because of these errors, a platform may "earn" a high-risk label unfairly. If a publicized risk ranking proves to be incorrect for a particular platform, then MMS would have to answer some tough questions. Because insurance companies would likely use the risk ranking to set rates, owners of identified "risky" platforms could face considerable additional expenses. At the same time, it is only by having the data publicly aired that errors are going to be identified and corrected.

What Are the Effects of Knowing that a Platform Is Ranked High or Low?

MMS Inspectors. Even if inspection priorities were not changed, once inspectors were told about the rankings, it would be difficult for them to do their jobs as previously. They likely would become more diligent on the high-risk platforms and less so on low-risk platforms. These unintentional changes would alter the underlying structure of the risk-ranking models, perhaps invalidating their continued use.

Operating and Owning Companies. The risk rankings might come as no surprise to the platform operators and owners, who probably already know how well they are doing and how "risky" their operation is. However, rankings could lead some owners to institute additional procedures and install safety equipment.

Despite the fact that the ranking is primarily intended as an MMS management tool, it could easily have financial consequences for the platform owners and operators. Insurance companies may react in ways that could increase operating costs. Owners could move personnel from low-ranked to high-ranked platforms. The effect on economic values of platforms could lead some owners to make buying and selling decisions on the basis of the platforms' risk rankings.

Environmental Organizations, Insurance Companies, and Lawyers. An accurate risk ranking would be of interest to many organizations not directly related to the operations of the platforms themselves. Environmental groups could use the rankings to target their efforts and protests to encourage improved safety. Legal groups could use the rankings to argue that proper precautions were not taken in the event of an accident or spill. Insurance companies could use the rankings as an input to their rate structure.

Two factors need to be considered. First, as currently done, the rankings are potentially flawed because of data error. Second, an above-average ranking does not necessarily imply poor management practice or unreasonable risk. Just as the airline with the most flights is more likely to have an accident than a small airline, a high-volume, complex platform with hundreds of workers is more likely to have an accident than a small, low-volume, unmanned platform. How these groups of interested stakeholders will view the ranking information should be considered by MMS. The general release of the rankings should be accompanied by a clear explanation of their meaning.

When Should the Ranking Be Updated, How Long Will the Results of This Study Remain Valid, and When Should New Models Be Calculated?

The greater the release of the ranking information (the larger the audience), the quicker the ranking models will become outdated and their value diminished. Widely publicized rankings could affect inspection protocols, platform operation and maintenance procedures, insurance rates, and public interest. They would affect the safety of platform systems and, in turn, the inputs to the models, the underlying structure of the models, and the accuracy of the models' predictions.

Does the "Risk Ranking" Take into Account the Consequences of a Spill in a Sensitive Area?

Our study did not evaluate the consequences of a spill from a platform adjacent to a "sensitive area" (beach, wetlands, or a shoreline community). In fact, the models do not predict the severity of an accident or the size of a spill. Accidents and spills were treated as binary variables (that is, 1,000-gallon and 10,000-gallon spills were treated equally as "spill events"). The environmental damage caused by a 10,000-gallon spill 400 miles from shore could be less than a 1,000-gallon spill near a salt marsh during nesting season.

All of the inspection districts have "environmentally sensitive" areas that could be severely affected by even a small spill. Therefore, even a moderately ranked platform could warrant increased study if it were near one of these sensitive areas. A follow-on study could geographically plot the rankings and their proximity to environmentally sensitive areas, taking into account likely wind and current information. Combining one of the more sophisticated fate and transport models with an environmental impact model might allow for a "consequence minimization" strategy to be determined analytically.

How Could Our Approach Be Applied to Similar Problems that Exist in the Government?

Because MMS collected platform data without a clear decision model in mind, data fields that were important to our study were not maintained as well as they might have been (for example, the lack of platform identification numbers for some of the accident data). However, the existence of the data allowed this preliminary study to be completed. Other federal, state, and local agencies also have large data sets, collected over many years, that are underutilized. By applying the latest data-mining techniques to these data sets, critical policy insights are possible that could help reduce risks to human health, safety, and the environment. Studies of problems from water quality (see Chapter 17) to air pollution caused by international shipping (see Chapter 13) have benefited from such serendipitous data exploitation.

Conclusions

Models to reduce risks can be constructed in many ways. For offshore oil production platforms, both statistical data analysis and expert decision models were successful in predicting the likelihood of future accidents or spills for the 4,000 platforms on the OCS. These models can be very valuable to the controlling federal agency as they set policy to allocate limited resources to reduce risks. However, the problem is quickly complicated. Data errors may limit a model's accuracy for individual platforms, results may be misused or distorted, and reactions to the model may negate its future use. Building an analytical model is only the first step in the process. Knowing where an accident is likely to occur does not mean that the accident can be easily prevented.

Notes

1. The work that serves as the basis for this chapter is a three-year study that was supported by the U.S. Department of Interior, Minerals Management Service; Carnegie Mellon University, Department of Engineering and Public Policy; and the U.S. Department of Energy, National Energy Technology Laboratory. The full results of the study can be obtained from the University of Minnesota Dissertation Reprint Service (Shultz 1999).
2. In general, the outer continental shelf is three or more miles from shore.
3. The United States' inspection program has a good reputation for safety internationally. Other countries routinely request information regarding how the U.S. Minerals Management Service conducts the inspection program. In addition, in the United States, all of the inspectors are government employees. In some other countries, the oil companies hire the inspectors.
4. An Oracle database called the Technical Information Management System (TIMS) is used to keep track of all information. In earlier years, accident and spill information was maintained separately from inspection records in a database called OPAC or GOPAC.
5. This review was in accordance with the three risk-factor screening surveys that had been conducted by the U.S. Minerals Management Service in 1995 and 1996.
6. For a more in-depth discussion of principal components, see Hatcher 1994.

References

Hatcher, L. 1994. *A Step-by-Step Approach to Using the SAS System for Factor Analysis and Structural Equation Modeling.* SAS, Inc., Cary, NC.

Langaas, M. 1995. *Discrimination and Classification.* Technical Report Statistics 1/ 1995. Department of Mathematical Sciences, The Norwegian Institute of Technology, The University of Trondheim, Norway.

MMS (Minerals Management Service). 1990. *Investigation of March 19, 1989. Fire, South Pass Block 60, Platform B, Lease OCS-G 1608.* OCS Report MMS 90-0016. MMS, U.S. Department of the Interior, Washington, DC.

———. 1992. *Accidents Associated with Oil and Gas Operations, Outer Continental Shelf, 1956–1990.* OCS Report MMS 92-0058. MMS, U.S. Department of the Interior, Washington, DC.

———. 1995. *Accidents Associated with Oil and Gas Operations, Outer Continental Shelf, 1991–1994.* OCS Report MMS 95-0052. MMS, U.S. Department of the Interior, Washington, DC.

———. 1998. *Accidents Associated with Oil and Gas Operations, Outer Continental Shelf, 1995–1996.* OCS Report MMS 98-0030. MMS, U.S. Department of the Interior, Washington, DC.

Paragon Engineering. 1996. Panel on Evaluation of Case Studies of Risk Assessment of Offshore Platforms. Draft working paper. September 30. Paragon Engineering Services Inc., Houston, TX.

Pitblado, R., and Turney, R. 1996. *Risk Assessment in the Process Industries.* Institution of Chemical Engineers, Washington, DC.

Shultz, John R. 1999. *The Risk of Accidents and Spills at Offshore Production Platforms.* Doctoral Dissertation. May. The Department of Engineering and Public Policy, Carnegie Mellon University, Pittsburgh, PA.

Swets, J. 1988. Measuring the Accuracy of Diagnostic Systems. *Science* 240 : 1285–1293.

U.S. Congress, OTA (Office of Technology Assessment). 1985. *Oil and Gas Technologies for the Arctic and Deepwater.* OTA-O-270. May. OTA, Washington, DC.

Wadsworth, H. 1990. *Handbook of Statistical Methods for Engineers and Scientists,* edited by H.M. Wadsworth. McGraw-Hill, New York, NY.

16

Evaluating Occupational Safety Costs and Policy in an Input–Output Framework

DEANNA H. MATTHEWS AND LESTER B. LAVE

Improving occupational safety has been a significant policy goal in state-enacted workers' compensation laws throughout this century. By 1965, the U.S. Congress judged state regulation to be insufficient to protect workers and so created the federal Occupational Safety and Health Administration (OSHA). The annual OSHA budget for the 1980s and 1990s averaged around \$300 million. During that time, the number of workplace inspections per year ranged from almost 50,000 in 1988 to around 30,000 in 1998. To alleviate hazards in specific facilities, and thus reduce the need for inspections, the budget for compliance assistance rose from about \$12 million before 1996 to more than \$40 million in 1998.

Despite the efforts of employers and regulators, workplaces inherently have risk. OSHA and its state counterparts continue to reduce workplace injuries and illnesses. The current safety regulations involve detailed descriptions of personal protective equipment, protective equipment for machines and facilities, and training requirements. Inspections target facilities with poor safety or poor compliance records. Fines are levied to punish employers and to force actions to eliminate or at least reduce hazards. Major incidents that involve death or severe injury make the news headlines and win an immediate investigation by OSHA, generally followed by fines and orders to remedy the situation.

One problem with current OSHA regulation and data collection is the reliance on tabulated data for lost workday injuries, disability days, and fatalities. The data collected by OSHA identifies industries with high rates of incidents, not necessarily identifying where the costs of incidents are high. These three statistics are imperfectly related to the worker or social costs of injury. Assembling a single index of the cost of worker injury would focus the attention of companies, unions, and regulations and would provide better information to industrial and consumer purchasers.

A concern for cost via workers' compensation and OSHA fines has led almost all companies to improve their safety performance by internalizing the cost of hazards. However, companies have exhibited little concern for the safety records of the companies from which they purchase goods and services. In contrast, the concern for quality and the acceptance of standards such as International Organization for Standardization (ISO) 9000 for quality management and quality assurance have led companies to insist their suppliers maintain and document their concern for quality and level of performance. To a lesser extent, environmental concerns as expressed in life-cycle analysis and ISO 14000 for environmental management have led firms to choose suppliers with satisfactory environmental performance. Consumers reacted to their presumption of the environmental damages done by plastic grocery bags by demanding paper bags. Ford Motor Company and other major manufacturers demand that their suppliers document the environmental effects of producing the ordered components.

We hypothesize that safety concerns have not reached up into the supply chain because no estimates of the occupational injuries are embodied in a product or service. We expect that manufacturing companies and consumers would react to data on the occupational injuries embodied in a product or service by seeking alternatives that cause fewer injuries and deaths to workers. For example, in this chapter, we examine the OSHA data on injury rates for coal and iron ore mining, transportation of these minerals, iron and steel making, and automobile manufacture. The occupational injuries embodied in a car are not only those associated with assembling the vehicle. Rather, the injuries accumulate from the mining and transport of coal and iron ore; making, rolling, and casting steel; and manufacturing the components that are assembled into the automobile. The occupational injuries associated with assembling a car are a small fraction of the total occupational injuries from the entire supply chain.

In this chapter, we focus on the performance of sectors with respect to occupational safety. First, we consider the direct injuries, illnesses, and fatalities that occur within each sector, as reported to OSHA. We translate these "direct" injuries, illnesses, and fatalities into the occupational injuries associated with each of 38 sectors through their own safety records, the safety records of the suppliers to the sectors, and the use of an input–output model. The next step is to translate the injuries, illnesses, and fatalities into monetary values. This necessarily controversial step involves weighting minor injuries, severe injuries, occupational disease, and fatalities. However, the result is a single indicator of occupational safety. This index, when normalized by the number of workers or the industry output, provides a way of comparing safety records between industries. Finally, we discuss the policy implications of these calculations and new approaches for data analysis from regulations in the future.

Current Design of Safety Data and Regulations

OSHA collects data on occupational safety on a regular basis. The data are aggregated by Standard Industrial Classification (SIC) code and published by the U.S. Bureau of Labor Statistics (BLS). Individual companies and facilities report the data to OSHA; they include such information as the type of injury or illness that occurred, the cause of the incident, and the resulting work restrictions on the injured or ill employee.

Annually, businesses submit a summary of incidents based on the rate of injury and illness experienced per 100 full-time workers. (Nonfatal injuries are reported separately from fatalities.) The overall injury and illness rates for private industry has fluctuated between 6.7 and 8.9 cases per 100 full-time workers between 1980 and 1998 (U.S. BLS 1999b). These rates are a relative estimate of safety and are somewhat comparable across industries and facilities of different scope and size. However, the rates do not distinguish between job types within an industry (within the mining industry, for example, a machine operator and a clerical worker have very different safety risks). Approximately 6 million incidents are reported annually.

Although the cause of an industrial incident that results in injury or illness is reported, the severity of the injury or illness is difficult to assess. A fall could result in an injury as mild as a minor bruise or as serious as quadriplegia. One means to assess the severity of safety incidents is via the lost workday data. Lost workday cases include incidents that resulted in an employee missing one or more days of work (excluding the day of the injury or illness) or an employee being transferred to a restricted work activity, or both. An injury or illness recorded as having lost workdays or restricted work activity can be assumed to be more severe than an incident that did not result in a lost or restricted workday. The rate of lost workday cases per 100 full-time workers has fallen from 4.0 in the late 1980s to 3.1 in 1998, with about 3 million individual incidents per year (U.S. BLS 1999b). The median days away from work for all injuries resulting in lost work days was 6 days in 1992; the category that averaged 2 days was Heat and Chemical Burns, whereas the

Carpal Tunnel Syndrome category averaged 32 days (U.S. BLS 1994). Because the rate data are comparable across industries, these data are used in this analysis with annual average employee figures to calculate the total number of lost workday cases for each industry sector.

Data on incidents involving fatal injuries are published separately in the Census of Fatal Occupational Injuries. Deaths in private industry caused by workplace hazards have hovered around 6,000 per year for the past decade (U.S. BLS 1999a). Unlike injuries or illnesses, the occurrence of a fatality is extremely rare in some sectors. Records of incidents that result in fatalities are somewhat more consistent than nonfatal injury and illness data, but the data have some anomalies. For example, the Tobacco sector (SIC 21) has a low recorded rate of lost workday injuries, no reported fatalities between 1992 and 1997, and then five fatalities in 1998. We question whether these data could be accurate. Similarly, the Leather and Leather Products sector (SIC 31) has no recorded fatalities between 1992 and 1998, but the lost workday rate of the sector ranged from 5.5 to 4.3 incidents per 100 full-time workers during this time, ranking it as one of the 10 most hazardous industries.

Although the data on safety incidents are collected and reported regularly, they have some limitations. The data are self-reported, which is likely to lead to underreporting. Because the rate data are the basis for OSHA auditing, facilities have some incentive to underreport to avoid scrutiny by regulators. An increase in the lost workday rate between 1988 and 1992 appears to be due to several large industries being fined in 1988 for not reporting incidents. Audits completed by OSHA estimate that total injury and illness cases are underreported by about 10% and that lost workday cases are underreported by about 25% (Conway and Svenson 1998). Also, illnesses caused by occupational exposure are difficult to identify. The cause of an illness initially may not be connected to workplace conditions or may have a delayed onset and not be recognized as being caused by the workplace. These unidentified illnesses do not affect the rate of incidence for a facility but do affect the costs sustained by companies.

The rate data collected by OSHA have been used to focus on reducing incidents in industries where rates are high and ignore industries where rates are generally low. The stagnant overall rates of both nonfatal and fatal injury and illness over the past decade may be due to the limitation of the regulatory design and how performance of industries is measured. The focus on high rates of injuries has led to regulatory control tailored to situations in a specific industry. Only recently has OSHA begun collecting the cause of injuries or illness from the annual reports of businesses (Abraham, Weber, and Personick 1996). By analyzing the root causes of incidents across industries, OSHA is now attempting to address problems that are common in different workplaces. If OSHA estimated the cost of safety incidents to companies, it could increase awareness of the true costs of hazards in the workplace among employers and consumers.

The Economic Costs of Safety

The Value of Safety

Health and safety incidents have a cost to employers and employees. When an employee suffers an injury or illness, the employer incurs several costs. On-site treatment of the employee, medical expenses, and workers' compensation claims must be paid. In addition, productivity is decreased if any damage to machinery or facilities must be repaired, if adjustments must be made to eliminate or reduce the hazard, or if an untrained temporary worker must replace the injured or ill employee. In the long run, workers' compensation premiums may rise if insurers continue to pay for injuries and illnesses that occur at a facility.

Aside from these direct employer costs, the injured worker may lose wages while absent from work or assigned to different work and may suffer from pain and inconvenience (perhaps permanent, including disability) caused by the injury or illness. Because of the range of severity of incidents, placing an average cost on injuries and illnesses is difficult. The risk–dollar trade-off has been widely studied, but little consistency has been found across studies.

Studies have used various approaches to value occupational injuries and illnesses (Viscusi 1992). The method used most commonly in the past was the "human capital" approach, in which lost wages were added to medical treatment costs to estimate the value. A more satisfactory approach estimates the "willingness to pay" for workers. These studies examine the wage premium that is associated with more dangerous jobs, calculating the implicit value that the worker places on safety. The range of estimates for the value of nonfatal injuries and illnesses is almost a factor of 10, from $14,000 to $131,000 for a single employee and a single incident. Table 16-1 is a list of injury and illness valuations culled from a survey of studies.

The range in valuations is primarily a result of the method used and differences in the nature of the incidents. The human capital method gives valuations much smaller than the revealed willingness-to-pay method. For example, the U.S. Environmental Protection Agency (EPA) used $83 as the amount workers were willing to pay to avoid a day of lost work in determining the costs and benefits of the Clean Air Act, reasoning that this was the median daily wage across all workers in 1992 (U.S. EPA 1997). Butler (shown

TABLE 16-1. Survey of Injury and Illness Valuation Estimates

Reference	*Value ($)*
Butler 1983	730 per day
Viscusi and O'Connor 1984	14,000–18,000
Viscusi and Evans 1990	18,000–29,000
Olson 1981	19,000–25,000
Viscusi 1978, 1979	20,000–39,000
Garen 1988	21,000
V. K. Smith 1983	28,000
French and Kendall 1991	38,000
Viscusi 1981	46,000
Kniesner and Leeth 1991	47,000
Viscusi 1978b	48,000–49,000
Viscusi and Moore 1987	55,000
Hersch and Viscusi 1990	57,000
Leigh and Folsom 1984	77,000–89,000
Biddle and Zarkin 1988	131,000
Mean	43,500
Median	38,000

Note: All values are in 1992 dollars, per incident except where noted. References are cited in the source document.

Source: Viscusi 1992.

in Table 16-1) used a willingness-to-pay approach to estimate a value of $730 per day, nine times greater (Viscusi 1992). Another source of variation in the estimates in Table 16-1 is differences in the nature of the incident, because they do not all involve the same number of lost workdays or levels of potential harm. The estimates cover costs that might be experienced across the scope of possible nonfatal injuries and illnesses (that is, a bone fracture would lie at the lower end of the cost scale, whereas loss of a limb would lie at the higher end of the cost scale). Employees are assumed to demand a higher wage premium to accept employment in a risky industry compared with employment in a safe industry. All of these factors help to explain the range of values from $14,000 to $131,000.

For fatalities, the studies attempt to estimate the implicit willingness to pay to prevent premature mortality. Again, these data sources include wage rates and risk levels associated with specific jobs. The estimates for the willingness to pay to avoid premature mortality range from a low of $600,000 to a high of $13.5 million—more than an order of magnitude. The range of valuations of a fatality attempts to control for factors other than risk of fatality that would influence wage rates. The range also depends on the likelihood of

TABLE 16-2. Survey of Mortality Valuation Estimates

Reference	*Value (million $)*
Kniesner and Leeth 1991 (United States)	0.6
Smith and Gilbert 1984	0.7
Dillingham 1985	0.9
Butler 1983	1.1
Miller and Guria 1991	1.2
Moore and Viscusi 1988a	2.5
Viscusi, Magat, and Huber 1991b	2.7
Marin and Psacharopoulos 1982	2.8
Gegax and others 1985	3.3
Kniesner and Leeth 1991 (Australia)	3.3
Gerking, de Haan, and Schulze 1988	3.4
Cousineau, Lacroix, and Girard 1988	3.6
Jones-Lee 1989	3.8
Dillingham 1985	3.9
Viscusi 1978, 1979	4.1
R.S. Smith 1976	4.6
V. K. Smith 1976	4.7
Olson 1981	5.2
Viscusi 1981	6.5
R S. Smith 1974	7.2
Moore and Viscusi 1988a	7.3
Kniesner and Leeth 1991 (Japan)	7.6
Herzog and Schlottman 1987	9.1
Leigh and Folsom 1984	9.7
Leigh 1987	10.4
Garen 1988	13.5
Mean	4.76
Median	3.85

Note: All values are in 1990 dollars per fatality. References are cited in the source document.
Source: Viscusi 1992.

fatality occurring. Workers demand a different wage premium for a high risk (1/100 per year) than for a low risk (1/100,000) per year. Valuations are summarized in Table 16-2.

Arguments can be made concerning the validity of any one of these estimates; no single value is likely to gain consensus. The variation within the costs associated with injury, illness, and death leads to a wide range of dollar estimates. One approach to handle this uncertainty is to work backward, by calculating the cost per injury or premature death averted. If this calculated "cost-effectiveness" estimate is within the range of values implicit in past legislation and regulatory decisions, then the intervention might be justified (Viscusi 1992). In the following analysis, the median values for injuries and illnesses and for fatalities ($38,000 and $4.1 million, respectively, adjusted to 1992 dollars) were used in comparing the total costs of injuries with industry output.

Economic Input–Output

A model of industrial interdependencies is necessary to trace the implications up and down the supply chain to connect each occupational injury to a good or service purchased by a company or a consumer. We use the economic input–output (EIO) framework developed by Leontief (1986). The model considers the economic interdependencies of producing goods and services as a linear system. The underlying assumption of the model is that an increase in demand for the goods and services from any sector will require a proportional increase in goods and services from all sectors providing inputs to that sector. The result is an EIO matrix of total requirements across all sectors that allows the examination of the economy-wide effects, including the supply chain effects of changes in one sector.

Leontief (1986) also proposed to augment this economic model with other industry data, such as external environmental or safety impacts. Augmentation of the EIO model with external data allows estimation of the total external burden caused by all sectors from a change in demand of only one sector. Augmenting the model in this fashion gives a life-cycle perspective of the impacts of the interdependencies in the economy. Such a model has been developed (Lave and others 1995), and analyses have been performed that include resource inputs such as electricity, fuels, ores, and fertilizers and environmental outputs such as toxic and conventional emissions by media, hazardous waste generation and management, global warming potential, and ozone-depleting substances (Hendrickson and others 1998; Matthews and Lave 2000). In this chapter, the EIO framework analysis is augmented with occupational lost workday cases and fatalities to estimate the total occupational risk across the economy associated with increasing demand for a given sector.

The approach of the EIO model (see Appendix at the end of this chapter for a complete derivation of the EIO model) can be explained as follows by using the petroleum and coal sector as an example. The petroleum and coal sector produces goods as demanded by its customers, which are the sector's *direct output.* To produce these goods, the sector requires supplies of materials and energy from various other sectors, such as coal mining; oil and gas extraction; electric, gas, and sanitary services; and ultimately, the suppliers to those sectors. This is considered the *indirect output* for the petroleum and coal sector. If the sector has an increased demand for its final product, then the direct output of the sector increases, as does the indirect output from the supplier sectors, because the petroleum and coal sector now requires additional supplies for its production. The sum of the direct output and the indirect outputs of the chain of supplier sectors is the *total economic output.*

Similarly, the method can be used with safety incidents. The *direct safety incidents* are simply the number of lost workdays or fatalities within the petroleum and coal sector divided by the total output of the sector. The *indirect safety incidents* are the portion of safety incidents of supplier sectors that occur from producing goods and services for the petroleum and coal sector. When the demand for goods from the petroleum and coal sector increases, as production increases, the safety incidents within the sector will increase. Similarly, the safety incidents in supplier sectors will increase. Again, the total occurrences of safety incidents is the sum of the direct safety incidents from the petroleum and coal sector itself plus the indirect safety incidents from the chain of supplier sectors. Thus, given the risk of each industry sector and the relationships between sectors of the economy, the total (direct plus indirect) risk resulting from an increment of demand in a particular sector can be determined. The *indirect risk*—that risk attributable to other industries—then can be calculated using simple subtraction: total risk minus direct risk.

This EIO model form assumes that the relationships within the economy and its associated external impacts are proportional. For example, a 20% increase in output from the petroleum and coal sector will result in a 20% increase in lost workday cases or fatalities from that sector. Likewise, supplier industries will experience similar proportional increases in both industry output and safety incidents. This proportionality may not fully describe the relationship between safety incidents and economic output, but in most cases, it should be a sufficient first approximation. Other best available estimates for changes in safety incidents may be only linear extrapolations as well. The U.S. Department of Commerce collects information from a sample of firms in the nation every five years to calculate the input–output model for the U.S. economy (U.S. DOC 1994). The total requirements matrix at the 80 × 80 EIO sector level for 1992 was used in this analysis. The matrix was aggregated to a 38 × 38 SIC sector level to match the sector divisions used with the published safety data. The aggregation was based on the proportion of output of each of the industry sectors included in the major sector categories.

Injuries and Illnesses

Table 16-3 provides results of the analysis using lost workday case data and economic output of the various industries. Data for the direct lost workday

TABLE 16-3. Lost Workday Risk per Million Dollars of Output

SIC	Industry	Direct rate/100 workers	Cases/million $ output			Indirect-to-direct ratio	Value of cases (% of output)	
			Indirect	*Direct*	*Total*		*Direct*	*Total*
1,2,7–9	Agriculture, forestry, and fishery	5.4	0.363	0.278	0.641	1.305	1.06	2.44
10	Metal mining	3.3	0.327	0.163	0.490	2.001	0.62	1.86
12	Coal mining	7.7	0.264	0.359	0.623	0.735	1.36	2.37
13	Oil and gas extraction	3.0	0.180	0.100	0.280	1.805	0.38	1.06
14	Nonmetallic materials	3.7	0.246	0.274	0.521	0.897	1.04	1.98
15–17	Construction	5.8	0.407	0.382	0.789	1.066	1.45	3.00
20	Food and kindred products	9.5	0.532	0.385	0.917	1.383	1.46	3.49
21	Tobacco	2.4	0.174	0.029	0.203	5.953	0.11	0.77
22	Textiles	4.2	0.477	0.369	0.846	1.295	1.40	3.21
23	Apparel	4.0	0.481	0.567	1.048	0.849	2.15	3.98
24	Lumber and wood products	7.6	0.564	0.590	1.154	0.956	2.24	4.39
25	Furniture and fixtures	6.6	0.435	0.731	1.166	0.595	2.78	4.43
26	Paper and allied products	5.0	0.430	0.261	0.692	1.647	0.99	2.63
27	Printing and publishing	3.2	0.273	0.288	0.561	0.949	1.09	2.13
28	Chemicals and allied products	2.8	0.330	0.104	0.434	3.183	0.39	1.65
29	Petroleum and coal products	2.8	0.327	0.031	0.357	10.695	0.12	1.36
30	Rubber and miscellaneous products	6.8	0.339	0.532	0.872	0.637	2.02	3.31
31	Leather and leather products	5.4	0.558	0.688	1.246	0.811	2.62	4.74

32	Stone, clay, and glass products	6.1	0.353	0.514	0.867	0.686	1.95	3.30
33	Primary metal industries	7.1	0.463	0.351	0.814	1.320	1.33	3.09
34	Fabricated metal industries	6.6	0.408	0.532	0.940	0.768	2.02	3.57
35	Industrial machinery and equipment	4.2	0.402	0.325	0.727	1.236	1.24	2.76
36	Electronic and electric equipment	3.6	0.368	0.264	0.632	1.392	1.00	2.40
37	Transportation equipment	7.1	0.530	0.344	0.875	1.539	1.31	3.32
38	Instruments and related products	2.7	0.286	0.193	0.479	1.486	0.73	1.82
39	Miscellaneous manufacturing	5.0	0.403	0.440	0.843	0.916	1.67	3.20
40–47	Transportation	6.7	0.293	0.612	0.904	0.479	2.32	3.44
48	Communications	1.8	0.212	0.096	0.308	2.202	0.37	1.17
49	Electric, gas, and sanitary services	3.6	0.247	0.121	0.368	2.033	0.46	1.40
50–57,59	Wholesale and retail trade	3.6	0.145	0.621	0.766	0.233	2.36	2.91
58	Eating and drinking	3.1	0.365	0.729	1.094	0.501	2.77	4.16
60–64,67	Finance and insurance	1.9	0.202	0.224	0.425	0.901	0.85	1.62
65	Real estate	3.1	0.096	0.039	0.135	2.432	0.15	0.51
70,72,76	Services	3.8	0.226	0.767	0.992	0.295	2.91	3.77
73	Business services	2.6	0.164	0.284	0.449	0.578	1.08	1.70
75	Auto repair, services, and parking	3.3	0.291	0.209	0.500	1.388	0.80	1.90
79	Amusement and recreation services	4.4	0.243	0.424	0.667	0.574	1.61	2.54
80–83	Health, legal, education, and social services	3.4	0.175	0.427	0.602	0.411	1.62	2.29

Note: SIC = Standard Industrial Classification code.

rate per 100 full-time workers range from a high of 9.5 in Food and Kindred Products (SIC 20) to a low of 1.8 for Communications (SIC 48). Other sectors with high rates are those usually thought to be risky, including Coal Mining (SIC 12) at 7.7 and Lumber and Wood Products (SIC 24) at 7.6. As indicated earlier, these rate data are comparable across sectors and give an indication of the risk involved for all employees in an industry. Again, because only a small proportion of workers have the dangerous mining or logging jobs, the rate for an industry is decreased by averaging the much safer jobs of, for example, clerks.

The safety impact per dollar of output tells a different story. The direct lost workday cases per million dollars of output (Table 16-3) measure risk based on the products and processes of an industry. The Services sector (SIC 70, 72, 76) now tops the list with 0.77 lost workday cases per million dollars of output, followed by Furniture and Fixtures (SIC 25) with 0.73 cases and Eating and Drinking establishments (SIC 58) with 0.73 cases. Using the EIO model to calculate supply chain effects across all industry sectors, the total (direct plus indirect) lost workday cases per million dollars of output for an additional $1 million in final demand is calculated. Leather and Leather Products tops the list with 1.25 lost workday cases throughout the economy per $1 million of the sector output.

The lost workday cases directly attributed to an industry sector can be compared with the indirect lost workday cases—those occurring in other industry sectors. In Table 16-3, the ratio of indirect to direct lost workday cases gives the relative risk in other industries due to production in an industry. A value of one indicates an equal number of lost workday cases inside and outside the industry; numbers greater than one indicate higher numbers of lost workday cases in supplier industries. Petroleum and Coal Products (SIC 29) is significantly higher in indirect lost workday cases than other industries, with a ratio of 10.7. We infer that the high ratio means that great care is taken to protect worker health and safety in this sector, although suppliers are notably unsafe. The Tobacco industry (SIC 21) has about half that at 5.9, and Chemical and Allied Products (SIC 28) follows with a ratio of 3.2. These results indicate that although an industry may be perceived as fairly safe based on reported data, like these three sectors are with less than 3.0 lost workday rates per 100 full-time employees, the overall risk created throughout the economy can be very high. In the case of Real Estate (SIC 65), which ranks fourth with a ratio of 2.4, the transfer of property involves generally safe clerical office work; however, sales include the development of property by the Construction sector (SIC 15–17), a high-risk industry.

A third way to compare occupational safety across industries is to examine the cost incurred for injuries and illnesses (Table 16-3). The direct and total lost workday cases for each industry were valued at $38,000 each, then this estimated valuation was divided by the value of output. Considering direct lost workday cases, safety costs ranged from almost 3% (Services, Furniture and Fixtures, and Eating and Drinking) to 0.1% (Tobacco, Petroleum and Coal Products, and Real Estate) of output for lost workday cases. Values between 5% and just under 1% are seen for the costs of total lost workday cases. These results indicate the relative cost of safety imbedded in products and services that eventually are passed on to the consumer.

The same analysis was completed using dollars of value added rather than dollars of output as the normalizing factor. Results are similar, although the percentage of value added for lost workday costs is approximately double across the industry sectors, because value added is only a fraction of the value of total output for most industries. The correlation between direct and total lost workday cases per dollar of output and per dollar of value added is shown in Table 16-4. These four measures are highly correlated, as shown by the correlation coefficients exceeding 0.9. The lost workday rate per 100 workers is positively correlated with the other measures, but the relationship is much smaller. Thus, whereas the difference between output and value added is important for some industries, it is not important in general. When we compute the correlations among the direct and total lost workday valuations per dollar of output or per dollar of value added, we find similarly high correlations.

Fatalities

The analysis was repeated using the data on fatalities in the workplace. Results per dollar of output are given in Table 16-5. The fatality rate per 1,000 employees ranges from a high of 0.68 for Agriculture, Forestry, and Fisheries (SIC 1, 2, 7–9) to a low of 0.01 for six different sectors. The Leather and Leather Products sector had no recorded fatalities from 1992 to 1998. When considering the number of fatalities normalized to $100 million of output, Agriculture, Forestry, and Fisheries; Lumber and Wood Products; and Services rank as the top three sectors for both direct and total fatalities.

The allocation of direct and indirect fatalities among the different categories reveals important differences among industries. Examining the ratio of indirect to direct fatalities, three sectors (Petroleum and Coal Products, Food and Kindred Products, and Tobacco) have a much higher proportion of fatalities occurring in the supply chain rather than directly within the industry. For every fatality within the Petroleum and Coal Products sector, a dozen fatalities occur in other sectors that supply it with input materials. Not surprisingly, sectors normally considered as having high risk—Agriculture, Forestry, and Fishery; Mining (SIC 10, 12); and Construction—are at the bot-

TABLE 16-4. Correlation Matrix for Direct and Total Lost Workday Cases by Output and Value Added

	Lost workday rate/100 workers	*Lost workday cases/ million $ output*		*Lost workday cases/ million $ value added*	
		Direct	*Total*	*Direct*	*Total*
Lost workday rate/ 100 workers	1				
Lost workday cases/ million $ output					
Direct	0.487	1			
Total	0.645	0.915	1		
Lost workday cases/ million $ value added					
Direct	0.640	0.912	0.985	1	
Total	0.685	0.770	0.955	0.958	1

tom of this list. These sectors are the most dangerous, so the sectors they buy from are necessarily less dangerous. In fact, the 10 most risky sectors by fatality rate are in the bottom 15 in the ratio of indirect to direct fatalities, and all have ratios less than one. Many fatalities within these sectors are due to providing inputs to sectors farther down the supply chain.

Finally, the value of fatalities as a percentage of output was determined as shown in Table 16-5. Each fatality was valued at $4.1 million. The costs range between 0.02% and 1.50% for all sectors. As a percentage of output, the Agriculture, Forestry, and Fisheries sector ranks first, and Services ranks second; only these two sectors have costs exceeding 1% of output. The value of total fatalities (direct plus indirect) ranges from 2% to 0.1%. As with lost workday cases, the value of fatalities as a percentage of value added is about two times higher. The correlation between direct and total fatalities is similar to that for lost workdays. One difference is the increased relationship between the fatality rate for the various sectors and the number of fatalities divided by dollar of output, both direct and total.

Injuries and Illnesses versus Fatalities

The ratios of indirect to direct safety impacts exhibit significant differences between the lost workday and fatality results. As shown in Tables 16-3 and 16-5, the ratios of indirect to direct lost workdays are very different from those for fatalities. Only 7 of the 38 sectors have ratios greater than two for lost workdays, but 21 have ratios greater than two for fatalities. These results indicate that lost workdays are a result of activity within a sector, but fatalities are caused by demands from sectors farther up the supply chain. Also, some sectors have considerable differences between the ratios for lost workdays versus fatalities. For example, Food and Kindred Products has ratios of 1.4 for lost workdays but 11.2 for fatalities. Only 11 sectors have higher ratios for lost workdays than for fatalities.

The cost estimates in the last two columns of Tables 16-3 and 16-5 also reveal a surprising result. In contrast to regulatory studies of air pollution, the percentage of costs for lost workday injuries and illnesses is more than for fatalities across all industry sectors. The cost percentages for output range between 5% and 0.5% for lost workday cases and from 2% to 0.1% for fatalities (Tables 16-3 and 16-5). These results generally agree with work by Leigh and Miller (1998) and by Viscusi and Zeckhauser (1994), but this analysis provides some advantages. Leigh and Miller do not account for supply chain impacts in estimates of total costs and do not include the agriculture sector. Viscusi and Zeckhauser use an EIO analysis and estimate direct plus indirect impacts. However, they aggregate fatalities into only eight major categories, whereas our analysis examines fatality impacts in 38 sectors. This dissagregation provides more detail in comparing the impacts of lost workdays versus fatalities within the same sector.

A New Approach to Safety Regulations

The overall costs of safety incidents are not trivial. Summing across all industries, roughly 3.7 million workday cases and 6,000 fatalities occur annually. Using our valuations, the annual social costs are then $130 billion for injuries and illnesses, and $23 billion for fatalities. The total cost of safety incidents, $150 billion, was 2.4% of the gross domestic product for the year. The annual OSHA budget of approximately $300 million seems incommensurate with the $150-billion loss. Our analysis does not estimate the reduction in injuries and disease that would result from a 1% increase in OSHA's enforcement budget, but the value of reducing injuries and disease likely would be more than $3 million per year.

A more surprising result of this analysis is the difference between the cost of lost workday injuries and illnesses versus the cost of fatalities. For all industries, the cost of injuries and illnesses is 2.22% of total output, whereas the cost of fatalities is only 0.68% of total output. The indirect cost of injuries and illnesses to the economy is substantial. This disparity in costs is a direct reflection of regulatory approach to focus on reducing fatalities. One measure of regulatory success is the cost-effectiveness of the program—for safety

TABLE 16-5. Fatality Risk per $100 Million of Output

SIC	Industry	*Direct fatality rate/ 1,000 workers*	*Fatalities/$100 million output*			*Indirect-to-direct ratio*	*Value of fatalities (% of output)*	
			Indirect	*Direct*	*Total*		*Direct*	*Total*
1,2,7–9	Agriculture, forestry, and fishery	0.68	0.160	0.348	0.508	0.459	1.43	2.08
10	Metal mining	0.25	0.086	0.125	0.211	0.689	0.51	0.87
12	Coal mining	0.32	0.060	0.151	0.211	0.396	0.62	0.87
13	Oil and gas extraction	0.24	0.052	0.081	0.133	0.640	0.33	0.54
14	Nonmetallic materials	0.24	0.044	0.181	0.225	0.244	0.74	0.92
15–17	Construction	0.23	0.074	0.153	0.226	0.483	0.63	0.93
20	Food and kindred products	0.05	0.216	0.019	0.235	11.176	0.08	0.97
21	Tobacco	0.05	0.056	0.006	0.062	8.975	0.03	0.25
22	Textiles	0.02	0.082	0.015	0.097	5.403	0.06	0.40
23	Apparel	0.01	0.065	0.011	0.077	5.683	0.05	0.31
24	Lumber and wood products	0.29	0.195	0.228	0.423	0.853	0.94	1.73
25	Furniture and fixtures	0.02	0.083	0.025	0.108	3.288	0.10	0.44
26	Paper and allied products	0.03	0.089	0.018	0.107	5.029	0.07	0.44
27	Printing and publishing	0.02	0.043	0.020	0.063	2.149	0.08	0.26
28	Chemicals and allied products	0.03	0.065	0.013	0.078	5.193	0.05	0.32
29	Petroleum and coal products	0.09	0.114	0.010	0.124	12.011	0.04	0.51
30	Rubber and miscellaneous products	0.03	0.063	0.023	0.086	2.780	0.09	0.35
31	Leather and leather products	0.00	0.078	0.000	0.078	NA	0.00	0.32

32	Stone, clay, and glass products	0.09	0.081	0.077	0.158	1.051	0.32	0.65
33	Primary metal industries	0.09	0.086	0.043	0.130	1.998	0.18	0.53
34	Fabricated metal industries	0.04	0.065	0.029	0.094	2.206	0.12	0.38
35	Industrial machinery and equipment	0.03	0.053	0.022	0.075	2.406	0.09	0.31
36	Electronic and electric equipment	0.01	0.048	0.008	0.056	5.960	0.03	0.23
37	Transportation equipment	0.03	0.065	0.014	0.080	4.598	0.06	0.33
38	Instruments and related products	0.01	0.037	0.006	0.043	5.645	0.03	0.18
39	Miscellaneous manufacturing	0.03	0.061	0.027	0.087	2.274	0.11	0.36
40–47	Transportation	0.23	0.068	0.213	0.281	0.321	0.87	1.15
48	Communications	0.02	0.035	0.012	0.046	3.012	0.05	0.19
49	Electric, gas, and sanitary services	0.09	0.075	0.030	0.105	2.470	0.12	0.43
50–57,59	Wholesale and retail trade	0.04	0.025	0.073	0.097	0.341	0.30	0.40
58	Eating and drinking	0.03	0.085	0.059	0.145	1.433	0.24	0.59
60–64,67	Finance and insurance	0.01	0.025	0.008	0.033	3.181	0.03	0.14
65	Real estate	0.04	0.024	0.006	0.029	4.248	0.02	0.12
70,72,76	Services	0.13	0.041	0.256	0.297	0.159	1.05	1.22
73	Business services	0.04	0.023	0.042	0.065	0.561	0.17	0.27
75	Auto repair, services, and parking	0.12	0.030	0.078	0.107	0.381	0.32	0.44
79	Amusement and recreation services	0.06	0.044	0.060	0.105	0.739	0.25	0.43
80–83	Health, legal, education, and social services	0.01	0.028	0.014	0.042	2.023	0.06	0.17

regulations, the cost per avoided injury or illness or fatality. With six million injuries and illnesses annually, which cost businesses three times as much as fatalities, perhaps policymakers should consider a level of action that targets preventing these lesser incidents.

Workers' compensation and tort law suits internalize most of the social costs of occupational health and safety, but a portion of the costs are borne by the injured worker and society more generally—pain and suffering, health care costs, and so forth. This internalization of costs has given companies incentive to maintain safe workplaces. However, the indirect costs of hazards may not be fully addressed. For example, the cost of lost productivity or machine repair may not be attributed to a safety incident but instead would be reflected in other accounting categories. Dorman (1998) discusses the difficulties in assessing the true costs of safety within a firm. As with environmental accounting projects, safety accounting exercises and information on the cost estimates of safety throughout the supply chain would assist in estimating the indirect or hidden costs of safety to further encourage changes in facility practices.

From the regulator's point of view, an emphasis on cost over the rate of injury or illness per employee could cause a shift in thinking. Two prominent sectors—Services, and Eating and Drinking—fall into the lower range when considering only lost workday rate per 100 employees, with rates of 3.8 and 3.1, respectively. But the sectors rank first and second when examining costs as a percentage of output. OSHA gives little attention to small retail establishments, such as restaurants. In addition to their small size, facilities in these sectors have low direct injury rates per employee, giving OSHA reason to focus inspections elsewhere. However, these data suggest that eating and drinking establishments should receive a much higher priority. Evaluation of these workplaces to reduce common hazards could be quite beneficial in reducing overall expenditures on injuries. By targeting industries with high rates per employee, the government may be missing opportunities to demonstrate to industry leaders how costly hazards may be. This awareness of safety costs may potentially increase the willingness of employers to seek out compliance assistance and voluntarily promote safety in the workplace—further alleviating the need for OSHA oversight and inspection.

Regulators target their efforts at workplaces where they believe their limited resources can have the most influence. Thus, they tend to focus on large workplaces and plants with the greatest safety and health burden. The calculation of indirect safety costs adds a new dimension to setting priorities that might be used within the framework of the Government Performance and Results Act. With the analysis of safety data in the EIO framework, OSHA regulators can examine the total health costs for a product or service and the effectiveness of their interventions on this cost. A large total cost implies that somewhere in the supply chain, some employer is either being lax or operating a highly dangerous process. For suppliers with large plants and extremely dangerous operations, OSHA will have already made its presence felt. This analysis may provide a new way of identifying workplaces that were not recognized as unsafe previously because they are small.

Total safety risk also would add a new dimension to assessing how well an industry is managing safety in its own workplace. Some measures to cut the costs of safety may lower the industry-reported incidents but increase risk overall. Contractor employment is a classic example (Levine 1997). An industry can subcontract a high-risk job to an outside industry, and any safety incidents will be attributed to the subcontracting firm. Because a subcontractor employee may not be familiar with a facility and its hazards (or may be new or temporarily employed and not aware of safety hazards), the subcontractor employee may experience higher risk exposure than a permanent full-time industry employee. As a result, the direct safety problems and direct costs of safety are lower than realized. Use of the EIO analysis in this framework helps to identify industries with high indirect safety issues in the supply chain. Analysis of total safety risk across the economy also would provide regulators with a measurement of whether safety risks were being reduced overall rather than merely shifted from one industry to another.

Beyond what regulators may accomplish, identifying the total safety cost of a product or service may encourage consumers to influence industries. For half of the industry sectors, the total lost workdays are more than double the direct lost workdays. For example, in the Chemical and Allied Products sector, total lost workdays are four times the direct lost workdays. Obviously, this sector buys raw materials and intermediate products that involve many lost workdays. In contrast, for Wholesale and Retail Trade (SIC 50–57, 59), total lost workdays adds only 25% to direct lost workdays. These major differences in lost workdays for each sector's supply chain could influence the way that consumers view individual sectors. Buyers exert considerable influence on sellers to improve their performance with respect to the quality and price of the product or service. They also could exert considerable influence with respect to the occupational injuries and illnesses embodied in the service or product.

Conclusions

Even though state and federal regulators have successfully reduced rates of occupational injury and disease, the toll remains high. After decades of federal regulation and almost a century of state regulation, trying harder or increasing budgets marginally is unlikely to have much effect. Instead, exposing the costs of hazards throughout the supply chain to both employers and consumers may make a difference.

We present a method for translating the OSHA data on occupational illness, injury, and death for specific industries into direct and indirect occupational injuries for each sector. It calculates the expected increase in occupational injuries, illness, and death throughout the entire supply chain, from minerals extraction to energy supply, components manufacture, and final assembly. In doing so, it reveals that many service sectors (such as restaurants) are responsible for few direct occupational injury problems but, through the goods and services they purchase, are responsible for many occupational injuries throughout the economy.

Our study does not comment on whether OSHA's regulation of the most dangerous jobs is sufficiently stringent; we have not examined the costs and benefits of more stringent regulation. However, by providing companies and consumers with information about the occupational injuries associated with their purchases, we provide the data they need to change the system, if they desire to do that. As businesses and consumers have become more concerned about quality and the environmental implications of their actions, they have reached up the supply chain to demand better performance. An assembler can no longer hide behind the excuse that the lack of quality or unacceptable environmental performance comes from a supplier. The assembler is responsible for ensuring the quality of the assembled product, not only the assembly, and the environmental performance of the supply chain, not only the assembler. We think it possible that the assembler and customer will care about the occupational hazards associated with all aspects of the final good or service. If so, responsible companies and consumers will open a new route to improving occupational safety that is independent of state and federal regulation.

Appendix

Economic Input–Output Analysis

Mathematically, the development of the economic input–output (EIO) model to include safety can be explained as follows. (The reader is directed to Miller and Blair 1985 for additional reading on EIO analysis.)

First, consider the direct requirements matrix (**D**) of intersectoral relationships. Each sector requires some final product output from various other sectors to complete its own production of output. For example, the petroleum and coal products sector requires output from the coal mining, oil and gas extraction, and fabricated metals industries. In matrix form, the rows of

D would indicate the amount of output from industry *i* required to produce one unit of output of industry *j*.

Second, consider the final demand vector (F) of goods in the economy, such as the demand for petroleum. The sector in consideration must produce F units of output to meet this demand. At the same time, $\mathbf{D} \times \mathbf{F}$ units of output are produced in all other sectors. So, the result is more than the demand for the petroleum and coal products sector but also the demand for its direct supplier sectors. The resulting total output vector ($\mathbf{X}_{\text{direct}}$) of the entire economy can be written

$$\mathbf{X}_{\text{direct}} = (\mathbf{I} + \mathbf{D})\mathbf{F}$$

where **I** is an identity matrix to represent the output of each sector itself, and F is a vector of final demand for the sectors. $\mathbf{X}_{\text{direct}}$ is then a vector of total output from each sector.

This relationship takes into account only one level of suppliers, however. The demand of output from the first-tier of suppliers creates a demand for output from their direct suppliers (that is, the second-tier suppliers of the sector in consideration). For example, the demand for petroleum results in a demand from the oil and gas extraction sector (first tier), which in turn results in a demand from the electric generators (second tier) to operate the gas extraction facilities. The second-tier supplier requirements could be calculated by further multiplication of the direct requirements matrix by the final demand, or **DDF**. In many cases, third and fourth or more tiers of suppliers exist, resulting in a summation of many of these factors so that the total output can be calculated as

$$\mathbf{X} = (\mathbf{I} + \mathbf{D} + \mathbf{DD} + \mathbf{DDD} + \ldots)\mathbf{F}$$

where **X** (no subscript) is a vector including all supplier outputs, direct and indirect.

The expression $(\mathbf{I} + \mathbf{D} + \mathbf{DD} + \mathbf{DDD} + \ldots)$ can be shown to be equivalent to $(\mathbf{I} - \mathbf{D})^{-1}$, which is called the *total requirements matrix*, or the Leontief inverse. The relationship between final demand and total output can be expressed compactly as

$$\mathbf{X} = (\mathbf{I} - \mathbf{D})^{-1}\mathbf{F} \text{ or } \Delta\mathbf{X} = (\mathbf{I} - \mathbf{D})^{-1}\Delta\mathbf{F}$$

where the latter expression indicates that the EIO framework can be used to determine relative changes in total output based on an incremental change in final demand. Typically, the values in the matrices and vectors are expressed in dollar figures (that is, in the direct requirements matrix [**D**], the dollar value of output from industry i used to produce one dollar of output from industry j). This puts all items in the economy—petroleum, coal, or electricity—into comparable units. As such, the model also can be used with figures for value added rather than total output.

The EIO analysis then can be augmented with additional noneconomic data. The total external outputs associated with each dollar of economic output can be determined by adding information, in this analysis safety data, to the EIO framework. First, the safety impact per dollar of output is calculated via

$$R_i = \frac{\text{Lost workday cases}}{X_i} \quad \text{or} \quad R_i = \frac{\text{Fatalities}}{X_i}$$

where R_i is used to denote the risk in sector i, and X_i is the total output for sector i. This measure of risk, based on reported safety data, is equivalent to the direct risk of a sector. The risk measures form the vector **R**.

To determine the total (direct plus indirect) risk throughout the economy, the direct risk value is used with the EIO model. A vector of the total external outputs (**B**) can be obtained by multiplying the total economic output at each stage by the risk:

$$\Delta\mathbf{B} = \hat{\mathbf{R}} \times \Delta\mathbf{X} \quad \text{or} \quad \Delta\mathbf{B} = \hat{\mathbf{R}}(\mathbf{I} - \mathbf{D})^{-1}\Delta\mathbf{F}$$

where $\hat{\mathbf{R}}$ is a matrix with the elements of the vector **R** along the diagonal and zeros elsewhere, and $\Delta\mathbf{X}$ is the vector of relative change in total output based on an incremental change in final demand. For this analysis, an increment of final demand ($\Delta\mathbf{F}$) of $1 million is used. The indirect safety risk is then calculated by subtracting the direct impact (**R**) from the total impact ($\Delta\mathbf{B}$).

References

Abraham, K.G., W.L. Weber, and M.E. Personick. 1996. Improvements in the BLS Safety and Health Statistical System. *Monthly Labor Review* 119(4): 3–12.

Conway, H., and J. Svenson. 1998. Occupational Injury and Illness Rates, 1992–1996: Why They Fell. *Monthly Labor Review* 121(11): 36–58.

Dorman, P. 1998. Cost Internalization in Occupational Safety and Health: Prospects and Limitations. In *Research in Human Capital and Development* (Volume 12), edited by I. Farquhar and A. Sorkin. Stamford, CT: JAI Press, 99–121.

Hendrickson, C.T., A. Horvath, S. Joshi, and L.B. Lave. 1998. Economic Input–Output Models for Environmental Life Cycle Assessment. *Environmental Science and Technology* 32(7): 184A–191A.

Lave, L.B., E. Cobas, C.T. Hendrickson, and F.C. McMichael. 1995. Using Input–Output Analysis to Estimate Economy-Wide Discharges. *Environmental Science and Technology* 29(9): 420A–426A.

Leigh, J.P., and T.R. Miller. 1998. Ranking Industries Based upon the Costs of Job-Related Injuries and Diseases. In *Research in Human Capital and Development* (Volume 12), edited by I. Farquhar and A. Sorkin. Stamford, CT: JAI Press, 47–72.

Leontief, W. 1986. *Input–Output Economics.* New York: Oxford University Press.

Levine, D. 1997. They Should Solve Their Own Problems: Reinventing Workplace Regulation. In *Government Regulation of the Employment Relationship,* edited by B.E. Kaufman. Madison, WI: Industrial Relations Research Association.

Matthews, H.S., and L.B. Lave. 2000. Applications of Environmental Valuation of Determining Externality Costs. *Environmental Science and Technology* 34(8): 1390–1395.

Miller, R.E., and P.D. Blair. 1985. *Input–Output Analysis: Foundations and Extensions.* Englewood Cliffs, NJ: Prentice-Hall.

U.S. BLS (Bureau of Labor Statistics). 1994. *Survey of Occupational Injuries and Illnesses 1992.* Washington, DC: U.S. Department of Labor.

———. 1999a. *Census of Fatal Occupational Injuries.* Washington, DC: U.S. Department of Labor.

———. 1999b. *Survey of Occupational Injuries and Illnesses 1998.* Washington, DC: U.S. Department of Labor.

U.S. DOC (Department of Commerce). 1994. *Input–Output Accounts of the U.S. Economy, 1992 Benchmark* (computer diskettes). Washington, DC: U.S. DOC.

U.S. EPA (Environmental Protection Agency). 1997. *Benefits and Costs of the Clean Air Act.* EPA 410-R-97-002. Washington, DC: U.S. EPA.

Viscusi, W.K. 1992. *Fatal Tradeoffs: Public and Private Responsibilities for Risk.* New York: Oxford University Press.

Viscusi, W.K., and R.J. Zeckhauser. 1994. The Fatality and Injury Costs of Expenditures. *Journal of Risk and Uncertainty* 8(1): 19–41.

17

Integrating Performance in the Design of a Water Pollution Trading Program

MARTIN T. SCHULTZ AND MITCHELL J. SMALL

In many areas of the United States, environmental managers are faced with the challenge of improving water quality where administrative and legal barriers prevent control of pollution at many important sources. A source of water pollution can be characterized as a point source or a nonpoint source. *Point sources* are sources associated with a pipe that discharges wastewater into streams, rivers, or lakes. Examples of point sources are municipal wastewater treatment plants or industrial manufacturing facilities. The U.S. Environmental Protection Agency (EPA) regulates these sources by issuing permits that specify the amount and types of pollutants each point source may discharge. In contrast, pollution from *nonpoint sources* (NPSs) can be difficult to control and regulate. These diffuse sources of pollution include agricultural fields, highways, and urban areas. The migration of pollutants from these sources to surface waters is less direct because there is no single discharge point. For example, rainfall and irrigation runoff from agricultural fields transport toxic chemicals over land to rivers and streams. The amount of pollutant that is carried in runoff from an agricultural field to a stream is a difficult quantity to measure. Once in the stream, these pollutants mix with others, so it is difficult to determine their origin. Malik, Larson, and Ribaudo (1994) discuss efforts to address the NPS problem through legislation and other means.

The Clean Water Act, enacted in 1972, requires point source dischargers to install and operate pollution controls to satisfy technology-based effluent limitations that restrict the quantity of pollutants in wastewater discharge. If, after the installation of the required pollution controls, water quality in the receiving stream does not meet water quality standards, regulators may require that point sources meet more stringent limits on the allowable concentration of pollutants in wastewater discharge. These standards are known as *water quality–based effluent limitations.* Section 303 of the Clean Water Act prescribes total maximum daily load (TMDL) as the regulatory program under which revised effluent limitations will be determined. Under this program, the maximum daily load is the assimilative capacity at low flow, the amount of pollutant that can be assimilated in the receiving water body during a low-flow event without violating water quality criteria (U.S. EPA 1991a). Available assimilative capacity is the difference between TMDL and the ambient pollutant load. Regulators will determine the available assimilative capacity by accounting for pollution from background sources and NPSs first. *Background load* occurs naturally in the stream and is not the result of human activity. Residual assimilative capacity will be allocated among point sources while allowing a margin of safety in the stream. These effluent limitations are water quality–based because they reflect the assimilative capacity of the stream. The implication is that where no assimilative capacity is available, no point source discharges will be allowed. However, the contribution of pollutants from NPSs will continue, because they are unregulated. Under these conditions, water quality–based effluent limitations imposed at point sources that discharge into TMDL-regulated watersheds and river basins could be very restrictive and extremely expensive.

As of 2000, EPA had drafted regulations regarding the implementation of TMDLs, but specific requirements and procedures remained a contentious issue. In general, municipal and industrial point sources argue for regulations that distribute the costs of achieving environmental improvements more evenly among pollution sources. Agricultural and other unregulated NPSs argue against any attempts to develop new regulations that involve them. Under the current system that fails to regulate selected sources, pollution offsets are one solution to the problem of extremely restrictive and expensive pollution controls at some point sources. Offsets allow point sources that discharge more pollutants than their allocation to compensate for downstream environmental impacts by reducing pollution from other sources (U.S. EPA 1991b). *Point source/nonpoint source trading* (PSNPST) is a possible standardized program under which such offsets might be obtained. Because it is often cheaper to reduce pollutant load at NPSs than at point sources, PSNPST can benefit dischargers by reducing compliance costs and, under an appropriate set of trading rules, can benefit regulators and society by generating net improvements in water quality. However, this solution ignores concerns about whether the resulting distribution of costs is equitable.

Regulatory programs can impose high costs on society, and the decision to pursue one approach rather than another can result in long-term implications for society. Therefore, these decisions justify a careful analysis of alternatives and an evaluation of the proposed regulatory designs. Quantitative analysis assists in both design and evaluation. For example, analysis helps force the analyst to formalize performance metrics and decision criteria. In addition, collection and analysis of data to support the evaluation provides an opportunity to better understand the technical problems and to test or support key assumptions. These activities should lead to a better understanding of the problem and could lead to a restructuring of the regulatory analysis. Because analysis is an iterative process, such changes generally improve the quality of study results. When multiple stakeholder groups participate in the decisionmaking process, evaluations should not only highlight the implications of differences in beliefs and values among the participants but also encourage negotiation and the eventual collaborative selection of performance metrics to implement the model (Diwekar and Small 1999). In this way, managers and regulators can use analysis as a consensus-building tool.

In this chapter, we illustrate how the evaluation of performance can inform regulatory design through study of a proposed set of trading rules for PSNPST. The trading designs evaluated here are based on target zone trading (TZT), an approach to trading developed for Clear Creek Watershed near Denver, Colorado. In TZT, trades would involve different water quality attributes at different places within the watershed. Dischargers that use assimilative capacity in the watershed would offset these contributions to pollutant load by reducing excess pollutant load at polluted sites. The amount of assimilative capacity a discharger might use would be determined by the amount of excess load that it reduced (Hydrosphere Resource Consultants 1997). TZT implementation would occur as the state implements a TMDL program that allows point source loads at designated polluted sites to be offset by reducing pollutant loads from NPSs.

Several factors distinguish NPSs from point sources, and trading ratios can help account for these differences. NPSs are diffuse pollution sources characterized by a large variance in emissions, large uncertainty in the effectiveness of water quality controls, and large uncertainty in the effectiveness of monitoring and enforcement programs (Harrington, Krupnick, and Peskin 1985; Milon 1987; Dunn and Shortle 1988; Segerson 1988; Shortle 1990; Letson 1992b; Letson, Crutchfield, and Malik 1993). Trading ratios compensate for the differences in loading variability and reliability of controls so that reductions in pollutant load originating at an NPS yield equivalent water quality benefits as reductions in pollutant load originating at the point source (Letson, Crutchfield, and Malik 1993; Malik, Letson, and Crutchfield 1993). One example of a trading program that uses trading ratios is the PSNPST program at Lake Dillon, Colorado. It requires a two-to-one decrease in load at an NPS to offset increases in load at a point source (U.S. EPA

1996a; Powers 1998). In another example, Rahr Malting Company of Minnesota agreed to offset its daily discharge of 150 pounds of five-day carbonaceous biological oxygen demand (CBOD5) to the Minnesota River by reducing upstream NPS phosphorus load. It used a trading ratio of one pound of total phosphorus upstream being equivalent to eight pounds of CBOD5 discharged in the TMDL reach (Senjem 1997). The rationale for the trading ratio is different in each example.

In TZT, the trading ratio differs from that described in the preceding paragraph. It is the ratio of the fraction of the available assimilative capacity used at unpolluted sites (where the level of pollution is below the threshold amount) to the fraction of excess load reduced at polluted sites (Hydrosphere Resource Consultants 1997). A TZT trading ratio may be thought of as an allowable ratio of environmental loss (at unpolluted sites) to environmental gain (at polluted sites). Thus, even for trades among like sources in different locations in the watershed (for example, point source–point source trades), TZT trading ratios different from 1.0 are used to procure net benefits to the environment as a result of the trade. When TZT is combined with PSNPST, factors for both the location differences and the source (point versus nonpoint) type differences can be incorporated into the trading ratio. In the analysis presented in this chapter, TZT trading ratios account for both differences in the variability of point source and NPS loading rates and an allowable ratio of environmental loss to environmental gain at sites within the watershed.

The issues addressed in this analysis are the character and value of the trading ratio. Specifically, should multiple values of the trading ratio be assigned for different locations within the watershed, and what should the value of the trading ratio be? We also estimate the compliance cost and illustrate the difficulties associated with implementing TMDLs in a watershed.

The organization of this chapter is as follows. First, we discuss current topics in trading design in a literature review. This discussion is followed by background information about Clear Creek watershed, TZT, and prospective options for pollution control. In these sections, emphasis is placed on providing an appreciation for the problem without the burden of technical details that are available (Schultz, Farrow, and Small 1999). Discussions of the optimization procedure and study results conclude.

Topics in Water Pollution Control, Regulation, and Trading

The use of optimization methods to identify least-cost pollution control strategies began in the 1960s. Loucks, Revelle, and Lynn (1968) and Revelle, Loucks, and Lynn (1968) show how water quality criteria can be met at minimum total cost by selecting an appropriate set of controls at treatment locations on the Willamette River in Oregon. However, early evaluations such as these provided little insight into the policy levers that could or should be used to reach a desired allocation, nor did they evaluate distributive and equity implications associated with pollution control decisions.

The idea that markets could be used as a tool in environmental regulation is generally attributed to Crocker (1966) and Dales (1968). Markets can take many forms. Baumol and Oates (1971) suggested emissions taxes that would force dischargers to incorporate environmental damages or costs as factors in production and pollution control decisions. Montgomery (1972) suggested transferable discharge permits. In its most basic form, a permit market operates by initially distributing permits among pollution sources. Because permits specify the maximum amount of pollutant that may be released at each source, they establish the maximum load. Participants then buy and sell permits until marginal pollution control costs are equal at all sources, thus eliminating incentives to buy or sell. A regulatory agency or an environmental group can intervene after the initial distribution to reduce pollutant load by purchasing permits from dischargers. Critics of trading systems who argue that tradable permits institutionalize the right to pollute have valid concerns. Standard permits also institutionalize the right to pollute, but they are granted by the regulatory agency and may be revoked or amended during periodic administrative reviews. Depending on the regulatory design, tradable permits might be recovered only by purchasing them from a polluter. Trading systems may also yield cost savings while encouraging the degradation of environmental quality at unimpaired locations (Atkinson and Tietenberg 1987). Variants of the basic trading scheme have been devised to address some of these concerns and are discussed by Hanley, Shogren, and White (1997).

Theory suggests that economic efficiency is achieved in a pollution trading program when each polluter controls emissions at a marginal cost equal to the public's marginal willingness to pay for environmental quality maintained by these controls. Because marginal social costs are difficult or impossible to determine, analysts typically ignore economic efficiency in evaluation of trades and instead evaluate the cost-effectiveness of emissions controls given ambient environmental quality standards set by regulators. A typical method for identifying a cost-effective water quality trading program is to minimize treatment cost subject to ambient water quality standards. For pollutants that are uniformly mixed in the environment, the optimal level of control at each source is that at which marginal control costs are equal among all sources. However, many pollutants are not uniformly mixed in the environment. For example, organic wastes decay in the water column, and therefore these concentrations decrease downstream, as the distance from the source increases. For pollutants that are not uniformly mixed, the optimal level of control at each source is that at which the marginal control cost equals the weighted average cost of reducing downstream loads (Tietenberg 1985). One drawback to this approach is that it allows the degradation of water quality at downstream sites up to the standard. Although additional constraints have been formulated to prevent this result, differences in *compliance cost* (defined as the sum of all treatment costs at all point sources and NPSs in the watershed) rather than differences in water quality are the primary criteria for distinguishing among alternative constraint formulations representing trading program designs (Burn and McBean 1985; Tietenberg 1985; Hanley, Shogren, and White 1997).

Trading programs for air emissions have been analyzed using this framework (Tietenberg 1980; Atkinson and Tietenberg 1982; Krupnick, Oates, and Van De Verg 1983). Other studies have applied the principles of trading to reduce water pollution control costs among point sources. O'Neil (1983) and O'Neil and others (1983) study trading in the Fox River of Wisconsin. Burn and McBean (1985) consider the mathematical programming problem in the context of a hypothetical river segment. Letson (1992a) looks at the trading of conventional pollutants in the Lower Colorado River, Texas. Schleich, White, and Stephenson (1996) and Schleich and White (1997) discuss trading in the Fox-Wolf river basin of Wisconsin and assess the cost implications of alternative nutrient reduction targets in different locations within the river basin.

The Clean Water Act does not now regulate NPSs, and no other regulatory mechanism reduces pollutant load from these sources. Achieving water quality criteria in many waterways will require reducing unregulated NPS contributions to load. Trading targets NPSs indirectly. Dunn and Shortle (1988) discuss proposed trading between point sources and agricultural NPSs in the Chesapeake Bay region, and Crutchfield, Letson, and Malik (1994) consider the potential for PSNPST in coastal watersheds. Milon (1987) evaluates water quality impacts and compliance costs for a hypothetical PSNPST program for the Honey Creek Basin, in Ohio. His analysis controls for differences in the reliability between point source and NPS controls.

Far more analyses describing potential benefits to trading exist than do actual trading programs. Some authors attempt to explain why trading programs are not yet common. Hanley, Hallet, and Moffat (1990) and Farrow (1993) cite institutional barriers such as a lack of provisions for trading in the Clean Water Act as one reason such programs are rare. They also suggest that regulators tend to place a higher priority on legislative mandates to enforce pollution controls than on reducing regulatory compliance costs. Stavins (1999) describes how incentive structures facing regulatory agencies, industry, environmental groups, and legislators can impede the development and implementation of effective pollution trading programs. EPA's *Draft Framework for Watershed Based Trading* (1996a) lists 26 current and potential water pollution trading programs in the United States. Among these programs, a common feature is that they fail to generate the activity and cost savings projected by analysts. For example, the Dillon Reservoir trading program established in 1984 has yielded one PSNPST, and only three trades have occurred among NPSs (Powers 1998).

Reasons for this underperformance relative to projections are discussed in the literature. Stavins (1995) considers that transaction costs may impede trading activity. Transaction costs may arise for several reasons. For example, a paucity of buyers and sellers in the market can make it difficult and costly to find a suitable trading partner when one is needed (Atkinson and Tietenberg 1991; Crutchfield, Letson, and Malik 1994). Administrative complexity also imposes costs that impede trading activity (Hahn 1989). Complex administrative requirements create informational burdens and make it more difficult to predict the outcome of regulatory decisions. The sequential process of trading in practice has been identified as a possible reason (Atkinson and Tietenberg 1991). Analysts who use optimization to estimate trading benefits assume that trades occur simultaneously and multilaterally under conditions of full information. In practice, however, trades are sequential and bilateral, with only partial and uncertain cost information available. Hanley, Shogren, and White (1997) list principal–agent problems as another reason firms might not engage in trades. A principal–agent relationship is one in which an agent makes decisions for a principal. Problems arise when the agent's profit- or utility-maximizing choices are at odds with those of the principal. For example, a treatment plant manager may benefit more from increases in staff, budget, and status achieved by increasing the scale of waste treatment operations than from enhanced profits achieved by permit trading.

Overview of Clear Creek Watershed Study Area

Clear Creek is a 570-square-mile, high-elevation watershed near Denver, Colorado, where regulatory agencies and stakeholders have considered proposals for trading programs to address the problem of acid mine drainage (Figure 17-1). Land cover in the watershed is 68% forest and range, 14% alpine, 13% urban, 3% agricultural, and 2% other (U.S. EPA 1996b). The study area, the 400-square-mile upper watershed above the town of Golden, is characterized by a steep and rugged landscape. Clear Creek drops 3,600 feet as it traverses the 40 miles between the upper reaches of the watershed and Golden.

This area has been heavily mined for gold, silver, copper, zinc, and other minerals. Although the number of abandoned mines is uncertain, Colorado's

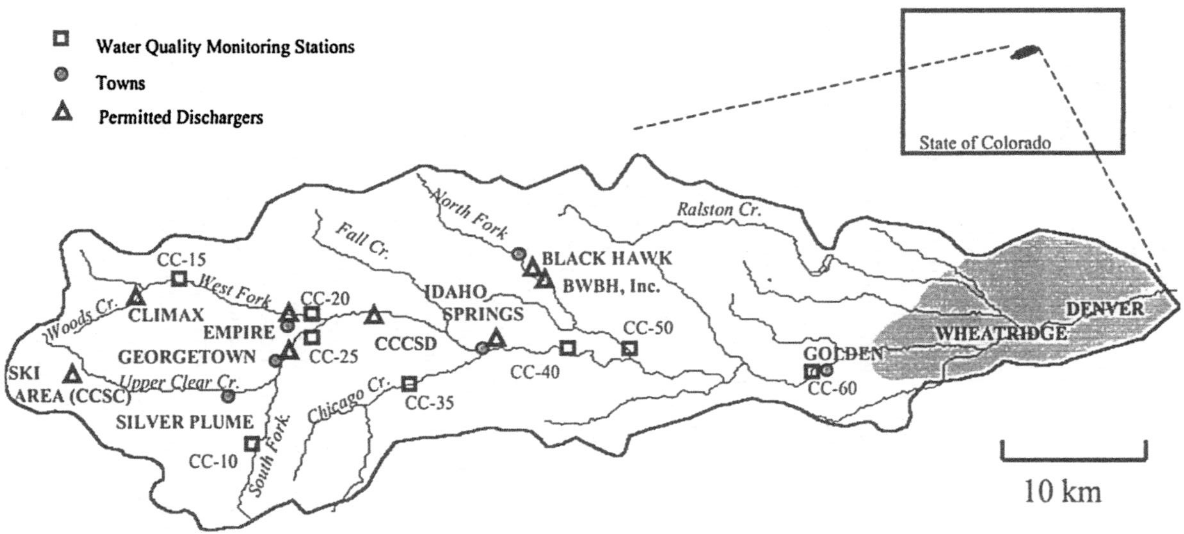

FIGURE 17-1. Clear Creek Watershed Study Area

Notes: Shown are Clear Creek (unlabeled, but extends across the map from where Upper Clear Creek meets South Fork on the left to past Denver on the right) and tributaries, eight water quality monitoring stations (labeled CC-10 through CC-60), six towns, and eight permitted dischargers. Where town names coincide with permitted municipal wastewater treatment plants (for example, Idaho Springs, Black Hawk, Georgetown, and Empire), a single label is used to indicate the name of both the town and the municipal wastewater treatment plant. The shaded area around Denver and Wheatridge indicates the extent of the Denver metropolitan area within the watershed.

Inactive Mine Program Survey has documented 1,998 sites in Clear Creek, Gilpin, and Jefferson Counties, of which 1,350 are located in the Clear Creek Watershed (USGS 1996). Large and heavily mined sites are well-known. However, most sites are smaller than one acre, and little information about them is available. Regulators have not identified legally and financially responsible parties to clean up these sites. In aggregate, these abandoned mines are an unregulated and diffuse source of water pollution that can be considered an NPS. Other NPSs contributing to water quality impairments in Clear Creek are timber harvesting and grazing areas, highways, and developed areas.

Backed by an active stakeholder community, the Clear Creek Watershed has been the site of proactive water quality programs in the past. In 1991, EPA named Clear Creek as one pilot area in which to implement its watershed approach to water quality management. The state of Colorado, local businesses, and federal land management agencies have undertaken voluntary mine cleanups and riparian area restorations. Local governments in the upper watershed have agreed to enforce ordinances that dictate best management practices for earth-moving, silviculture, and construction activities (UCCWA 1994). In 1994, the National Forum on Non-Point Source Pollution sponsored by the National Geographic Society and the Conservation Fund supported a multiyear study to develop locally designed approaches to NPS pollution trading. This study led to a TZT proposal (Hydrosphere Resource Consultants 1997), described in detail in the next section. Despite these efforts, the state of Colorado has determined that water quality in some areas does not support designated uses and has identified seven stream segments for TMDLs. TMDLs have been recommended for one or more metals, including copper, zinc, radium-226 and -228, iron, manganese, and cadmium (CDPHE 1998a). The state plans to begin implementing TMDLs as early as 2001.

TMDLs could provide an incentive for PSNPST if effluent limitations become more restrictive. The Clean Water Act and Colorado Department of Public Health and Environment regulations require regulators to first allocate assimilative capacity to background and NPSs. Any remaining capacity then will be allocated among point sources while maintaining a margin of safety in the stream segment. A margin of safety prevents uses of assimilative capacity from exceeding 85% of the TMDL. If naturally occurring background loads and unregulated NPS loads exceed 85% of the TMDL, then reductions in NPS loads—in the form of offsets or by some other means—will be required before point sources are permitted to discharge the pollutant at issue. Where unregulated NPSs are the dominant source of pollution, point sources may bear high costs of pollution control.

Target Zone Trading

TZT (Hydrosphere Resource Consultants 1997) would allow permitted dischargers to offset their pollution by implementing NPS pollution controls at abandoned mining sites. The proposal is distinguished from others because it permits an increase of one pollutant (for example, copper) at one site in exchange for control of another pollutant (for example, zinc) at a different site. Rules of the program are not finalized, but one suggestion is to issue a credit that is equal to the percentage of excess load reduced at polluted sites that are designated by the regulatory agency. That credit would be used to allow the reduction of available assimilative capacity at other designated sites by a percentage equal to or less than the credit amount.

Designated sites likely would correspond to sites for which water quality monitoring data exists. To illustrate, Figure 17-2 shows three designated sites in relation to the TMDL. Load at designated site A is 33% of the TMDL. Load at designated site B is 1.6 times the TMDL, and the load at designated

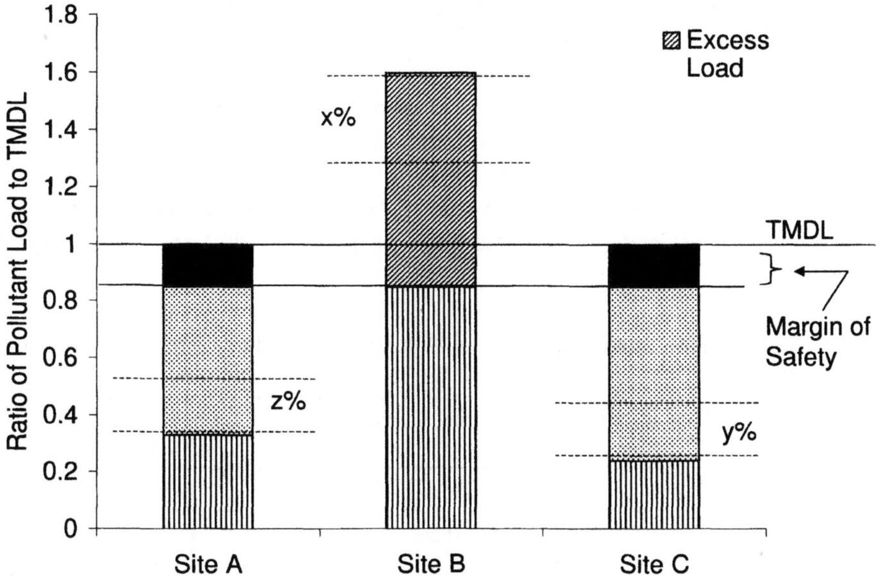

FIGURE 17-2. Target Zone Trading Involving Three Sites

Notes: Reductions in excess load at Site B, expressed as a percentage of the excess of the total maximum daily load (TMDL; *x*%), are obtained from pollutant traders in exchange for use of available assimilative capacity, expressed as a percentage of available load below the TMDL (*y*%, *z*%). One additional condition of the trade is that the mass of pollutant load increased at Sites A and C is always less than or equal to the mass of pollutant load reduced at Site B.

site C is 67% of the TMDL. Designated sites A and C are less than the target zone (85%). An offset that reduces excess load at B, for example, by $x = 40\%$ (that is, 40% of the load that is above the TMDL) can be used as credit to use assimilative capacity at site A or site C. The percentage of assimilative capacity (y and z) used at the unpolluted sites must be less than or equal to the percent decrease in excess load at the polluted site. (With reference to Figure 17-2, $x \geq y + z$.) A simple trading ratio would reflect the ratio of the load offset to the load reduced: $0 < (y + z)/x < 1$.

An example in which one discharger contributes load to one polluted site and one unpolluted site will help illustrate the trading ratio. Assume that the available capacity at the unpolluted site is 1 kilogram (kg) per day and the excess load at the polluted site is 5 kg/day. Also assume that the discharger proposes to use 0.5 kg/day of assimilative capacity at each of the two sites. At the unpolluted site, that amount would be 50% of the assimilative capacity. If a trading ratio of 0.7 were mandated, the discharger would be required to reduce the excess load at the polluted site by 4.07 kg/day. The net reduction in load at the polluted site would be 3.57 kg/day because the discharger would be adding 0.5 kg/day at that site. For this example, Table 17-1 lists the load reductions associated with several trading ratios. In this example, the entire excess load is eliminated when the trading ratio is 0.5 or less.

The TZT trading ratio has a value between 0 and 1 for each target pollutant at each polluted site. A value of 1 imposes the requirement that point sources using 1% of available assimilative capacity at unpolluted sites reduce excess load at the polluted site by 1%. Similarly, a value of 0.5 requires a 2% reduction in excess load, and a value of 0.25 requires a 4% reduction in excess load. One strategy that has been proposed is that the TZT trading

TABLE 17-1. Sample Calculations of Load Reduction at a Polluted Site with Different Trading Ratios

Trading ratio	*NPS load reduction required*		*Point source additions to load*		*Net reduction in load*	
	kg/day	*%*	*kg/day*	*%*	*kg/day*	*%*
0.4	6.75	135.0	0.50	10.0	6.25	125.0
0.5	5.50	110.0	0.50	10.0	5.00	100.0
0.6	4.67	93.3	0.50	10.0	4.17	83.3
0.7	4.07	81.4	0.50	10.0	3.57	71.4
0.8	3.63	72.5	0.50	10.0	3.13	62.5
0.9	3.28	65.5	0.50	10.0	2.78	55.5
1	3.00	60.0	0.50	10.0	2.50	50.0

Note: % = percent of excess load at the polluted site.

ratio could take multiple values within the watershed and that the value could be lower where the level of pollution is greater (Hydrosphere Resource Consultants 1997). Lower trading ratios require greater reductions in excess load at polluted sites.

The expression for the trading ratio becomes more complex when more than two sites and one pollutant are considered (Schultz, Farrow, and Small 1999). The trading ratio cannot be calculated directly because the level of neither point source control nor NPS control is known for any particular trading ratio. The optimization procedure that is described later resolves this problem. The program proposes a trading ratio value and then solves a linear program to determine the cost-minimizing levels of pollution control for that trading ratio. The optimization is constrained so that the solution satisfies the regulatory inequality on available assimilative capacity. An optimal trading ratio is then found by iterative testing of trading ratio values. Testing continues until additional improvements in water quality are difficult to obtain.

Water Quality in Clear Creek

Stream flow in relatively fast-flowing Clear Creek is highly variable. For example, flows at Golden increase from a low of around 63,000 cubic meters (m^3) per day in February to a peak of 1.2 million m^3/day in June or July. Relatively high concentrations of toxic metals are observed during both low-flow and high-flow periods. During low flow, not much pollution enters the stream because of dry weather conditions, but concentrations are high because less water is available to dilute the pollutant load. At high flow, loading rates are very high as a result of the increased mine drainage, snow melt, and rainfall runoff, but concentrations are low because a large volume of water is available to dilute the pollutant load. Violations of the water quality criteria can occur under both conditions. Although this variation complicates the analysis and makes Clear Creek a likely candidate for a seasonal TMDL, for simplicity of analysis and methodological illustration, this study uses a single TMDL that is based on low-flow conditions. The TMDL is calculated at the three-year low flow using acute water quality criteria as directed by the state of Colorado (CDPHE 1998b). An acute water quality criterion is a limit that the ambient pollutant concentration may not exceed more than once in any three-year period. It differs from a chronic water quality criterion, which specifies limits for average pollutant concentrations over a particular averaging period.

Pollutants originate from background sources, point sources, and NPSs. Naturally occurring background loads are estimated from observed concentrations at a pristine location in upstream Chicago Creek. This location and the locations of point source dischargers considered in the analysis are shown in Figure 17-1. Seven wastewater treatment plants discharge into Clear Creek: Climax Mine, Empire, Black Hawk, Georgetown, Central Clear Creek Sanitation District, Clear Creek Ski Corporation, and Idaho Springs. An eighth point source, Bullwhackers Black Hawk, Inc. (BWBH, Inc.), is a small discharger associated with gambling interests near Black Hawk. Discharge volumes and point source loading rates used in this study are estimated from data contained in EPA's Permit Compliance System database and EPA documents (U.S. EPA 1982).

A loading rate, expressed in kilograms per day, gives the mass of pollutant released into the environment over a specified time period. We divided Clear Creek watershed into eight subwatersheds, and NPS loading rates in these eight areas were characterized by using empirically derived lognormal probability distributions based on regression analysis of observed flows and concentration data. NPS area loading rates are expressed on a unit area basis of kilograms per square kilometer per day ($kg/km^2/day$). For copper, the expected value of the NPS area loading rate ranges from 0.0001 to 0.026 $kg/km^2/day$, and coefficients of variation range from 0.73 to 1.67. For zinc, the expected values of the NPS area loading rates range from 0.008 to 0.278 $kg/km^2/day$, and coefficients of variation range from 0.69 to 1.66. The coefficient of variation was calculated by dividing the standard deviation by the mean. These lognormal distributions reflect uncertainty and variability in the NPS area loading rate given low-flow conditions.

The responses of pollutant concentrations to changes in treatment strategies at low flow were modeled using a one-dimensional fate and transport model. The NPS loading rates were simulated, and then the weighted sum of point source and NPS loads were added to background loads at each polluted site. It was assumed that there was no net settling or resuspension, and because metals do not decay in the water column, calculation of downstream concentrations was straightforward. Figure 17-3 shows the resulting expected value and 95% confidence bounds for the ratio of estimated pre-trade load to allowable load (TMDL) at the eight designated sites considered in this study of Clear Creek (CC-10 through CC-60). These sites were chosen because they correspond to long-term stream flow and water quality monitoring locations in the watershed. Reading from left to right in Figure 17-3 corresponds to moving downstream, although some monitoring locations are located along tributaries to the main stem (see Figure 17-1). Because this analysis was conducted at low flow, the load ratio is synonymous with the ratio of the expected value of the modeled ambient concentration to the acute criteria concentration. The two dashed lines correspond to 1 and 0.85 on the y-axis. These sites are in compliance if the ratio of the expected value to acute crite-

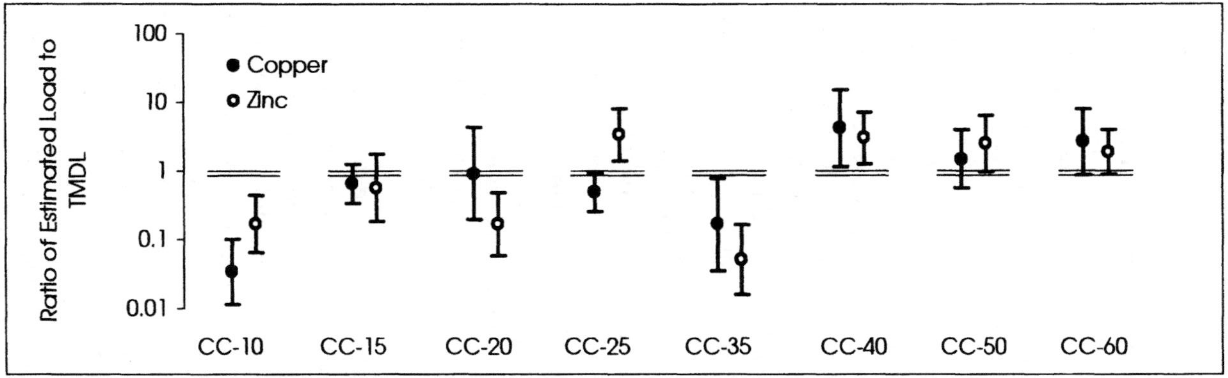

FIGURE 17-3. Pre-Trade Water Quality in Clear Creek

Notes: Sites CC-10 through CC-60 correspond to long-term monitoring locations in the Clear Creek watershed. Stations are numbered approximately in the order they occur as one moves downstream, from west to east, but some stations are located on tributaries to the main stem. See Figure 17-1 for clarification on the location of each station. Points are expected values and error bars are 95% confidence bounds for the ratios of estimated pre-trade load to allowable load (TMDL). The horizontal line at $y = 1$ indicates the TMDL, and the horizontal line at $y = 0.85$ indicates the TMDL with a margin of safety. Monitoring locations where the expected value exceeds 85% of the TMDL are identified as polluted sites. CC-15 is not identified as a polluted site for the purpose of this analysis because the expected value is below 85% of TMDL.

ria concentration is ≤0.85. Where the confidence intervals extend above the upper dashed line, the probability that load exceeds the TMDL is >0.025.

Treatment Costs

Municipal wastewater treatment plants remove most of the metals in their influent through the passive effects of primary and secondary treatment designed to remove conventional pollutants. Under revised effluent limitations, dischargers may need to install additional pollution controls. Although several processes are potentially available to dischargers, our study assumed that treatment plants would adopt precipitation, the least expensive treatment option currently available (Eckenfelder 1989; Krishnan and others 1991). The assumed average daily amortized cost of constructing and operating a system to remove copper and zinc from wastewater was based on capital and operating cost equations for sodium hydroxide precipitation to remove metal from industrial wastewater (U.S. EPA 1998a, 1998b).

Treating secondary effluent to achieve an effluent copper concentration of 0.02 milligrams per liter (mg/l) has an estimated cost of \$370/day plus \$0.10/m^3. For example, the capital and operating costs associated with this process at a 1,200-m^3/day wastewater treatment facility would be \$490/day. Treating secondary effluent to achieve an effluent zinc concentration of 0.1 mg/l has an estimated cost of \$310/day plus \$0.11/m^3. Because minimum effluent concentrations are achieved for different metals at different levels of pH, achieving minimum effluent concentrations for both metals at one plant would require the installation of two separate processes. The total cost for these two treatment processes operating together would be \$680 per day plus \$0.21/m^3 of wastewater discharged. A 10-year period was used for amortization of capital costs.

The cost and nature of NPS controls was one of the more uncertain variables used in our analysis. NPS controls to reduce metal loads vary from site to site. Controls could include construction and maintenance of sediment traps, removal or capping of mine tailings and rock piles, or construction and operation of plants to treat mine discharge. We assumed that NPS control costs could be represented by using an average daily amortized cost per unit area. This assumption was necessary because of a lack of information on the cost and effectiveness of controls at specific sites in the watershed. As with the treatment technology considered for point sources, the average daily cost of control was calculated by assuming a capital investment amortized over 10 years with a 7% discount rate. Lyon and Farrow (1995) report that \$5–10/acre/year would cover most NPS control practices in agriculture. We assumed that this amount provided a lower bound for the amortized NPS control cost. Because higher costs are likely at abandoned mining sites, we used a cost of \$50/km^2/day.

Optimization Algorithm

A two-stage optimization algorithm using simulated annealing (SA) and a mixed-integer linear program (MILP) was used to select trading ratios that resulted in the best water quality given the combined actions of water treatment facilities acting independently to minimize treatment costs. SA is an approach to stochastic optimization attributed to Metropolis and others (1953) and further developed by Kirkpatrick, Gelatt, and Vecchi (1983), Cerny (1985), and Aarts, Korst, and Van Laarhoven (1997).

The first step was to initialize the SA decision variables, the trading ratios. MILP identifies the level of treatment effort at point sources and NPSs that minimizes the aggregate cost of compliance with the trading program at the given trading ratio. Compliance is achieved when all pollutant loads discharged from point sources are offset at each downstream-polluted site according to the applicable trading ratio. When trades involve multiple pollutants, a weight is used to account for differences in the toxicity of pollutants and the tolerance for pollutants at each site. Weighting is necessary because some pollutants are more toxic than others. Reducing one unit of zinc load, for example, results in less environmental improvement than reducing one unit of copper load. In addition, ambient water quality criteria differ among sites according to the designated water use. Thus, the tolerance for a pollutant differs among sites. (This weight is calculated $w_{jr} = \bar{c}_{ref,r} / \bar{c}_{jr}$. In this equation, w is the weight on pollutant j at site r and $\bar{c}_{ref,r}$ is the ambient criteria concentration for the reference pollutant, copper, at the site. The variable $\bar{c}_{jr}$ is the site-specific ambient criteria concentration of the pollutant for which the weight is to be calculated.)

The water quality impact of point source and NPS controls was simulated, and this information was used to calculate the SA regulatory objective function to be minimized. This objective function could take one of many forms depending on the chosen objectives. In our study, the regulatory objective function measures the deviation of the simulated ambient concentrations from acute criteria concentrations. The regulatory objective function is

$$\min \sum_{\substack{\text{Target}\\ \text{pollutants}}} \sum_{\substack{\text{Polluted}\\ \text{sites}}} \left|\bar{F} - F_{0.975}\right|$$

where $\bar{F}$ is the acute criteria concentration for the pollutant at a site and $F_{0.975}$ is the upper bound of the 95% confidence interval on the estimated pollutant concentration. The regulatory objective function takes units of micrograms per liter (μg/l) because it measures the deviation of ambient pollutant concentration both above and below the acute water quality criteria. In implementation, the summation in the regulatory objective function is over target pollutants at polluted sites where the ratio of the sum of background and NPS load to TMDL is >0.85.

This regulatory objective minimizes excess load rather than total load and therefore reflects a policy choice that does not encourage overcontrol of pollution in the watershed. Operationally, this regulatory objective function helps us to find the combination of trading ratios that satisfies acute water quality criteria at the lowest cost. Note, however, that it implies a penalty for reducing loads below the TMDL. Although reducing pollutant loads below the TMDL might produce environmental benefits, these reductions are not called for in regulations. Because $F_{0.975}$ is the 97.5th percentile of the probability distribution that reflects uncertainty in the pollutant loads at a site, the expected value of loads is in many cases well below the TMDL. Other objective functions also might be considered. For example, rather than treating all loads equally, the differences might be weighted in acute criteria concentrations by a toxicity weight. Deviations in water quality then would be expressed in a copper toxicity equivalent, which would give greater importance to a 1-μg/l decrease in copper concentration than to zinc concentration and greater importance to decreases in concentration where ambient water quality criteria are more restrictive.

Using the 97.5th percentile of the probability distribution rather than an expected value in the objective function is a means of incorporating uncertainty in NPS loading rates into the trading ratios that are the solution to the SA optimization. Some authors have suggested that these trading ratios also should compensate for not only a greater variance in loadings at NPSs but also less accuracy in monitoring, greater difficulty in enforcement, and increased uncertainty in treatment costs (Malik, Letson, and Crutchfield 1993). Uncertainty in enforcement and monitoring impose costs on society because a regulatory agency must expend more effort to ensure compliance with a trading agreement—or settle for less assurance of compliance. These uncertainties in compliance can reduce a regulator's willingness to allow a PSNPST. Lower trading ratios designed to compensate for these uncertainties would require more cleanup at NPSs and could help reduce a regulatory agency's reluctance to allow a trade. Conversely, treatment or cleanup cost is probably less predictable at NPSs than at point sources. This uncertainty in cleanup cost can reduce a point source's willingness to participate in a trading program and therefore suggests that higher trading ratios may be needed to maintain the attractiveness of trades among point sources.

Results and Discussion

Insights into the design of the trading program can be gained by evaluating program outcomes under alternative regulatory designs. The specific issue to be addressed is whether the trading ratio should take different values at different sites. The analysis accomplishes this by comparing baseline and scenario optimization results. The baseline version of the optimization imposes the condition that trading ratios take an identical value at each site. This is the baseline scenario against which alternative trading designs should be compared. The scenario version of the optimization relaxes the requirement that trading ratios take an identical value at all polluted sites. In this scenario, trading ratios are allowed to take multiple values. Compliance costs and water quality improvements are assessed against the baseline scenario.

Results of this analysis also can be compared with similar metrics under both pre-trade conditions and alternative command-and-control conditions. The pre-trade value of the regulatory objective function is 8,102 μg/l, and the pre-trade compliance cost is 0. This information can be combined with scenario results to calculate the cost-effectiveness of scenarios. Under an alternative command-and-control program that requires the implementation of the tertiary treatment technology at all point sources while ignoring NPSs, the regulatory objective function is 8,077 μg/l and compliance cost is approximately $7,000 per day. This command-and-control scenario was not used as a baseline in our study because there was no comparable mechanism to address that portion of the loads that could not be removed from the wastewater stream by using the tertiary treatment technology.

Results of the two optimizations are provided in Table 17-2. Under a uniform trading ratio, the optimal trading ratio for pollutant j at polluted site r (δ_{jr}) that minimizes the regulatory objective function is 0.093 (Case 1). This means that the required reduction in pollutant at the polluted site is almost 10 times that of the increase in load at the unpolluted site, resulting in a compliance cost of $68,800 per day and a regulatory objective function of 984 μg/l. As would be expected, relaxing constraints by allowing multiple trading ratios reduces compliance costs while allowing greater progress

TABLE 17-2. Compliance Costs and Objective Function Values for Two Optimizations

Case	*Compliance cost ($/day)*	*Objective value (μg/l)*	*Trading ratio characteristics*
1	68,800	984	Uniform, $\delta_{jr} = 0.093$
2	68,500	517	Multiple values of δ_{jr}

Note: δ_{jr} = optimal trading ratio for pollutant j at polluted site r.

toward the goal established for the watershed. Allowing multiple trading ratios reduces compliance costs by \$300 to \$68,500 per day and achieves a sum of absolute deviations from acute criteria of 517 µg/l (Case 2). This value of the regulatory objective is 47% lower than under a uniform trading ratio. The difference in compliance cost is quite small and, because uncertainties in the costs are large, can be considered insignificant.

Water quality under the two trading regimes is compared with pre-trade conditions in Figure 17-4, which shows the ratio of estimated pre- and post-trade load to TMDL for copper and zinc. Loads at CC-10 and CC-15 remain unchanged under both scenarios. At CC-15, the 97.5th percentile of the load ratio remains above the threshold. This site was not initially identified as a polluted site because the expected values were below 85% of the TMDL.

The effects of each trading program can be observed at monitoring locations CC-20 to CC-60. Where the expected value of loads was above the TMDL in the pre-trade evaluation, expected values are brought below the standard. Under a uniform trading ratio, the 97.5th percentile of zinc load exceeds the TMDL at CC-25, and the 97.5th percentile of copper load exceeds the TMDL at CC-60. Under a variable trading ratio, the 97.5th percentile of zinc loads is brought below the TMDL at CC-25, but there is a trade-off in that the expected value of the copper loads are somewhat higher at CC-25, CC-40, and CC-50. The expected value of zinc load also is higher at CC-50.

Table 17-3 lists optimal values of the trading ratio at polluted sites for Case 2, in which trading ratios can take different values at different locations and for different pollutants. For example, the trading ratio for copper is 0.046 at CC-20, but 0.163 at CC-40. Also listed in Table 17-3 is the load ratio, the ratio of estimated pre-trade load to TMDL. At CC-20, the expected value of the ratio of estimated copper load to copper TMDL ($E[x]$) is 0.93. This value is >0.85, which is the limit for this ratio after allowing for a margin of safety equal to 15% of the TMDL. Table 17-3 also shows the 97.5th percentile of the pre-trade ratios, $F_{0.975}$. This distribution represents uncertainty in the

TABLE 17-3. Trading Ratios and Pre-Trade Load Ratios for Two Pollutants

	Copper				*Zinc*			
Receptor	CC-20	CC-40	CC-50	CC-60	CC-25	CC-40	CC-50	CC-60
Trading ratio	0.046	0.163	0.022	0.145	0.183	0.159	0.066	0.089
Load ratio								
$E[x]$	0.93	4.26	1.50	2.67	3.42	3.04	2.51	1.89
$F_{0.975}$	4.35	15.59	3.98	8.11	8.15	7.24	6.39	3.98

Notes: $E[x]$ = expected value of the estimated load ratio; $F_{0.975}$ = upper bound of the 95% confidence interval on the estimated load ratio.

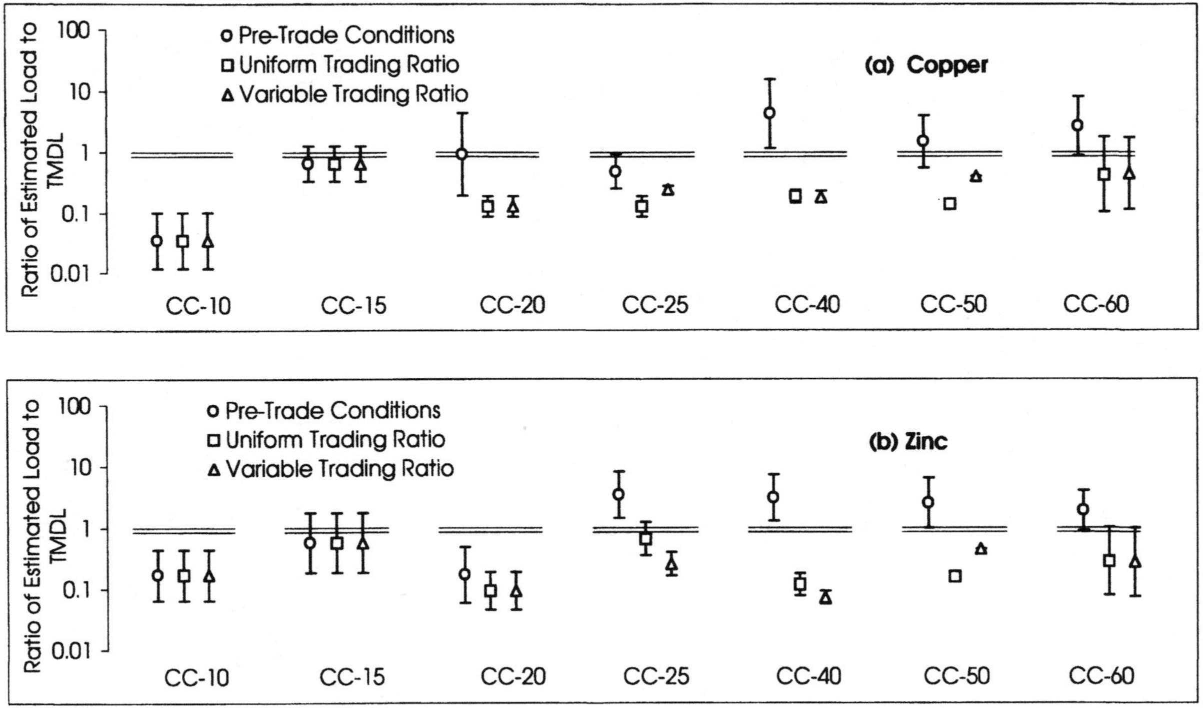

FIGURE 17-4. Effect of Two Trading Designs on Water Quality

Notes: Plots show the ratios of estimated copper (top) and zinc (bottom) loads to TMDL within their 95% confidence bounds. Both trading designs reduce the variance of the pollutant loads. A variable trading ratio achieves compliance at a greater number of monitoring locations, but the expected value of the load ratios is higher at some sites.

expected value of the ratio of estimated load to TMDL at this site; these pre-trade ratios are illustrated in Figure 17-3.

These results suggest advantages to using multiple trading ratios within the watershed. The original TZT proposal suggested that one strategy for determining the value of the trading ratio at a polluted site might be to relate them to the degree of impairment at that site (Hydrosphere Resource Consultants 1997). The effect would be to force greater reductions in excess load where impairments were greatest. However, results of our analysis suggest that, in the optimal combination of trading ratios, this strategy might not be good for the watershed as a whole; lower values of the trading ratio are not associated with sites where the degree of impairment is highest.

Of course, our evaluation is specific to one watershed and assumes that the objective function used in this analysis is indeed the appropriate objective for the watershed. A different regulatory objective might weight the deviations from acute criteria to account for differences in toxicity among pollutants and differences in water quality criteria among sites. In that case, optimization would identify a different set of trading ratios. It is up to stakeholders to determine the objectives and identify the best set of trading ratios.

In this chapter, we demonstrate that quantitative analysis can provide an explicit basis for choosing trading ratios or similar parameters. It is conceivable that regulators in such a situation might have adopted trading ratios without doing an analysis first. If those trading ratios were too high or too low, the result could be very little trading, suboptimal compliance costs, and few of the impacts on water quality achievable using an alternative design for the regulatory program.

Acknowledgements

Funding for this project was provided by the Center for Study and Improvement of Regulation and the Department of Engineering and Public Policy at Carnegie Mellon University.

Appreciation is extended to Liz Casman, Urmila Diwekar, Scott Farrow, and Tim Johnson. The authors appreciate the information obtained from and helpful discussions with those working locally on the Clear Creek Watershed problem, including Holly Fliniau, Paul Frowhart, R.L. Jones, Allen Medine, Carl Mount, Carl Norbeck, Tom Settle, and Tim Steele. They thank R.L. Jones of the Upper Clear Creek Watershed Advisory Group for providing the water quality data used in this study.

The authors alone are responsible for all errors, interpretations, and omissions.

References

Aarts, Emile, Jan Korst, and Peter J.M. Van Laarhoven. 1997. Simulated Annealing. In *Local Search in Combinatorial Optimization*, edited by E. Aarts and J.K. Lenstra. Chichester, U.K.: Wiley.

Atkinson, Scott, and Tom Tietenberg. 1982. The Empirical Properties of Two Classes of Designs for Transferable Discharge Permit Markets. *Journal of Environmental Economics and Management* 1: 237–250.

———. 1987. Economic Implications of Emissions Trading Rules. *Canadian Journal of Economics* 20: 370–386.

———. 1991. Market Failure in Incentive-Based Regulation: The Case of Emission Trading. *Journal of Environmental Economics and Management* 21: 17–31.

Baumol, William J., and Wallace E. Oates. 1971. The Use of Standards and Prices for the Protection of the Environment. *Swedish Journal of Economics* 73: 42–54.

Burn, Donald H., and Edward A. McBean. 1985. Optimization Modeling of Water Quality in an Uncertain Environment. *Water Resources Research* 21(7): 934–940.

CDPHE (Colorado Department of Public Health and Environment). 1998a. *Water Quality Limited Segments: Colorado's 1998 303(d) List* (Draft). January. Denver, CO: Water Quality Control Division.

———. 1998b. Regulation No. 31: Basic Standards and Methodologies for Surface Water. *Code of Colorado Regulations* 5(March): 1002–31.

Cerny, V. 1985. A Thermodynamical Approach to the Travelling Salesman Problem. *Journal of Optimization Theory and Applications* 45: 41–51.

Crocker, Thomas D. 1966. The Structuring of Atmospheric Pollution Control Systems. In *The Economics of Air Pollution*, edited by H. Wolozin. New York: W.W. Norton and Co.

Crutchfield, Stephen R., David Letson, and Arun S. Malik. 1994. Feasibility of Point and Non-Point Source Pollution Control. *Water Resources Research* 30(10): 2825–2836.

Dales, John H. 1968. Land, Water, and Ownership. *Canadian Journal of Economics* 1: 791–804.

Diwekar, Urmila, and Mitchell J. Small. 1999. Industrial Ecology and Process Optimization. *Journal of Industrial Ecology* 2(3): 11–13.

Dunn, James W., and James S. Shortle. 1988. Agricultural Non-Point Source Pollution Control in Theory and Practice. *Marine Resource Economics* 5: 259–270.

Eckenfelder, W. Wesley. 1989. *Industrial Water Pollution Control.* New York: McGraw-Hill.

Farrow, Scott. 1993. The Existing Basis and Potential for Damage Fees and Tradable Allowances. In *Clean Water and the American Economy Proceedings*, edited by U.S. Environmental Protection Agency (EPA). EPA 800-R-93-001a. Washington, DC: U.S. EPA, Office of Water.

Hahn, Robert. 1989. Economic Prescriptions for Environmental Problems: How the Patient Followed the Doctor's Orders. *Journal of Economic Perspectives* 3(2): 95–114.

Hanley, Nick, S. Hallet, and I. Moffat. 1990. Why Is More Notice Not Taken of Economists' Prescriptions for the Control of Pollution? *Environment and Planning* 22: 1421–1439.

Hanley, Nick, Jason F. Shogren, and Ben White. 1997. *Environmental Economics in Theory and Practice.* New York: Oxford University Press.

Harrington, Winston, Alan J. Krupnick, and Henry M. Peskin. 1985. Policies for Non-Point Source Pollution Control. *Journal of Soil and Water Conservation* 40: 27–32.

Hydrosphere Resource Consultants, Inc. (Barbara J.B. Green, Charles W. Howe, Lawrence J. MacDonnell, and Ann S. Maest). 1997. *Possible Approaches to Evaluating and Implementing Unlike Transactions: Orphan Site Feasibility Study, Phase III, Task 3.* Report to the Conservation Fund. Boulder, CO: Hydrosphere Resources Consultants, Inc.

Kirkpatrick, S., C.D. Gelatt Jr., and M.P. Vecchi. 1983. Optimization by Simulated Annealing. *Science* 220: 671–680.

Krishnan, E.R., P.W. Utrecht, A.N. Patkar, J.S. Davis, S.G. Pour, and M.E. Foerst. 1991. *Recovery of Metals from Sludges and Wastewaters.* Park Ridge, NJ: Noyes Data Corporation.

Krupnick, A.J., W.E. Oates, and E. Van De Verg. 1983. On Marketable Air Pollution Permits: The Case for a System of Pollution Offsets. *Journal of Environmental Economics and Management* 10: 233–247.

Letson, David. 1992a. Simulation of Two-Pollutant Two-Season Pollutant Offset System for the Colorado River of Texas Below Austin. *Water Resources Research* 28: 1311–1318.

———. 1992b. Point/Non-Point Source Pollution Reduction Trading: An Interpretive Survey. *Natural Resources Journal* 32: 219–232.

Letson, David, Stephen R. Crutchfield, and Arun S. Malik. 1993. Point Non-Point Source Trading for Controlling Pollutant Loadings to Coastal Waters: A Feasibility Study. In *Theory, Modeling, and Experience in the Management of Nonpoint Source Pollution Control,* edited by C.S. Russell and J.S. Shogren. Dordrecht, the Netherlands: Kluwer Academic Publishers.

Loucks, Daniel P., Charles S. Revelle, and Walter R. Lynn. 1968. Linear Programming Models for Water Pollution Control. *Management Science* 14(4): B166–B181.

Lyon, Randolph M., and Scott Farrow. 1995. An Economic Analysis of Clean Water Act Issues. *Water Resources Research* 31(1): 213–223.

Malik, Arun S., David Letson, and Stephen R. Crutchfield. 1993. Point/Non-Point Source Trading of Pollution Abatement: Choosing the Right Trading Ratio. *American Journal of Agricultural Economics* 75: 959–967.

Malik, Arun S., Bruce A. Larson, and Marc Ribaudo. 1994. Economic Incentives for Agricultural Non-Point Source Pollution Control. *Water Resources Bulletin* 30(3): 471–480.

Metropolis, N., A. Rosenbluth, M. Rosenbluth, A. Teller, and E. Teller. 1953. Equations of State Calculations by Fast Computing Machines. *Journal of Chemical Physics* 21: 1087–1092.

Milon, J. Walter. 1987. Optimizing Non-Point Source Controls in Water Quality Regulation. *Water Resources Bulletin* 23(3): 387–396.

Montgomery, W. David. 1972. Markets in Licences and Efficient Pollution Control Programmes. *Journal of Economic Theory* 5: 395–418.

O'Neil, William B. 1983. Transferable Discharge Permit Trading under Varying Stream Conditions: A Simulation of Multiperiod Permit Market Performance on the Fox River, Wisconsin. *Water Resources Research* 19(3): 608–612.

O'Neil, William B., David Martin, Christina Moore, and Erhard Joeres. 1983. Transferable Discharge Permits and Economic Efficiency: The Fox River. *Journal of Environmental Economics and Management* 10: 346–355.

Powers, Ann. 1998. Reducing Nitrogen Pollution on Long Island Sound: Is There a Place for Pollutant Trading? *Columbia Journal of Environment* 23(2): 137–216.

Revelle, Charles S., Daniel P. Loucks, and Walter R. Lynn. 1968. Linear Programming Applied to Water Quality Management. *Water Resources Research* 4(1): 1–9.

Schleich, Joachim, and David White. 1997. Cost Minimization of Nutrient Reduction in Watershed Management Using Linear Programming. *Journal of the American Water Resources Association* 33(1): 135–142.

Schleich, Joachim, David White, and Kurt Stephenson. 1996. Cost Implications in Achieving Alternative Water Quality Targets. *Water Resources Research* 32(9): 2879–2884.

Schultz, Martin T., Scott Farrow, and Mitchell J. Small. 1999. Evaluating a Proposed System of Point Non-Point Source Offsets. Working paper. Pittsburgh, PA: Carnegie Mellon University, Center for Study and Improvement of Regulation.

Segerson, Kathleen. 1988. Uncertainty and Incentives for Non-Point Pollution Control. *Journal of Environmental Economics and Management* 15: 87–98.

Senjem, Norman. 1997. *Case Study: Minnesota–Pollutant Trading at Rahr Malting Co.* http://www. pca.state.mn.us/hot/es-mn-r (accessed June 12, 1999).

Shortle, James S. 1990. The Allocative Efficiency Implications of Water Pollution Abatement Cost Comparisons. *Water Resources Research* 26(5): 793–797.

Stavins, Robert N. 1995. Transaction Costs and Tradeable Permits. *Journal of Environmental Economics and Management* 29: 133–148.

———. 1999. What Can We Learn from the Grand Policy Experiment? Lessons from SO_2 Allowance Trading. *Journal of Environmental Economics and Management* 12(3): 69–88.

Tietenberg, Tom H. 1980. Transferable Discharge Permits and the Control of Stationary Source Air Pollution: A Survey and Synthesis. *Land Economics* 56: 391–416.

———. 1985. *Emissions Trading: An Exercise in Reforming Pollution Policy.* Washington, DC: Resources for the Future.

UCCWA (Upper Clear Creek Watershed Association). 1994. *Guide to Water Quality Protection and Erosion Control.* Idaho Springs, CO: Upper Clear Creek Watershed Association.

U.S. EPA (Environmental Protection Agency). 1982. *Fate of Priority Pollutants in Publicly Owned Treatment Works.* EPA 440/1-2/303. Washington, DC: U.S. EPA, Office of Water.

———. 1991a. *Technical Support Document for Water Quality-Based Toxics Control.* EPA 505-90-001. Washington, DC: U.S. EPA, Office of Water.

———. 1991b. *Guidance for Water Quality-Based Decisions: The TMDL Process.* EPA 440/4-91-001. Washington, DC: U.S. EPA, Office of Water.

———. 1996a. *Draft Framework for Watershed Based Trading*. EPA 800-R-96-001. Washington, DC: U.S. EPA, Office of Water.

———. 1996b. *Basins Version 1 CD-ROM Database: Region 8*. EPA 823-C-96-008. Washington, DC: U.S. EPA, Office of Water and Office of Science and Technology.

———. 1998a. *Development Document for Proposed Effluent Limitations Guidelines and Standards for Industrial Waste Combustors*. EPA 821-B-97-009. Washington, DC: U.S. EPA, Office of Water.

———. 1998b. *Economic Analysis and Cost Effectiveness Analysis of Proposed Effluent Limitations Guidelines and Standards for Industrial Waste Combustors*. EPA 821-B-97-010. Washington, DC: U.S. EPA, Office of Water.

USGS (U.S. Geological Survey). 1996. *National Overview of Abandoned Mine Land Sites Utilizing the Minerals Availability System (MAS) and Geographic Information System (GIS) Technology*. Open file report 96-0549. Washington, DC: U.S. Geological Survey.

18

Analysis of the Benefits and Costs of Clean Air

H. Scott Matthews

The United States entered a new era of environmental protection in 1970 with the passage of the Clean Air Act (CAA), which redefined the national effort toward achieving and maintaining healthy air quality levels in the United States. Since then, it has been twice amended, in 1977 and in 1990. The CAA had several goals:

- to establish national air quality standards for the primary criteria air pollutants (states were tasked with developing and implementing plans to achieve these standards over time),
- to improve the enforcement of meeting the standards,
- to force new sources to meet standards using best available technologies, and
- to deal with hazardous pollutants and emissions from automobiles.

The 1977 amendments tightened standards for areas already in attainment. Areas not in compliance were given provisions toward meeting the deadlines, such as permit requirements for new sources. The 1990 amendments further strengthened the act and recognized the existence of complex problems in certain areas. They listed 189 hazardous air pollutants and programs for controlling these emissions and established programs to reduce releases of acid rain precursors and ozone-depleting substances. Again, penalties and enforcement were strengthened. Finally, Section 812 of the 1990 amendments required the U.S. Environmental Protection Agency (EPA) to perform a "retrospective" benefit–cost analysis of the CAA and its amendments up to 1990 as well as "prospective" analyses every two years afterward. These benefit–cost analyses were required to consider impacts on public health, economy, and the environment of the legislation. Specific effects mentioned for inclusion were employment, productivity, cost of living, and economic growth.

The retrospective analysis (also known as the 812 report) was completed in 1997. The first draft of the prospective analysis for 1990–2010 was completed in 1999. Both studies determine a net economic benefit to society of improved air quality as a result of CAA legislation. Actual emissions over time were compared with levels of emissions estimated as if legislation had never been enacted (the "no control" scenario), and the resulting reductions in human health effects from lower levels of pollution were valuated. In short, this calculation sought to determine the "net reduced amount of emissions" resulting from the act. This exercise also assumed no changes in the geographic distribution of the population and no changes in the economy as a result of cleaner air. These assumptions themselves are somewhat problematic, because geographic distributions did change appreciably over the period, and pollution controls led to higher costs, unemployment, and so forth. However, in this chapter, we make no attempt to correct for scenarios with alternative assumptions.

The differences in pollutant emissions in raw terms are only one part of a complex process undertaken to determine the overall net economic benefit of clean air. The reports go through a detailed process of characterizing the major parts of the problem, which entails the following steps:

1. Direct cost estimation
2. Macroeconomic modeling
3. Emissions modeling
4. Air quality modeling
5. Health and environmental effects modeling
6. Economic valuation
7. Results aggregation and uncertainty characterization

These steps were completed sequentially, and the output of each step was an input into the next one. The direct costs of complying with CAA legislation were estimated across public and private entities since 1970. These costs (mostly derived from U.S. Census survey data about compliance expenditures) were fed into a macroeconomic model to estimate economic conditions such as production, employment, and price levels over time under the control and no-control scenarios. The resulting economic indicators were fed into emissions models to estimate emissions levels under the two scenarios. These emissions models estimated releases of total suspended particulates (TSPs), sulfur dioxide (SO_2), nitrogen oxides (NO_x), carbon monoxide (CO), volatile organic compounds (VOCs), and lead. Ozone (O_3), a criteria air pollutant, is formed in the atmosphere through interactions of NO_x and VOCs. The models used benchmarks of emissions levels in 1975, 1980, 1985, and 1990. Linear interpolation was used when needed to estimate releases in intermediate years. Emissions in 1970 were assumed to be equal in both scenarios.

Generally, air quality modeling requires large-scale computing and funding systems. However, the reports generally relied on simple linear scaling approaches but were based on actual air quality data from 1970 to 1990. One exception was the modeling of ozone-related air quality, in which more sophisticated methods were undertaken to yield more detailed results. The retrospective analysis, for example, modeled ozone air quality over 147 urban areas in the continental United States.

The results of the air quality modeling exercise were then used to estimate physical effects as a result of releases in specific areas (for example, emissions in a particular urban area). Physical effects, health-based outcomes such as morbidity and mortality resulting from exposure to emissions, are found by using concentration–response functions. No new analysis was carried out for concentration–response functions; existing scientific literature was surveyed. Differential incidence rates of effects such as premature death, bronchitis, and lost workdays as a result of emissions were estimated between the control and no-control scenarios. An example result of this step is an estimate of decreased cases of chronic bronchitis as a result of cleaner air.

After the various effects of importance were estimated, economic valuation studies were surveyed to determine the economic benefits of the estimated fewer health effects. For example, 26 studies on consumer willingness to pay to avoid premature death (also known as "value of life" studies) were collected as a proxy for the economic benefit of preventing mortality. Five of the studies are contingent valuation studies, and the others estimate increased wages for incurring risk. The retrospective analysis estimated a mean of $4.8 million per avoided death through this method (in adjusted 1990 dollars). Surveys of similar studies were used for other health effects.

Finally, the economic benefits of reduced health effects were monetized and compared with the direct costs of complying with clean air legislation. The stream of benefits from 1970 to 1990 was adjusted to 1990 dollars. In the end, the retrospective report estimated the central net benefit of clean air to be $22 trillion in 1990 dollars. It also estimated the 5th-percentile benefits at $5.6 trillion and the 95th-percentile benefits at $49.4 trillion. However, the vast majority of the benefit came from reduced deaths as a result of lower levels of airborne particulates.

Despite the considerable amount of resources expended on the retrospective and prospective analyses, the reports fail to address several basic but important issues related to the net benefits of clean air. First, valuations of benefits from reduced emissions of the various pollutants are not reported. Instead, only aggregate net benefits are reported (for example, the $22 trillion figure referenced above). This omission makes validating or comparing these results with previous studies difficult, because one of the preferred analytical tools of policy analysts is cost-effectiveness analysis rather than merely benefit–cost analysis. Cost-effectiveness analysis does not value the benefits of externalities such as emissions or lives saved, and it estimates a measure in dollars per ton.

Second, although the report is based on emissions data disaggregated by source (for example, mobile or stationary), resulting benefit–cost calculations were not done across these source categories. This distinction is important because previous studies have suggested that mobile sources have high relative costs and low relative benefits, whereas stationary sources generally have high benefits and low costs. If this result were to hold for EPA's efforts, it would create some resistance to additional levels of reductions based on equity issues.

In this chapter, we seek to remedy some of these shortcomings to provide a greater level of detail and insight into the United States' efficiency in achieving clean air over the past 30 years as well as into the future. To achieve this result, we present existing levels of pollution and the implied benefits of reduced pollution from the retrospective analysis. Then, we compare the implied benefits with existing studies on the benefits of reduced pollution to validate the results.

Air Pollution Costs of Industrial Activity

Current environmental regulation does not eliminate all pollution discharges or even all environmental damage from the discharges. Table 18-1 lists estimates for several pollutants. EPA estimated in a report other than the CAA report that 84 million metric tons of total criteria air pollutant emissions and 4.5 billion metric tons of carbon dioxide equivalents were emitted as a result of industrial activity in 1992. The majority of emissions resulted from the manufacture and transport of goods, and the remainder came from private transportation.

Since the 1970 CAA, opponents of environmental legislation have contended that air pollution abatement costs are large, costing jobs and disrupting the economy. However, EPA's 812 report concluded that the social benefits of abating air pollutants since 1970 exceed costs by a large margin. The EPA analysis does not attempt to justify individual regulations. In particular, the majority of benefits come from reducing the concentration of small particles and eliminating lead from gasoline. Most costs come from reducing the

TABLE 18-1. Total U.S. Emissions and Estimated External Costs of Various Pollutants, 1992

Pollutant	*Emissions (metric tons)*	*Estimated external cost (billion $)*
CO	24 million	13
NO_x	17 million	18
PM (PM_{10})	11 million	31
SO_2	20 million	36
VOCs	12 million	17
GWP[a]	4.5 billion	63
Total	—[b]	178

Notes: Value for global warming potential (GWP) was converted from 1.22 billion MMTCE (million metric tons of carbon equivalent) to metric tons of carbon dioxide (CO_2) equivalents by multiplying by the weight fraction of carbon in CO_2, 44/12. CO = carbon monoxide; NO_x = nitrogen oxides; PM = particulate matter; PM_{10} = PM with diameter smaller than 10 μm; SO_2 = sulfur dioxide.

[a] In CO_2 equivalents.

[b] Cannot add emissions (different units).

Sources: U.S. EPA 1997a, 1997b, 1997c, 1998.

emissions of pollutants that cause urban smog (ozone). In the remainder of this chapter, we break down the sources of pollution and valuations (listed in Table 18-1) into their components and analyze them separately.

External Environmental Costs

The estimated external costs in Table 18-1 depend on damages per unit of pollution. These damages result from market incentives that motivate businesses to provide the desired goods and services at least cost, even if it means creating an externality by discharging pollution to the environment. More generally, an externality results whenever private costs (total cost to the firm) are different from social costs (total cost to society). For example, banning tetraethyl lead from being added to gasoline created an estimated $2 trillion of social benefit (U.S. EPA 1997a, 1997b, 1997c). Isolating products with large external costs and estimating their economy-wide environmental impacts helps make environmental regulation more effective and efficient.

The environmental costs estimated here do not include the industrial expenditures for emissions control. These expenditures are private costs and already are embedded in the prices of goods and services. In particular, the 1990 CAA mandated almost a 50% reduction in SO_2 emissions and allowed firms to trade emissions permits (Schmalensee and others 1998). Both the additional control costs and the costs for emissions permits are already included in the price of electricity and other goods. However, the damages resulting from such emissions in terms of human health and mortality are not included, and no regulation explicitly makes decisionmakers consider these costs.

One way to internalize pollution externalities is to levy an effluent fee each time there are emissions. The effect would start by increasing the cost of extracting raw materials, which would raise the costs of making intermediate goods, along with the effluent fees for their emissions. By the time commodities were offered for consumption, their costs and prices would reflect their cumulative environmental discharges. Collecting effluent fees would be administratively difficult. It would be far easier to collect the fee at the point of final sale, but incentives for miners, hazardous waste handlers, and so forth are not so clear.

Even if the U.S. Congress were to mandate the economically correct quantity reduction in SO_2 and other pollutants—that is, the quantity at which the marginal social cost of abatement equals the marginal social benefit of abatement—externalities still would have to be internalized. The producers have played their role by abating to the level at which marginal private benefit equals marginal private cost. However, consumers still have to decide how much of each good or service to purchase. Only if consumers make consumption decisions on the basis of prices that reflect all the externalities will social efficiency be attained. The need to charge firms for the marginal social loss from emissions is discussed in Joskow 1992, Freeman and Krupnick 1992, and Freeman and others 1992.

A Summary of the Valuations Derived from the Retrospective Report

EPA's retrospective benefit–cost analysis (U.S. EPA 1997c) of the CAA provides a comprehensive summary of the estimated costs and benefits achieved in the 20 years of environmental regulation up to 1990. Tables I-3 through I-5 of that report (not presented here) summarize that \$22 trillion of benefits in 1990 dollars accrued. The following discussion is provided to facilitate comparisons of EPA's analysis with our expanded analysis.

One of the line items in the benefit calculation of EPA's Table I-3 is mortality related to fine particulate exposure. The number of cases of mortality prevented per year as a result of the CAA was determined and multiplied by a mean benefit per prevented death of \$4.8 million (in 1990 dollars). This same procedure was performed for each of the 20 years, and the stream of values was adjusted to 1990 dollars using a 5% discount rate. For this particular component of the overall analysis, a benefit of $16.6 trillion in 1990 dollars was found. Note that mortality from long-term exposure to particulates represents more than 75% of all benefits identified by the report. It is unclear whether the original purpose of the CAA was to primarily reduce mortality from particulates in line with the benefit findings of the analysis.

To find the values needed for the calculation summarized above, the report modeled point estimates with a distribution and used a Monte Carlo–type simulation to determine the expected number of cases or incidents for each of the health effects. Such a simulation depends on a complicated set of functions for each health effect, and the mean values of the resulting distribution are reported. The actual monetized benefit values reported in the analysis come from statistically applying the whole distribution to the value per incident avoided.

To facilitate comparison with EPA's model, a few simplifying assumptions are made here to simulate the reported results.[1] Instead of relying on EPA's distributions for incidences and valuations, only the reported means were used. First, for each of the health and mortality effects, the mean (expected) value of the distribution of incidences was assumed. Then, the mean dollar valuation of the particular effect was used. As an example, Table 18-2 shows our method for finding the benefit from avoided mortality from fine particulate exposure as reported above. For this particular health effect, the assumption of the mean value in 1990 dollars was $4.8 million per death avoided, which was found by looking at a range of studies of consumer willingness to pay to prevent premature mortality. The product of the values of mean cases and valuation per case is our estimate of the total yearly benefit, which was then adjusted (discounted) to 1990 dollars using a discount rate of 5%. Note that year 0 is 1990, and year 20 is 1970 (the factor powers needed for the adjustments to 1990 dollars). Finally, the data for reduced TSP emissions came from the EPA report (U.S. EPA 1997c, Table B-16) as the difference between baseline 1970 and actual emissions. Reduced emissions are differences between the "with and without CAA" scenarios and are not discounted. Estimates were listed for every fifth year beginning in 1975. Interpolation was used for in-between years and for 1970–1974.

Table 18-3 is a summary of our simplified approach to duplicating the results of the 812 report by using mean estimates for all categories of physical effects considered by EPA. Because of space constraints, only the particulate mortality results are presented here in detail. We compare our mean estimate values with those presented in Table I-3 of the EPA report and note the individual percent differences between methods. Because of differences yielded when using probability distributions and sampling, some individual categories have significant differences. However, the overall net difference is nearly zero (less than $50 billion out of $22 trillion).

TABLE 18-2. Time Line of Simplified Estimation of Health Effects Valuations from Reduced Fine Particulate Mortality, 1970–1990

Year	*Mean cases (thousands)*	*Mean valuation ($, unadjusted)*	*Mean valuation (1990 $)*	*Reduced TSP emissions (short tons)*
0 (1990)	183.539	880,987,200,000	880,987,200,000	15,840,000
1	180.7596	867,646,080,000	911,028,384,000	15,430,000
2	177.9802	854,304,960,000	941,871,218,400	15,020,000
3	175.2008	840,963,840,000	973,520,765,280	14,610,000
4	172.4214	827,622,720,000	1,005,980,588,802	14,200,000
5	169.642	814,281,600,000	1,039,252,592,763	13,790,000
6	164.8904	791,473,920,000	1,060,650,749,860	13,468,000
7	160.1388	768,666,240,000	1,081,590,591,186	13,146,000
8	155.3872	745,858,560,000	1,101,972,789,769	12,824,000
9	150.6356	723,050,880,000	1,121,689,231,732	12,502,000
10	145.884	700,243,200,000	1,140,622,385,917	12,180,000
11	128.46	616,608,000,000	1,054,608,930,929	11,212,000
12	111.036	532,972,800,000	957,142,574,478	10,244,000
13	93.612	449,337,600,000	847,293,060,054	9,276,000
14	76.188	365,702,400,000	724,065,737,751	8,308,000
15	58.764	282,067,200,000	586,397,450,568	7,340,000
16	47.0112	225,653,760,000	492,573,858,477	5,872,000
17	35.2584	169,240,320,000	387,901,913,551	4,404,000
18	23.5056	112,826,880,000	271,531,339,485	2,936,000
19	11.7528	56,413,440,000	142,553,953,230	1,468,000
20 (1970)	0	0	0	0
Total			16,723,235,316,231	214,070,000

Note: TSP = total suspended particulate.

The bottom row of Table 18-3 indicates that the simplified method of using mean estimates produces a result only negligibly different from that produced using EPA's complicated method of using distributions for both cases and damages. Once this was accomplished, the calculation of comparative estimates of benefits from reduced pollution (in dollars per ton) was straightforward. We then could consider the 1990 valuation of mortality due to fine particulate emissions in dollars per ton by dividing the total emission of TSPs avoided (214 million tons) by the total benefit of $16.7 trillion. This result alone is more than $78,000 per short ton (or $86,000 per metric ton). Note that mortality from particulate emissions was assumed to result from emissions of fine particulates—for example, particulates less than 10 microns in diameter (PM_{10}) or less than 2.5 microns in diameter ($PM_{2.5}$)—not TSPs.

TABLE 18-3. Comparison of EPA 812 Report Benefit Category Valuations and Estimates (1990$)

		Value (billion $)		
Incidence	*Pollutant*	*EPA*	*Mean estimate*	*Difference (%)*
Mortality				
Long-term PM_{10} exposure	PM	16,632	16,723	1
Lead exposure	Lead	1,339	1,288	–4
Chronic bronchitis	PM	3,313	3,211	–3
Other lead-induced ailments				
Lost IQ points	Lead	377	420	11
IQ < 70 points	Lead	22	25	11
Hypertension	Lead	98	107	9
Coronary heart disease	Lead	13	13	
Atherothrombotic brain infarction	Lead	10	9	–10
Initial cerebrovascular accident	Lead	16	13	–18
Subtotal		536	587	10
Hospital admissions				
All respiratory	PM/O_3	9	9	4
COPD and pneumonia		9	9	
Ischemic heart disease	PM	4	4	–8
Congestive heart failure		5	5	
Other respiratory-related ailments				
Child: Short breath days	PM	6	7	8
Child: Acute bronchitis	PM	7	7	–4
Child: Upper and lower respiratory	PM	2	2	20
Adult: Any of 19 acute conditions	PM/O_3	46	43	–6
All: Asthma attacks	PM/O_3	0	0	
All: Increase in respiratory illness	NO_2	2	2	
All: Any symptom	SO_2	0	0	
Restricted activity and work loss days				
MRAD	PM	85	145	70%
Work loss days	PM	34	35	2%
Totals		22,029	22,077	0%

Notes: O_3 = ground-level ozone; COPD = chronic obstructive pulmonary disease; NO_2 = nitrogen dioxide; MRAD = minor restricted activity day.

Sources: U.S. EPA 1997c, Table I-3; author's analysis.

Assuming a ratio of PM_{10} to TSPs of 0.5, reduced PM_{10} emissions would be only 107 million tons, increasing the valuation to $172,000 per metric ton.

First, all relative damages are summarized. Table 18-4 shows the breakdown of total damages ($22 trillion) across the studied pollutants. Recall that the "benefits" identified in Table 18-3 arise from reduced emissions from the Pollutants column. In general, each row of this table lists a health effect attributed to a particular kind of pollutant. The identified pollutant categories were aggregated into the table by using EPA's classification assumptions.

Results show that the vast majority (99%) of all benefits cited in the retrospective report are attributable to prevented deaths. This fact also sheds light on several of the assumptions used to create the report, namely, that most of the benefit comes from avoided deaths from exposure to particulate emissions (91%). In effect, EPA has put most of its eggs in the particulate mortality basket. As a result, the implication of the report is that future regulation should target only reduced emissions of particulate matter and would achieve 90% of the benefit seen by regulating all conventional pollutants at a much lower cost. This result is unfortunate but expected due to the classification done by EPA.

Table 18-3 shows how this result happened. The only sources of mortality in the study were lead and long-term exposure to particulates. Although the independence of lead in the environment as a cause of death is not questionable, it is difficult to designate the cause of mortality from breathing dirty air to a specific pollutant source, because the various pollutants are often released together, and several studies have been unable to show statistical differences in the causation of death from particular emissions. This being determined, EPA attributed all mortality from nonlead sources to particulate matter, a highly controversial assumption. To be fair, one of the reasons behind this assumption is that studies have inconsistently attributed the health effects to particular pollutants. In other words, it is unclear whether particulate matter, SO_2, or NO_x is the cause of certain effects, but it is known that adverse health effects occur in the presence of these pollutants.

Table 18-4 takes a step forward from the published analysis by comparing the implied per-pollutant benefits to reductions of that source. The result is a series of implied benefits (in dollars per ton) from each of the pollutants.[2] The retrospective analysis implies benefits on the order of $100,000 per ton of reduced particulates and nearly $750,000 per ton of lead. The benefits from reduced CO, SO_2, and NO_x are negligible. We show later in this chapter how these implied valuations compare with existing valuations of these pollutants.

To bound the dollar-per-ton estimates in Table 18-4, an allocation mechanism was generated to "spread out" the mortality from nonlead sources among the other pollutants. We took the simplified results from above, added the emissions of nonlead sources, and distributed the value for particulate mortality across the nonlead sources. For example, particulates are only 17% of total nonlead emissions (214 million out of a total of 1.2 billion tons), so only 17% of the original particulate mortality value was distributed to particulate matter. This assumption is not based on any health science and is used only for illustration. The result of the proportionally adjusted process is shown in Table 18-5. One additional adjustment was considered in Table 18-5, because of the lack of literature on serious health damages resulting from CO emissions. In this final adjustment, only particulate matter, NO_x, and SO_2 received distributions of the original particulate mortality value.

Alternative Damage Valuation of Emissions

Many studies have translated damage estimates caused by air pollution into dollar values. Table 18-6 summarizes some of the studies that have valued the damage caused by SO_2 emissions and adjusts the valuations to 1990 dollars (FRED 2000).

The studies presented thus far represent a summary of the relevant literature. The results are not in close agreement; the range in 1990-valued estimates (\$710–4,300 per metric ton) reflects the substantial uncertainties associated with the effects of SO_2 on health, state-to-state variations in exposure, and the valuation of these effects.

In general, the method of determining these valuations comes from a damage function approach. The studies translate emissions first into ambient air concentrations and then into damage to human health, materials, plants and animals, visibility and aesthetics, and ecology by using a dose–response relationship. These human and ecological impacts are translated into economic values, and the externalities are identified (Lee 1992). In practice, the mortality losses dominate the valuation. The studies by Cifuentes and Lave (1993), Zuckerman and Ackerman (1995), and CEC/United States (1993) all follow damage function approaches.

Four of the valuation studies were done by state public utility commissions (PUCs). The states wanted to make sure that utilities did not build polluting generators on the basis of inadequate state and federal environmental regulation. They sought values so that a utility planning a new generation unit could account for the external costs of each fuel and technology as well as private costs. The values were to be independent of the specific location of a generating plant. Using these values, utilities were told to choose the fuel and technology that minimized the sum of private and social costs. The studies in some states (for example, Nevada) were not based on a specific method for internalizing environmental damages and were the result of a political selection amongst alternatives to meet public goals (Wiel 1992). These PUC

TABLE 18-4. Breakdown of Damages in EPA Retrospective Report (billion 1990$, short tons)

		EPA				*Comparison (mean) estimates*			
		Benefits				*Benefits*			
Source	*Abated (million tons)*	*Mortality*	*Other*	*Total*	*Implied cost ($/ton)*	*Mortality*	*Other*	*Total*	*Implied cost ($/ton)*
TSP	214	19,945	205	20,150	94,126	19,934	264	20,198	94,350
Lead	2.517	1,339	536	1,875	744,934	1,288	587	1,876	745,153
CO	763.1	0	3	3	3	0	2	2	3
NO_x	72	0	2	2	28	0	2	2	25
SO_2	189.5	0	0	0	0	0	0	0	1

TABLE 18-5. Emission Fraction Adjustments to Cost Values (billion 1990$, short tons)

		Distributed across all nonlead sources				*Distributed across nonlead sources (not CO)*			
		Benefits				*Benefits*			
Source	*Abated (million tons)*	*Mortality*	*Other*	*Total*	*Implied cost ($/ton)*	*Mortality*	*Other*	*Total*	*Implied cost ($/ton)*
PM	214	3,445	264	3,709	17,325	8,974	264	9,237	43,150
Lead	2.517	1,288	587	1,876	745,153	1,288	587	1,876	745,153
CO	763.1	12,281	2	12,283	16,097	0	2	2	3
NO^x	72	1,158	2	1,160	16,119	3,017	2	3,019	41,944
SO_2	189.5	3,050	0	3,050	16,094	7,944	0	7,944	41,919

TABLE 18-6. Social Damages per Metric Ton of SO_2 Air Emissions

Estimate	*($)*	*(1990$)*
Zuckerman and Ackerman 1995	4,900	4,300
CEC/United States 1993	3,770	3,400
ECO Northwest and others 1987	1,950	2,230
EPRI 1990	1,730	1,980
Nevada PSC 1990	1,700	1,700
Elkins and others 1985	1,470	1,680
California 1989	1,200	1,260
Rowe 1995	780	900
Repetto 1990	700	800
Cifuentes and Lave 1993	790	710
Median of studies	1,700	1,690

Note: Values in 1990 dollars were converted from valuations by using the chain-weighted GDP deflator.

Sources: Ottinger 1992; CEC/United States 1993; Cifuentes and Lave 1993; U.S. Congress, OTA 1994; Zuckerman and Ackerman 1995; ECO Northwest and others 1987.

studies were explicitly public–political valuations of air pollution emissions. Although the estimates are informed by the research literature, they represent an interpretation of the valuations by a political body after holding public hearings. Thus, they provide perspectives on the damage associated with emissions of various pollutants, rooted in damage function approaches.

The use of these various studies provides a broad range of estimates of the impact of air emissions on health and ecology as well as a range of valuations of the resulting damage. The underlying uncertainties are real, as demonstrated by EPA's selection of new national ambient air quality standards for small particles and ozone and the storm of criticism it provoked.

The body of social damage valuations was converted into 1990 dollars using the chain-weighted gross domestic product (GDP) deflator (FRED 2000). The valuations for all six pollutants are shown in Table 18-7. In our judgment, the range and median estimates for SO_2, particulate matter, NO_x, and VOCs are the best estimates currently available.[3]

The U.S. Economy in 1992

The data in the last column of Table 18-1 are the result of multiplying the EPA estimates of total 1992 emissions of each pollutant by the median estimate of dollar damage from Table 18-7. The estimate of total annual environmental damage from the criteria air pollutants is $180 billion in 1992, or

TABLE 18-7. Unit Social Damage Estimates from Air Emissions of Environmental Externalities

		External cost (1990$/metric ton of air emissions)			
Pollutant	*No. of studies*	*Minimum*	*Median*	*Mean*	*Maximum*
CO	2	1	480	480	970
NO_x	9	190	1,010	2,630	8,800
SO_2	10	710	1,690	1,900	4,300
PM	12	900	2,650	4,100	15,400
VOCs	5	150	1,300	1,500	4,040
GWP[a]	4	2	13	12	22

[a] In CO_2 equivalents.

Sources: Ottinger 1992; CEC/United States 1993; Cifuentes and Lave 1993; Frankhauser 1994; U.S. Congress, OTA 1994, Zuckerman and Ackerman 1995; ECO Northwest and others 1987.

5% of 1992 GDP. This social damage is surprisingly large, given the 28 years of effort resulting from the 1970 CAA. The estimated annual expenditures on abatement of these pollutants were $20 billion to $25 billion in 1990 dollars (U.S. EPA 1990). In few cases is the marginal cost of abatement as large as the median valuation, and so it is not surprising that the benefit of abatement is much larger than the expenditures on abatement.

Congress and EPA have tended to focus on the industries with the greatest emissions. When that industry has a high ratio of direct emissions costs to output, EPA and Congress have tended to soften emissions regulations to prevent bankruptcy. We suggest that attention could be focused on the sectors that cause the greatest social damage, especially when that damage is due to suppliers of the sector. For example, eating and drinking establishments can be shown to merit greater attention, particularly in a deregulated era where they can select their electricity supplier (Matthews and Lave 2000).

EPA reports that the discharges of SO_2, particulate matter, nitrogen dioxide, and VOCs are large, despite decades of regulation. A handful of sectors, such as electricity production, are responsible for the majority of air emissions. The environmental impacts of other sectors are due to their selection of suppliers among the most polluting sectors. For the most polluting sectors, the key to improving environmental performance is to reduce their own emissions. For the other sectors, the key to improving their environmental performance is to purchase fewer goods and services from the most polluting sectors.

We caution that our estimates of social cost are uncertain, reflecting the substantial uncertainty in damages and their valuation from air pollutants. These estimates do not include damage from other (toxic) air pollutants or liquid or solid discharges. They show that in 1992, the social loss from the six air pollutants was more than $180 billion (more than 5% of GDP). The Council on Environmental Quality reports that annual air pollution control expenditures are on the order of $30 billion, suggesting that the median external costs (totaling almost $200 billion) are significantly higher and that there may be room for improvement to bring costs and benefits in line with socially optimal levels.

Comparison of Literature and EPA Valuation

Considering the three sets of dollars per ton values implied by EPA's analysis (Tables 18-4 and 18-5), it is clear that these values are significantly higher than even the maximum in the range of values from the external cost studies earlier in this chapter (that is, in Tables 18-6 and 18-7). The values presented here are in short tons, whereas prior results are in metric tons. Valuations in dollars per metric ton would be only 10% higher, a marginal change that has no qualitative effect. This result generally supports the claim that the previous external cost results of the damage associated with the air pollution resulting from industrial production are underestimated if EPA's valuations are accepted. Regardless, any of the valuations implied by EPA and shown above are clearly well beyond the low or high end of existing studies in Table 18-7.

The great discrepancy between EPA's implied valuations and those in the literature should cause concern. Although EPA did not set out to generate dollar-per-ton estimates, these values can be concluded from a reading of the analysis. They represent current EPA methods for valuing health effects, so in general it seems that the values being used by EPA are extraordinarily high. On the other hand, if we assume that EPA's values are more likely to be correct than those in the literature, then the implication is that levels of controls on pollution are well below socially optimal levels. Regardless, using either our robust analysis or EPA's method, net benefits remain positive, and there appears to be significant room to tighten standards in the future.

Valuations by Source Category

Having investigated the aggregate benefits, a similar analysis was performed to compare the net benefits of the various source categories in the report. As discussed earlier, the retrospective and prospective analyses failed to break out benefits and costs by source (for example, mobile versus stationary). To illustrate the importance of such a disaggregation, we provide an example for this purpose. This new analysis disaggregates emissions, valuates the respective benefits, and makes comparisons across programs.

TABLE 18-8. Net Emissions of CO with the Clean Air Act (thousands of short tons)

Year	*Overall*	*Mobile*	*Stationary*	*Commercial/ residential*
1970	0	0	0	0
1971	–1,710	–1,376	–334	0
1972	–3,420	–2,752	–668	0
1973	–5,130	–4,128	–1,002	0
1974	–6,840	–5,504	–1,336	0
1975	–8,550	–6,880	–1,670	0
1976	–12,382	–10,610	–1,764	–8
1977	–16,214	–14,340	–1,858	–16
1978	–20,046	–18,070	–1,952	–24
1979	–23,878	–21,800	–2,046	–32
1980	–27,710	–25,530	–2,140	–40
1981	–34,762	–32,210	–2,508	–44
1982	–41,814	–38,890	–2,876	–48
1983	–48,866	–45,570	–3,244	–52
1984	–55,918	–52,250	–3,612	–56
1985	–62,970	–58,930	–3,980	–60
1986	–68,168	–63,914	–4,194	–60
1987	–73,366	–68,898	–4,408	–60
1988	–78,564	–73,882	–4,622	–60
1989	–83,762	–78,866	–4,836	–60
1990	–88,960	–83,850	–5,050	–60

Source: U.S. EPA 1997c, Table B-20.

Table 18-8 shows the net reduced emissions of CO with and without CAA by source for the period 1970–1990. Estimates for 1970–1974 were not published, so these values were estimated by interpolation. The source category (for example, mobile or stationary) for the data and the total emissions are given in Table 18-8. Values for the commercial and residential sector are not aggregated with the other stationary sources because they are not necessarily subject to what would traditionally be called "stationary source" regulations. Regardless, the reductions in the Commercial/Residential sector are estimated to be small and make negligible differences in the remainder of the analysis.

Through similar methods, we could generate results for the remaining pollutants studied in the retrospective analysis. Table 18-9 summarizes these findings across all six pollutants. These results were converted to metric tons for comparison with our external cost valuations above, and emissions of PM_{10} were assumed to be 40% of TSPs (EPA did not specifically report PM_{10} emissions over time).

TABLE 18-9. Net Reduced Emissions of Pollutants Estimated in Retrospective Analysis across All Sources (in thousands of metric tons)

	Annual overall reduced emissions					
Year	*CO*	*VOC*	*NO_x*	*SO_2*	*TSP*	*PM_{10}*
1970	0	0	0	0	0	0
1971	1,556	479	111	639	1,335	534
1972	3,111	957	222	1,277	2,671	1,068
1973	4,667	1,436	333	1,916	4,006	1,603
1974	6,222	1,914	444	2,554	5,342	2,137
1975	7,778	2,393	555	3,193	6,677	2,671
1976	11,264	3,208	890	4,665	7,556	3,022
1977	14,750	4,023	1,224	6,137	8,435	3,374
1978	18,236	4,838	1,559	7,609	9,314	3,725
1979	21,722	5,653	1,894	9,081	10,192	4,077
1980	25,208	6,468	2,229	10,553	11,071	4,428
1981	31,623	7,301	2,853	10,873	11,364	4,546
1982	38,038	8,135	3,477	11,193	11,657	4,663
1983	44,453	8,968	4,101	11,513	11,950	4,780
1984	50,869	9,801	4,725	11,833	12,243	4,897
1985	57,284	10,634	5,349	12,154	12,536	5,014
1986	62,012	11,562	5,926	12,578	12,909	5,163
1987	66,741	12,490	6,503	13,001	13,282	5,313
1988	71,470	13,418	7,079	13,425	13,655	5,462
1989	76,198	14,346	7,656	13,849	14,028	5,611
1990	80,927	15,274	8,233	14,273	14,401	5,760

Notes: Conversion from short tons to metric tons is assumed to be 0.9097; TSP-to-PM_{10} conversion factor is 0.4.

Source: U.S. EPA 1997c, Tables B-16 through B-20; authors' calculations.

Using our 1990-valued external cost valuation ranges from Table 18-7, the reduced emissions from Table 18-9 were valued for their economic benefit to society and are presented in Table 18-10. Total values have not been discounted; the reduced emissions per year were valued at 1990 valuations. Then, a 5% discount rate was used to adjust all valuations to 1990, as done in the EPA report. Of course, this valuation method is completely different from that used by EPA, but it represents a valuation from published literature given the estimated reduced emissions from EPA.

Using the median estimates from Table 18-7, the adjusted central estimate of benefits of the CAA is only on the order of $1.5 trillion (instead of $20 trillion). This result is about 90% lower than EPA's estimates. Our analysis does not include the benefits from reduced lead, which was 10% of the total bene-

TABLE 18-10. Overall Benefits (million 1990$)

Year	*Minimum*	*Median*	*Maximum*
1970	0	0	0
1971	1,029	3,976	15,391
1972	2,057	7,951	30,782
1973	3,086	11,927	46,173
1974	4,114	15,903	61,564
1975	5,143	19,878	76,955
1976	6,694	26,368	105,280
1977	8,244	32,858	128,320
1978	9,795	39,348	151,359
1979	11,346	45,838	174,398
1980	12,897	52,328	197,438
1981	13,480	57,973	217,172
1982	14,062	63,617	236,906
1983	14,645	69,262	256,640
1984	15,228	74,906	276,374
1985	15,811	80,551	296,108
1986	16,499	85,721	315,044
1987	17,188	90,891	333,981
1988	17,877	96,062	352,917
1989	18,566	101,232	371,853
1990	19,254	106,402	390,789
Total	227,015	1,082,993	4,035,445
Total at 5%	322,389	1,493,612	5,591,508

Note: Total is in nondiscounted 1990 dollars.

fits, or about $2 trillion. Similarly, Table 18-11 shows the results of disaggregating the values above into benefits from mobile and stationary sources and then economically valuing the reduced emissions per year (Tables 18-12 and 18-13 show the detailed emissions by source). Table 18-11 also shows the totals discounted at 5%.

So far, we are only able to consider the benefits from the CAA as reported. However, it is more interesting to know the net benefits, inclusive of the significant costs needed to achieve the estimated benefits. Table 18-14 summarizes annualized cost data reported in the retrospective analysis. EPA's analysis does not include annualized cost detail before 1973. In the absence of costs, we ignore benefits and costs from 1970 to 1972.

We have disaggregated all relevant information needed to separately determine the discounted benefits and costs of mobile and stationary sources as a

TABLE 18-11. Annual Stationary and Mobile Benefits from Reduced Emissions (million 1990$)

	Stationary source			*Mobile*		
Year	*Minimum*	*Median*	*Maximum*	*Minimum*	*Median*	*Maximum*
1970	0	0	0	0	0	0
1971	946	2,728	11,754	82	1,247	3,637
1972	1,892	5,457	23,507	165	2,495	7,275
1973	2,838	8,185	35,261	247	3,742	10,912
1974	3,785	10,913	47,014	330	4,989	14,550
1975	4,731	13,641	58,768	412	6,237	18,187
1976	6,127	17,392	71,977	559	8,914	26,137
1977	7,523	21,142	85,186	705	11,592	34,087
1978	8,919	24,892	98,395	852	14,269	42,037
1979	10,315	28,642	111,604	998	16,946	49,987
1980	11,711	32,393	124,813	1,144	19,624	57,937
1981	12,054	33,470	129,087	1,381	24,174	71,873
1982	12,398	34,548	133,362	1,617	28,725	85,808
1983	12,742	35,626	137,637	1,853	33,276	99,744
1984	13,085	36,703	141,912	2,089	37,826	113,679
1985	13,429	37,781	146,186	2,325	42,377	127,615
1986	13,878	39,082	151,211	2,565	46,246	140,117
1987	14,326	40,383	156,235	2,806	50,115	152,619
1988	14,775	41,685	161,259	3,046	53,984	165,121
1989	15,224	42,986	166,284	3,287	57,853	177,623
1990	15,672	44,287	171,308	3,527	61,723	190,125
Total	196,369	551,937	2,162,759	29,989	526,356	1,589,071
Total at 5%	281,617	791,857	3,121,989	39,891	695,425	2,092,171

Note: These totals do not match overall totals due to rounding.

result of the CAA. Table 18-15 shows the year-by-year overall, mobile, and stationary source net benefits using the median results from Tables 18-10, 18-11, and 18-14. The total is the difference of the "adjusted" values from these tables (discounted at 5%). Even with our lower inherent valuation of pollutant releases, net benefits are still positive.

Regulatory Improvement

The disaggregated and valued results paint a different picture of the environmental effects of nearly 30 years of clean air legislation. First, the net benefits

TABLE 18-12. Annual Mobile Reduced Emissions by Pollutant (thousands of metric tons)

Year	*CO*	*VOC*	NO_x	SO_2	*TSP*	PM_{10}
1970	0	0	0	0	0	0
1971	1,252	438	67	–2	11	4
1972	2,503	877	135	–4	22	9
1973	3,755	1,315	202	–5	33	13
1974	5,007	1,754	269	–7	44	17
1975	6,259	2,192	337	–9	55	22
1976	9,652	2,798	577	–7	69	28
1977	13,045	3,404	817	–5	84	33
1978	16,438	4,010	1,057	–4	98	39
1979	19,831	4,616	1,297	–2	113	45
1980	23,225	5,222	1,537	0	127	51
1981	29,301	6,066	2,050	–2	147	59
1982	35,378	6,910	2,564	–4	167	67
1983	41,455	7,754	3,077	–5	187	75
1984	47,532	8,598	3,590	–7	207	83
1985	53,609	9,443	4,103	–9	227	91
1986	58,143	10,347	4,596	–9	246	98
1987	62,677	11,251	5,089	–9	264	106
1988	67,210	12,155	5,582	–9	282	113
1989	71,744	13,060	6,075	–9	300	120
1990	76,278	13,964	6,568	–9	318	127

are about 95% less than EPA's estimates and fall below the minimum net benefits estimated in the retrospective analysis. Also, until 1982, net benefits from stationary source reductions far outweighed benefits from mobile sources. By 1990, annual net benefits from mobile source reductions were more than double the benefits from stationary sources.

This result may suggest that EPA should in the future concentrate similar air quality efforts on abating emissions from mobile sources because the "recent" benefits from these programs have produced large net benefits. It also does not preclude additional tightening of stationary sources, because net benefits have accrued there as well.

Annual benefits from stationary source reductions have remained largely unchanged since 1980, whereas benefits from mobile source reductions have increased dramatically (Tables 18-12 and 18-13). Unfortunately, the construction of EPA's analysis does not allow the evaluation of the underlying successes of EPA programs, only the overall result. Armed with the evidence

TABLE 18-13. Annual Stationary Reduced Emissions by Pollutant (thousands of metric tons)

Year	*CO*	*VOC*	NO_x	SO_2	*TSP*	PM_{10}
1970	0	0	0	0	0	0
1971	304	40	44	640	1,325	530
1972	608	80	87	1,281	2,649	1,060
1973	912	120	131	1,921	3,974	1,589
1974	1,215	160	175	2,562	5,298	2,119
1975	1,519	200	218	3,202	6,623	2,649
1976	1,605	369	311	4,670	7,485	2,994
1977	1,690	539	404	6,139	8,347	3,339
1978	1,776	708	497	7,607	9,210	3,684
1979	1,861	877	589	9,075	10,072	4,029
1980	1,947	1,046	682	10,543	10,935	4,374
1981	2,282	1,028	793	10,864	11,206	4,482
1982	2,616	1,010	904	11,184	11,477	4,591
1983	2,951	992	1,015	11,504	11,748	4,699
1984	3,286	973	1,126	11,824	12,019	4,808
1985	3,621	955	1,237	12,144	12,290	4,916
1986	3,815	979	1,319	12,568	12,647	5,059
1987	4,010	1,002	1,401	12,992	13,003	5,201
1988	4,205	1,026	1,483	13,416	13,360	5,344
1989	4,399	1,050	1,565	13,840	13,716	5,487
1990	4,594	1,073	1,647	14,264	14,073	5,629

of high net benefits from mobile sources, EPA might better justify future vehicle standards.

Conclusions

With the publication of the retrospective and prospective analyses, EPA was given the mandate and opportunity to justify the burden of air pollution regulation for the past 30 years and into the future. By officially reporting only the overall net benefits and excluding any validating analysis or disaggregation by source, the agency failed to distinguish between major programs of the current regulatory environment. EPA also missed an opportunity to further inform the public about the wide-ranging successes of its programs.

It is unclear whether the original intent of the CAA was to phase out emissions in such an inequitable fashion. Even though EPA's analysis and our own

TABLE 18-14. Annualized Costs at 5% (million 1990$)

Year	*Stationary capital cost*	*O&M*	*Recovery cost*	*Mobile capital cost*	*O&M*	*Other*	*Total expenses*	*Discounted total expenses at 5%*
1970	NA	NA	NA	NA	NA	NA	NA	0
1971	NA	NA	NA	NA	NA	NA	NA	0
1972	NA	NA	NA	NA	NA	NA	NA	0
1973	523	3,936	545	0	4,838	2,290	11,042	25,308
1974	1,194	4,778	746	98	5,927	2,184	13,435	29,327
1975	1,888	5,154	895	177	5,250	2,063	13,637	28,350
1976	2,630	5,768	1,074	645	4,459	2,183	14,611	28,929
1977	3,317	6,527	1,128	1,194	3,617	2,378	15,905	29,991
1978	3,968	6,991	1,158	1,784	1,705	2,487	15,777	28,333
1979	4,598	7,959	1,296	2,395	2,124	2,503	18,283	31,270
1980	5,277	8,791	1,361	3,053	2,826	2,226	20,812	33,901
1981	5,967	8,785	1,430	3,656	1,993	1,935	20,906	32,432
1982	6,610	7,855	1,158	4,313	750	1,755	20,125	29,734
1983	7,217	8,168	1,067	4,934	–201	1,684	20,735	29,176
1984	7,694	8,505	1,082	5,564	–406	1,634	21,909	29,360
1985	8,163	8,617	921	6,400	404	1,785	24,448	31,203
1986	8,593	8,477	1,013	6,924	–1,628	1,809	23,162	28,154
1987	9,005	8,602	1,117	7,416	–1,474	1,804	24,236	28,056
1988	9,410	8,143	1,206	7,831	–1,716	1,819	24,281	26,770
1989	9,804	8,259	1,171	8,237	–1,707	1,865	25,287	26,551
1990	10,222	8,842	1,256	8,531	–1,816	1,542	26,065	26,065
Total								522,910

Note: O&M = operations and management; NA = not applicable.
Source: U.S. EPA 1997c, Table A-9.

expanded analysis suggest that net benefits are positive, given the far higher burden on stationary sources in terms of costs, permitting, and management requirements, this analysis sheds some doubt on whether incremental effort to reduce stationary source emissions is worthwhile.

Notes

1. Chapter 19 provides an example of a more extended simulation for the prospective report.
2. The tons are as released, that is, not discounted.

TABLE 18-15. Overall, Mobile, and Stationary Source Net (Median) Benefits to Society from the Clean Air Act, 1975–1990 (million $)

Year	*Overall*	*Mobile*	*Stationary*
1973	2,028	–2,512	9,789
1974	5,386	–2,261	12,414
1975	12,975	1,683	15,581
1976	23,278	7,544	19,933
1977	31,968	12,786	23,431
1978	42,331	19,359	27,102
1979	47,129	21,255	29,728
1980	51,337	22,389	32,066
1981	57,503	28,739	31,257
1982	64,258	34,960	31,382
1983	68,282	40,163	29,982
1984	71,021	43,779	28,928
1985	71,603	45,401	27,978
1986	76,041	49,775	27,987
1987	77,162	51,136	27,660
1988	79,138	52,776	27,935
1989	79,742	53,890	27,399
1990	80,337	55,008	26,479
Total	941,520	535,869	457,030

3. The estimates for carbon monoxide are based on fewer studies and are less reliable; estimates for carbon dioxide implied by some estimates of the cost of meeting the Kyoto Agreement are larger than the values we use.

References

CEC (Commission of the European Communities)/United States. 1993. Externalities of the Fuel Cycle: Externe Project. Working documents 1, 2, 5, and 9. Brussels, Belgium: European Commission.

Cifuentes, Luis, and Lester B. Lave. 1993. Economic Valuation of Air Pollution Abatement: Benefits from Health Effects. *Annual Review of Energy and the Environment* 18: 319–342.

ECO Northwest and others. 1987. *Generic Coal Study: Quantification and Valuation of Environmental Impacts.* Report commissioned by Bonneville Power Administration. Portland, OR: ECO Northwest.

Frankhauser, Samuel. 1994. The Social Costs of Greenhouse Gas Emissions: An Expected Value Approach. *Energy Journal* 15(2): 157–170.

FRED (Federal Reserve Economic Data). 2000. *Chain-Weighted GDP Deflator.* http://www.stls.frb.org/fred/data/cpi/cpiaucsl (accessed Nov. 15, 2000).

Freeman, A.M., and Alan Krupnick. 1992. Externality Adders: A Response to Joskow. *Electricity Journal* August/September: 61–63.

Freeman, A.M., Dallas Burtraw, Winston Harrington, and Alan Krupnick. 1992. Weighing Environmental Externalities: How to Do It Right. *Electricity Journal* August/September: 18–25.

Joskow, Paul. 1992. Weighing Environmental Externalities: Let's Do It Right. *Electricity Journal* May: 53–67.

Lee, Russell. 1992. Estimating the Impacts, Damages, and Benefits of Fuel Cycles: Insights from an Ongoing Study. In *Social Costs of Energy, Proceedings of an International Conference Held at Racine, WI, Sept. 8–11, 1992,* edited by O. Hohmeyer and R.L. Ottinger. Berlin, Germany: Springer-Verlag.

Matthews, H. Scott, and Lester B. Lave. 2000. Applications of Environmental Valuation for Determining Externality Costs. *Environmental Science and Technology* 34: 1390–1395.

Ottinger, Richard L. 1992. Pollution Taxes: The Preferred Means of Incorporation of Environmental Externalities. In *Social Costs of Energy, Proceedings of an International Conference Held at Racine, WI, Sept. 8–11, 1992,* edited by O. Hohmeyer and R.L. Ottinger. Berlin, Germany: Springer-Verlag.

Schmalensee, Richard, Paul L. Joskow, A. Denny Ellerman, Juan P. Montero, and Elizabeth M. Bailey. 1998. An Interim Evaluation of Sulfur Dioxide Emissions Trading. *Journal of Economic Perspectives* 12(3): 53–68.

U.S. Congress, OTA (Office of Technology Assessment). 1994. *U.S. Congress Studies of the Environmental Costs of Electricity.* OTA ETI-134. Washington, DC: U.S. Government Printing Office.

U.S. EPA (Environmental Protection Agency). 1990. *Environmental Investments: The Cost of a Clean Environment.* Washington, DC: U.S. EPA.

———. 1997a. *Final Regulatory Impact Analysis: Control of Emissions of Air Pollution from Highway Heavy-Duty Engines.* Washington, DC: U.S. EPA.

———. 1997b. *National Air Pollutant Emission Trends Report, 1900–1996.* Washington, DC: U.S. EPA.

———. 1997c. *The Benefits and Costs of the Clean Air Act: 1970 to 1990.* Office of Air and Radiation. Washington, DC: U.S. EPA.

———. 1998. *Inventory of U.S. Greenhouse Gas Emissions and Sinks: 1990–1996* (1998 Draft). http://www.epa.gov/oppeoee1/globalwarming/inventory/1998-inv.html (accessed April 2000).

Wiel, Stephen. 1992. Why Utilities Should Incorporate Externalities. In *Social Costs of Energy, Proceedings of an International Conference Held at Racine, WI, Sept. 8–11, 1992,* edited by O. Hohmeyer and R.L. Ottinger. Berlin, Germany: Springer-Verlag.

Zuckerman, Brian, and Frank Ackerman. 1995. The 1994 Update of the Tellus Institute Packaging Study Impact Assessment Method. Presented at SETAC Impact Assessment Working Group Conference, Washington, DC, Jan. 25–26.

19

Facilitating Regulatory Design and Stakeholder Participation

The FERET Template with an Application to the Clean Air Act

R. SCOTT FARROW, EVA WONG, RAFAEL A. PONCE, ELAINE M. FAUSTMAN, AND RICHARD O. ZERBE

The federal government requires impact and benefit–cost analyses for major regulations, as do numerous states.[1] In addition to a role in regulatory development, such analyses are important to stakeholders of all persuasions who comment on regulations and participate in the public debate. However, the high cost and slow pace of conducting customized studies are barriers to their use early in regulatory development and in later public participation. The Fast Environmental Regulatory Evaluation Tool (FERET) is designed to provide a common platform reflecting best practice, lower costs, and faster delivery time in the expectation of improving regulatory development and stakeholder involvement.

As a risk assessment and benefit–cost tool, FERET integrates impact and benefit–cost analysis while incorporating uncertainty into both. In environmental policy analysis, uncertainty usually focuses on the risk assessment of health impacts. In contrast, benefit–cost evaluation typically involves a different group of researchers. This dichotomy also appears in separate policy suggestions to improve regulation through the use of risk and benefit–cost analyses.

In this chapter, we describe a computerized template for benefit–cost analysis that integrates aspects of risk and economic analysis. We illustrate its uses by evaluating alternative health and economics assumptions in the analysis of the future benefits and costs of the Clean Air Act (U.S. EPA 1997, 1999). The complete FERET model along with supporting documentation is provided on the CD-ROM provided with this book and also is available from the authors. (FERET operates using Microsoft Excel and Decisioneering Crystal Ball for full functionality.)

Regulatory Evaluation

How and when are regulatory evaluations produced? Various environmental, health, economic, and policy analysis professionals are called upon to produce impact and benefit–cost analyses. Analyses are typically carried out for only the small fraction of federal regulations that are considered major, about 1% in 1999 (OMB 2000), and the number for states is likely to be even lower. The cost of such analyses, in the few cases publicly known, ranges from $200,000 to $8 million and can often take years to complete (Morgenstern 1997). Could regulatory development benefit from application to the 99% of regulations that are not analyzed, or from a public debate that allows expanded access to experiment with alternative designs and scientific assumptions? Could the standard practice of focusing on a mean or average value with a few variations be better informed by a broader consideration of uncertainty throughout the analysis?

Believing that the answer to these questions is "yes," we set out to design a computerized template that would

- structure the basic integration of impacts and valuation;
- provide a core survey of the literature;[2]
- incorporate uncertainty through simulation methods; and
- deliver a bottom line benefit–cost analysis that reports quantitative impacts, economics values, and qualitative elements.

The resulting program is FERET.

Benefit–Cost Analysis and the Structure of FERET

Benefit–cost analysis is a long-established technique that is the subject of numerous books that address both estimation and conceptual issues (for example, Zerbe and Dively 1994; Hanley and Spash 1993). The conceptual underpinning of the benefit–cost approach is a field called *welfare economics,* which seeks to answer whether a society's well-being is improved after some action, such as a new regulation. Such analyses are problem-specific, but a common starting point is to assess changes in one or several markets while allowing for external nonmarket effects such as pollution damages. It is typically assumed that a competitive market exists (that is, there are many buyers and sellers[3]) and that welfare is increased as long as those who gain can potentially compensate those who lose (the Kaldor–Hicks criteria, as discussed in Zerbe and Dively 1994 and Farrow 1998, among others). FERET can be consistent with a welfare theoretic analysis because it computes net benefits as the difference between reductions in external costs and the costs of compliance. For example, the dominant benefits from the reduction of conventional air pollutants are health impacts avoided—the reduction in external impacts—while industry control costs represent the typical cost.

Some distributional issues related to income also can be evaluated in FERET, even if they receive little attention in standard benefit–cost analysis. As an example, compliance costs in FERET include taxes,[4] because tax transfers net out (are equal) among gaining and losing parties. However, analysts can segregate tax payments if they wish as a gain to one party and a loss to another. Other distributional issues involve who bears the costs and who receives the benefits. For instance, although the basic structure of FERET identifies the compliance cost as those of industry, such costs are wholly or partially passed on to consumers who bear the final costs. Newly emerged concerns under the name of environmental justice or equity investigate the impacts on minority or low-income groups who may bear a disproportionate burden of external costs. FERET can be used to assess these costs or benefits as well.

How FERET Facilitates Benefit–Cost Analysis

FERET was designed to not only improve on standard practice but also move to best practice. As support to the analyst, FERET provides current guidance from government and other sources such as that by the Office of Management and Budget in the Executive Office of the President, the U.S. Environmental Protection Agency (EPA), and the state of California. Outside critics also have developed various lists of what constitutes best practice.[5] Below, we list recent suggestions for improving benefit–cost analysis developed by a distinguished group of economists (Arrow and others 1996) and explain how FERET addresses these suggestions.

"Benefits and costs of proposed policies should be quantified wherever possible. Best estimates should be presented along with a description of the uncertainties." FERET provides the template to quantify costs and benefits. Best estimates, often interpreted as either the mean or the median value, are easily available in FERET, as is the entire statistical distribution as a representation of the uncertainty.

"A core set of economic assumptions should be used in calculating benefits and costs associated with environment, health, and safety regulation. Key variables include the social discount rate, the value of reducing risks of dying and accidents, and the value associated with other improvements in health." In FERET, the analyst selects the discount rate and other economic parameters with various Help files to assist in that choice. Furthermore, instead of depending on a single economic value, FERET provides a bibliography of values. The user has the opportunity to select one or many studies as the basis of key health and economic variables. Additional guidance is provided by identifying studies chosen by EPA in its work.

"Information should be presented clearly and succinctly in a regulatory impact (benefit–cost) analysis. Transparency is necessary if benefit–cost analysis is to inform decisionmaking." FERET, while not writing the text for a benefit–cost analysis, is an openly structured template in which the effects of different benefit or cost assumptions can be investigated. Written in Excel, it is (relatively) transparent in its form and operation. Users can use the built-in capabilities of commercial software such as Excel and Crystal Ball to track changes, prepare reports, and display results graphically.

"It is important to identify the incremental benefits and costs associated with different regulatory policies." FERET is designed to assess the impact of regulations through a "with and without" kind of analysis that is the basis for assessing incremental benefits. Although guidance is provided on this topic, only the analyst can establish the sequence of regulatory designs that identify various levels of "incremental" regulation. Once defined, the regulatory design becomes the basis for an incremental analysis.

"Whereas benefit–cost analysis should focus primarily on the overall relationship between benefits and costs, a good benefit–cost analysis will identify important distributional consequences of a policy." FERET can be used in several ways to identify the distributional aspects of a regulation. For instance, modifying the "exposed population" is one way to identify subsets of the overall population, perhaps based on income or race. Health and valuation studies also can be selected that focus on subsets of the population such as children. Finally, information about the direct incidence of costs among business, consumers, and the government can be identified.

"Not all impacts of a decision can be quantified or expressed in dollar terms. Care should be taken to ensure that quantitative factors do not dominate important qualitative factors in decisionmaking." FERET provides a special area to enter qualitative factors that may affect decisions. These factors are presented with the quantitative results. Also presented is a calculation showing how large the qualitative factors need to be to change the economic value of the regulation from positive to negative or vice versa.

"The more external review regulatory analyses receive, the better they are likely to be." FERET's default assumptions are based on EPA's peer-reviewed studies of the retrospective and prospective studies of the Clean Air Act. FERET has been circulated for testing among interested users. Most important, FERET provides a common structure for stakeholder participation in regulatory debate. If one stakeholder develops an analysis, his or her assumptions are readily apparent in FERET. Other stakeholders may conduct external reviews of the original analysis with analyses of their own.

Beyond Best Practice

FERET provides several additional capabilities. For instance, the most important assumptions can be quantitatively identified through the sensitivity analysis features of Crystal Ball, analysts can pick more than one value for key parameters and use the results of several studies, the optimal level of regulation can sometimes be calculated, and graphics and statistics are provided. Perhaps of most importance, FERET reports the physical and other impacts used in the production of the benefit estimates, which are of direct interest to various stakeholders.

The Structure of FERET When Data Are Known

The structure of FERET is relatively simple. First we discuss the structure as if key parameter values are known, and then we discuss uncertainty over those parameters. The end point of FERET's calculations are estimates of the *present value of net benefits,* the difference between the present value benefits (PVB) and present value costs (PVC). In regulatory applications, PVB are typically external (third-party) costs avoided, such as health impacts. The annual benefits of an external impact j (B_j) are

$$B_j = \Delta y_j \times V_j$$

where Δy_j is the change in impact j (such as health cases), which is valued at V_j. For example, impact j could be the number of premature deaths avoided through control of airborne particulate emissions. PVB is the sum over all impacts (J) and time periods (t, up to terminal time T):

$$\mathrm{PVB} = \sum_{t=0}^{T} \sum_{j=1}^{J} \frac{B_{jt}(1+i)^t}{(1+r)^t}$$

where i is any rate of growth in real benefits (say, due to increased population growth) and r is a real (excluding inflation) discount rate. B_{jt} may be zero in some early years if there is a latency period or a delay in implementation (the latter two delays are possible input choices for the user).

Health impacts in particular are measured (in the primary approach) through concentration–response functions, typically of the form

$$\Delta y_j = -y_j(e^{-\beta \Delta X} - 1)$$

where ΔX is the change in concentration of a pollutants and β is the concentration response (the percent change in cases caused by a unit change in concentration) estimated from epidemiological data.

On the cost side, PVC encompasses compliance costs by industry and any nonmarket costs (external costs) to consumers. These costs are context-specific to the regulatory design. Some analysts may have their own models of cost, in which case a cost distribution can be entered directly. Alternatively, FERET provides an EPA model, PROJECT, for cost estimation. PROJECT, which has been used in various forms in courts for more than a decade, computes the PVC on the basis of estimates of fixed, operating, and maintenance costs. The user is prompted for cost assumptions (with help available in documentation from EPA and provided with FERET) on the life of buildings and equipment associated with fixed costs. Taking into account the life of the project and any replacement cycle, the model computes the PVC.

Taxes also can be taken into account if the user wishes to identify some distributional impacts, such as those among industry, government, and the consumer. The basic FERET analysis sets the tax rate to zero on the basis of welfare economic results for the standard computation of social benefits and costs.

Intermediate information is obtained on all environmental impacts that are quantified. In addition, a *cost-effectiveness analysis* (that is, the least cost of achieving a given objective) can be carried out by setting the benefit to zero and studying costs of the regulatory alternative. Alternatively, benefits can be the sole focus of analysis. By finding the level of control that balances additional costs with additional benefits, the regulation can be optimized; this feature is not yet automated, but the underlying software is capable of performing the task.

Incorporating Uncertainty

Depending on the context, there can be uncertainty over several parameters, including the change in concentration, the concentration–response function, compliance costs, and the value of impacts (ΔX, β, C, and V). For concentration and cost, FERET uses a single user-defined distribution to represent the uncertainty. However, for health response and valuation, there is uncertainty as to which distribution among many should be used. Different studies report different distributions, and some studies report multiple distributions. To accommodate this surplus of information, FERET has a two-step procedure that first selects which distribution should be used in one trial and then selects a single measure from that distribution. The process is repeated so that on the next iteration, a different study (or the same one) is randomly selected and its distribution sampled. The resulting empirical "meta-distribution" takes into account the distribution over all the studies selected by the user.

FERET provides a flexible structure. For example, although the studies used by EPA (1997, 1999) are provided as a default, users can easily "turn off" any study from the bibliography so it will not be sampled; similarly, studies can be given different weights. Alternatively, users can select non-EPA studies or add their own studies. Documentation is provided for key characteristics of each study (demographics, location, peer review, source of data, and so on) as well as the full citation.

Figure 19-1 illustrates the steps to generate a single "trial" or "observation" for a single pollutant and a single impact. The first step is to sample from the change in concentration based on the user's regulatory design (Step 1). FERET then randomly selects one study from among the health studies chosen for use and samples from the distribution of the response function of that study (Steps 2 and 3). When the sample values are substituted into the concentration–response function, the number of cases changes. For most functions, the baseline number of cases is determined from information calculated using either EPA-provided incidence rates or national baseline incidence rates already integrated into FERET. Users can change the baseline if they have more local information. Valuing the change in cases follows similarly in Steps 4 and 5, in which a valuation study is randomly selected from among the potential set and then an observation is drawn from the distribution of the value from that study. When the value is multiplied times the change in the number of cases, the monetized benefits of the environmental improvement are recorded for that one trial. FERET then samples from the distribution of compliance cost.

The difference between the cost and the benefit, after putting both values in present value terms,[6] is the net present value of the action. The bottom line is a distribution of environmental and monetary impacts, including the present value of net benefits.

Output Reporting and Sensitivity

FERET provides a summary table showing the mean, 5th and 95th percentiles for the health impacts, and the present value. These are the data most

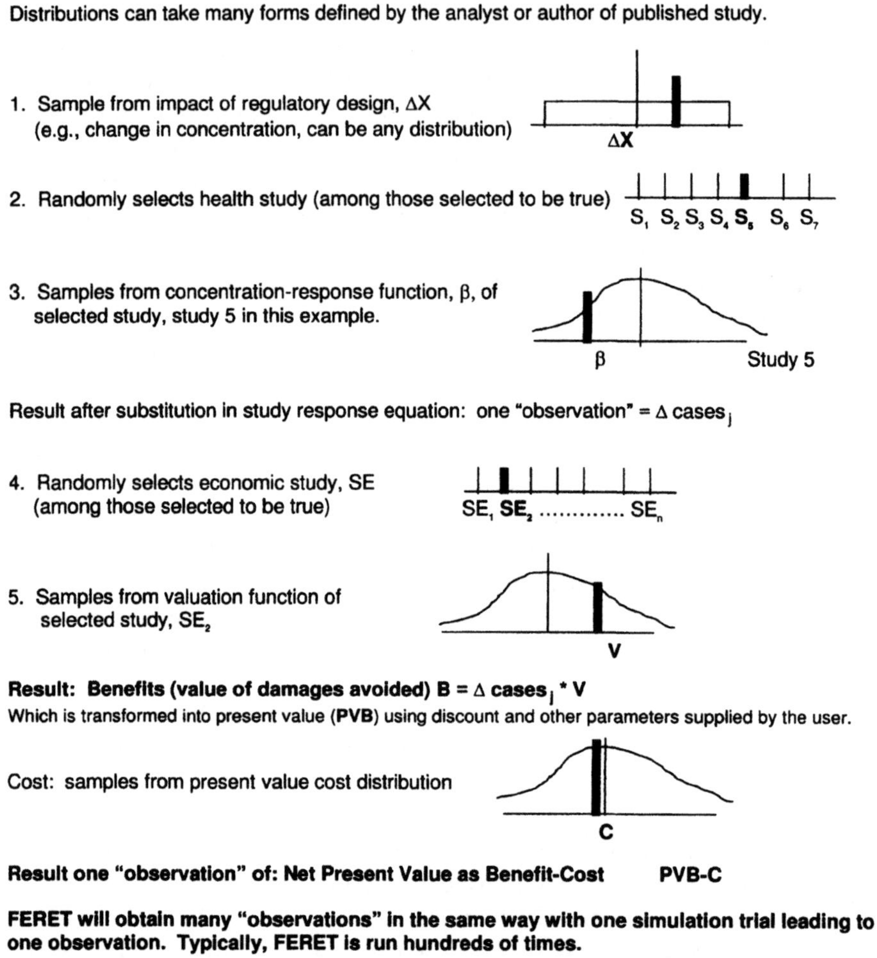

FIGURE 19-1. FERET Conceptual Flow Chart for One Impact of One Pollutant

frequently used in a benefit–cost analysis, although complete information on the statistical distribution for each intermediate and final output is available. Adjacent material provides qualitative information entered by the user from the regulatory design worksheet. Sensitivity analyses that identify the assumptions most highly correlated with a particular forecast can be automatically generated, as can sensitivity plots that show the outcome of varying a particular parameter. Data, including graphics of cumulative or frequency distributions, can be outputted in the form of a report.

At the end of an analysis, FERET produces the essentials of an economic risk assessment: a benefit–cost analysis with uncertainty, along with health and other impacts more traditionally associated with such studies. How such an analysis is used enters the territory of risk management. A decisionmaker applying standard economic analysis may suggest taking any or all of the actions whose mean (expected) present value is positive. In fact, distributional and other issues make the conclusion less clear. Typically, the analyst and the decisionmaker will want to know about the sensitivity of the results to changes in some assumptions, such as the discount rate or the exposed population, which is easily done in FERET. Others might feel that the direction of the qualitative impacts is clear and might, therefore, change a decision based on only the monetized impacts. Finally, some decisionmakers may use a modified decision criteria, perhaps due to budget constraints or some kind of precautionary principle. FERET assists in the risk and benefit–cost analysis, but it is not a risk manager or a decisionmaker.

Case Application: The Costs and Benefits of the Clean Air Act

EPA recently completed an important set of studies on the costs and benefits of the Clean Air Act (U.S. EPA 1997, 1999). EPA's analyses are based on an extensive review of the literature and the construction of elaborate air pollution, health, and valuation models that predict and value impacts on health and the environment. Numerous sensitivity and other studies were developed over the several-year period during which these studies were prepared. However, the public has little ability to investigate assumptions other than those used by EPA.

Because the structure of FERET is designed to allow analysts to follow the EPA approach but deviate from it when they wish, the EPA analyses provide a testing ground for both calibrating FERET and testing FERET's potential as a stakeholder participation tool. Consequently, in the last section, we report a calibration to EPA's results for 2010 and investigate several modifications of EPA's analysis.[7]

Calibration to EPA's Costs and Benefits

FERET is designed to follow the intent of and utilize the same statistical building blocks as EPA in its studies of the Clean Air Act. However, unavoidable differences result because the underlying EPA model is based on finer geographic resolution, and even the quite good documentation provided by EPA does not address all the technical questions that one might ask. The calibration efforts use the same health and economic studies, incidence rates, discount rate, latency periods, and so on (details are provided in the CD-

TABLE 19-1. Mean Values of Impacts, Benefits, and Costs: EPA and Calibration FERET

Comparison impact	*EPA 2010*	*FERET*
Present value net benefit (billion $)	83–86	83
Premature mortality (no. of deaths)	23,000	23,000
Total benefit (billion $)	110	110
Total cost (billion $)	27	27

ROM provided with this book). The impact of the Clean Air Act on national concentrations (the regulatory design) was calculated from EPA data to make those values consistent with key outcomes in the EPA report.

The driving factor in the costs and benefits of the Clean Air Act is the number of premature deaths avoided, which account for more than 90% of PVB. In Table 19-1, the EPA results for 2010 and a calibration result from FERET are presented for key economic and health dimensions.

Although any particular simulation can generate somewhat different outcomes, the results from FERET nearly replicated the EPA results. The benefits, costs, net present value, and premature mortality are quite accurate, but there is some ambiguity as to which present value one should calibrate. EPA variously reports the mean net present value as $83 billion (U.S. EPA 1999, iii), $86 billion (implied on p. 114), and $93 billion (p. 105). EPA tables suggest that $83 billion or $86 billion is intended. Some differences exist between EPA values and FERET predictions. For instance, FERET predicts somewhat higher incidence of respiratory and cardiovascular hospital admissions (due at least in part to a somewhat different aggregation process across studies) and a somewhat lower mean value per life saved.[8]

Figure 19-2 illustrates one of several of the visual outputs of FERET, showing a chart of the distribution of the present value of the net benefits. Because FERET's base-case calibration reasonably mimics the results of the EPA analysis, we use FERET to address several issues as if we were stakeholders interested in regulatory development or a fast turnaround and assessment of the regulatory analysis. The first question regards the importance of the mortality concentration–response study used by EPA. The second question is about the role of valuation studies that used contingent valuation survey techniques.

FERET is designed to allow the user to easily change the studies on which the analysis is based. Alternative FERET analyses were conducted that first use the short-run mortality studies considered by EPA in its draft retrospective study, followed by an alternative using two long-term studies instead of one. The use of long-term studies was advised by the EPA Clean Air Council. Two sets of changes were necessary to assess the importance of the shift from one long-term study to short-term studies. One change is the incidence rate and another the exposed population. Because the long-term study (Pope and

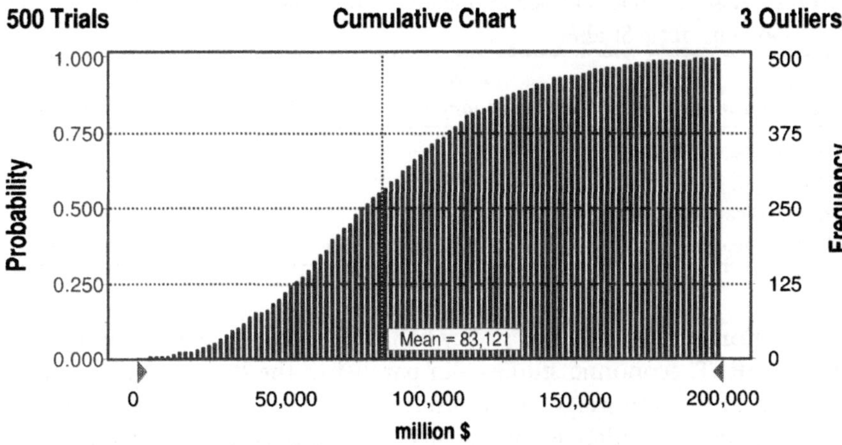

FIGURE 19-2. Cumulative Distribution of Net Present Value

others 1995) was based on premature deaths of people more than 30 years old, the incidence rate was changed from that used in the EPA study to correspond to an "all ages" incidence rate for the short-term studies, as was the exposed population. The result is presented in Table 19-2.

The difference between short and long-term studies is quite dramatic. The present value of net benefits drops by 98%, whereas deaths drop by almost 90%. There is a 50% chance that net benefits are negative using the short-term studies, with a daily improvement in concentration equal to the long-term mean. Alternatively, if a second long-term study (Dockery and others 1993) is used equally along with that of Pope and others (1995) (using the initial incidence rate and assuming the annual mean pollution change equals the median change), then net benefits increase to 136 billion and deaths averted increase to 36,000.

Does this result mean that the EPA analysis is flawed? No. It points out the importance of the decision to use the different types of studies. If long-term studies are the right choice, then the value of information is dramatic. It reveals that the reduction of pollution is much more important than previously thought, compared with short-term studies. Alternatively, it may suggest some caution in dropping one modeling approach, using short-term studies, and entirely depending on newer results.[9]

The second question investigated the importance of the method of estimating the value of a statistical life. Contingent valuation is a survey method (Farrow, Goldberg, and Small 2000) that is the subject of heated professional debate regarding its reliability and interpretation. Some analysts do not believe that such studies are consistent with sound economic science, whereas others believe that they can be consistent.

TABLE 19-2. Mean Values of Impacts, Benefits, and Costs: EPA and Calibration FERET Short- and Long-Term Studies

Comparison impact	*EPA 2010 (1 long term)*	*FERET 2010 (2 long term)*	*(short term)*
Present value net benefit (billion $)	83–86	137	2
Particulate-caused mortality (no. of deaths)	23,000	36,000	3,000

A solution as to its role in a policy debate (but not its science) is available. Within FERET, economic studies—in parallel to the health studies—can be turned on or off, and weighted in different ways. The importance of the contingent valuation studies is investigated by comparing the outcomes when they are and are not used. One other change was necessary to compare the valuation studies. EPA used a single distribution, fit to the mean estimates of the valuation studies, and drew values of a statistical life from that distribution. This method does not allow the user to delete studies selectively. When an alternative base case is estimated using individual valuation studies instead of EPA's distribution, the net benefits increase somewhat to more than $90 billion from the EPA results of $83 billion. When the contingent valuation studies are then removed from the study set, the median value per life saved changes by only $200,000 (from a mean of $4.4 million), and the net present value changes by only a few billion dollars (a few percent). This change is relatively small in the context of the broader variation of the outcome. We conclude that whether contingent valuation studies are used for the value of premature mortality is irrelevant to the overall outcome, therefore potentially removing a contentious issue from stakeholder discussions.

Finally, FERET automatically updates key data from the 1990 values used in the EPA report to more current dollars. These adjustments for the value of a dollar may have more of an effect on policy than many other modeling changes. EPA reported its basic results in 1990 dollars (as FERET does). When adjusted for inflation, as FERET can do for any year between 1990 and 2000 for key summary measures, the net benefits increase by 27%. In the perceptions of policymakers, this adjustment may be as real as any improvements in analysis.

Conclusions

FERET is a tool that integrates risk and policy analysis in a benefit–cost framework. Constructed around EPA's design and data for the costs and benefits of the Clean Air Act, FERET provides a quick tool for regulatory development and stakeholder participation. An application to the Clean Air Act illustrates its use as a stakeholder participation tool. It allows the quick investigation of alternative assumptions, such as which studies to include.

Using FERET, we found that the decision to move from short-term to long-term mortality studies is a key determinant of the benefits, whereas the distinction between economic studies that use or do not use contingent valuation is of relatively little importance. FERET appears to provide a flexible structure for use in regulatory development and stakeholder participation. Possible applications include state air quality regulations, as for older sites; screening to assess regulatory alternatives for federal regulation; support for company environmental management systems; environmental justice issues for different "exposed" populations; and health outcomes from health or safety regulation.

Notes

1. For a review of benefit–cost practice in the federal and state governments, see Farrow and Toman 1999 or Hahn 2000.

2. Morgenstern and Landy (1997, 461) state, "Contractors who conduct these studies argue that a remarkably large portion of the costs are associated with gathering information rather than from conducting actual analysis."

3. See, for instance, Varian 1992.

4. FERET includes a cost estimation program developed by the U.S. Environmental Protection Agency called PROJECT, which provides cost estimates that can remove the effect of taxation from the cost of the firm.

5. See, for example, Arrow and others 1996 and Farrow and Toman 1999.

6. The typical case will generate annual benefits that are put into present value terms using the regulatory design parameters of the timing of the action and Excel's present value formula. If the PROJECT model is used to generate costs, then the costs are already in present value terms.

7. However, FERET is being used to capture the nationwide output of the EPA analysis. The underlying EPA tools provide a finer geographic resolution, but this information has not been important to the debate, and the public is unlikely to have the ability to use the EPA model at such detail.

8. Run "FERET2010.xls" on the enclosed CD-ROM. You may wish to read calibration notes, also included.

9. For a different critique that might be quantified, see Lutter and Belzer 2000.

References

Arrow, K., M. Cropper, G. Eads, R. Hahn, L. Lave, R. Noll, P. Portney, M. Russell, R. Schmalensee, V.K. Smith, and R. Stavins. 1996. *Benefit–Cost Analysis in Environ-*

mental, Health, and Safety Regulation. AEI Press, Washington, DC. http://www.aeibrookings.org (accessed March 5, 2001).

Dockery, D.W., C. Pope, X. Xu, J. Spengler, J. Ware, M. Fay, B. Ferris, and F. Speizer. 1993. An Association between Air Pollution and Mortality in Six U.S. Cities. *New England Journal of Medicine* 329(24): 1753–1759.

Farrow, S. 1998. Environmental Equity and Sustainability: Rejecting the Kaldor Hicks Criteria. *Ecological Economics* 27(2): 183–188.

Farrow, S., and M. Toman. 1999. Using Environmental Benefit–Cost Analysis to Improve Government Performance. *Environment* 41(2): 12–15, 33–37.

Farrow, S., C. Goldburg, and M. Small (eds.). 2000. *Environmental Science and Technology* Special Edition (Economic Valuation), April 15.

Hahn, R. 2000. State and Federal Regulatory Reform: A Comparative Analysis. *The Journal of Legal Studies* 29(2): 873–912.

Hanley, N., and C. Spash. 1993. *Cost–Benefit Analysis and the Environment.* Edward Elgar: Hants, U.K.

Lutter, R., and R. Belzer. 2000. EPA Pats Itself on the Back. *Regulation* 23(3): 23–28.

Morgenstern, R. (ed). 1997. *Economic Analyses at the EPA.* Resources for the Future: Washington, DC.

Morgenstern, R., and M. Landy. 1997. Economic Analysis: Benefits, Costs, and Implications. In *Economic Analyses at the EPA,* edited by R. Morgenstern. Resources for the Future: Washington, DC, 455–478.

OMB (Office of Management and Budget). 2000. *Draft Report to Congress on the Costs and Benefits of Federal Regulations.* OMB: Washington, DC.

Pope, C.A., M. Thun, M. Namboodiri, D.W. Dockery, J. Evans, F. Speizer, and C. Heath. 1995. Particulate Air Pollution as a Predictor of Mortality in a Prospective Study of U.S. Adults. *American Journal of Critical Care Medicine* 151(3, Part 1): 669–674.

U.S. EPA (Environmental Protection Agency). 1997. *The Benefits and Costs of the Clean Air Act, 1970 to 1990.* U.S. EPA, Office of Air and Radiation: Washington, DC.

———. 1999. *The Benefits and Costs of the Clean Air Act, 1990 to 2010.* EPA-410-R-99-001. U.S. EPA, Office of Air and Radiation: Washington, DC.

Varian, H. 1992. *Microeconomics Analysis.* W.W. Norton and Company: New York.

Zerbe, R.O. Jr., and D.D. Dively. 1994. *Benefit–Cost Analysis in Theory and Practice.* HarperCollins College Publishers: New York.

20

Epilogue: The Challenge of Improving Regulation

Paul S. Fischbeck and R. Scott Farrow

The chapters in this book are devoted to improving regulation, recognizing that such improvement is an ongoing process. To paraphrase an advertising slogan, if good regulations were easy, then every agency would have them. Analysis of various kinds—behavioral, institutional, economic, technological, and quantitative—can contribute to improving regulation and need not cause paralysis.

As we peer into the uncertain regulatory mist, several themes seem likely to continue. Regulatory design may emerge as a more systematic tool but still subject to political laws of motion. For instance, interest in performance-based regulation brings to the fore concern about what our current regulations actually achieve and the performance that we would like them to achieve. Appropriately, debates will continue about balancing objectives that include effectiveness, equity, innovation, efficiency, and the ability to respond to new information, among others. New applications of market- and information-based regulatory designs appear destined for further investigation.

The rapid changes of recent years brought on in part by the evolution of computer technology have created new opportunities. At the analytical level, we believe that large gains are still to be made through the development of computer-based systems for decision support. Such systems might range from shared data access to tools that facilitate analysis. At a broader level, the rapid change has restored a concern for dynamic performance. In antitrust regulation, the computer age has reinforced the potential trade-off between static and dynamic performance. That concern is increasingly apparent in environmental, health, and safety regulation. The frequent first response of regulation—to stop an offending practice—is increasingly followed by concern about the dynamics of regulation: the unintended consequences, opportunities for innovation, and concern for uncertainty. In a world where many analysts pretended that the regulator was omniscient, there was little concern about such dynamics. As we have grown more sophisticated in appreciating the many aspects of uncertainty that underlie regulatory issues, future research is likely to investigate assessment, management criteria, and regulatory designs in light of uncertain science, human behavior, and institutions.

Although the focus of this book is the microevolutionary steps to improve regulation, such changes should be part of a package with more macro-oriented syntheses of how to improve regulation. We hope that a better understanding of the microfoundations of successful and unsuccessful regulation will help guide the investigation of policies such as organic environmental legislation to replace media-specific statutes and the broader application of risk-based, market-based, and other innovative mechanisms.

In this volume, we have sought to demonstrate the usefulness of confronting specific regulatory alternatives with evidence and analysis in a "bottom-up" approach to regulatory improvement. As a balance to discussions of sweeping changes, we offer incremental regulatory improvement and the development of tools that have multiple applications. What is presented here may not be the grand view from the mountaintop; instead, it provides the view from the toilers in the field, which has its own perspective and power. Much remains to be done on all fronts.

Index

For Product Safety Concerns and Information please contact our EU representative GPSR@taylorandfrancis.com Taylor & Francis Verlag GmbH, Kaufingerstraße 24, 80331 München, Germany

Batch number: 10725935

Printed by Printforce, the Netherlands